Communications in Computer and Information Science 2690

Series Editors

Rationale

The CCIS series is devoted to the publication of proceedings of computer science conferences. Its aim is to efficiently disseminate original research results in informatics in printed and electronic form. While the focus is on publication of peer-reviewed full papers presenting mature work, inclusion of reviewed short papers reporting on work in progress is welcome, too. Besides globally relevant meetings with internationally representative program committees guaranteeing a strict peer-reviewing and paper selection process, conferences run by societies or of high regional or national relevance are also considered for publication.

Topics

The topical scope of CCIS spans the entire spectrum of informatics ranging from foundational topics in the theory of computing to information and communications science and technology and a broad variety of interdisciplinary application fields.

Information for Volume Editors and Authors

Publication in CCIS is free of charge. No royalties are paid, however, we offer registered conference participants temporary free access to the online version of the conference proceedings on SpringerLink (http://link.springer.com) by means of an http referrer from the conference website and/or a number of complimentary printed copies, as specified in the official acceptance email of the event.

CCIS proceedings can be published in time for distribution at conferences or as post-proceedings, and delivered in the form of printed books and/or electronically as USBs and/or e-content licenses for accessing proceedings at SpringerLink. Furthermore, CCIS proceedings are included in the CCIS electronic book series hosted in the SpringerLink digital library at http://link.springer.com/bookseries/7899. Conferences publishing in CCIS are allowed to use our online conference service (Meteor) for managing the whole proceedings lifecycle (from submission and reviewing to preparing for publication) free of charge.

Publication process

The language of publication is exclusively English. Authors publishing in CCIS have to sign the Springer CCIS copyright transfer form, however, they are free to use their material published in CCIS for substantially changed, more elaborate subsequent publications elsewhere. For the preparation of the camera-ready papers/files, authors have to strictly adhere to the Springer CCIS Authors' Instructions and are strongly encouraged to use the CCIS LaTeX style files or templates.

Abstracting/Indexing

CCIS is abstracted/indexed in DBLP, Google Scholar, EI-Compendex, Mathematical Reviews, SCImago, Scopus. CCIS volumes are also submitted for the inclusion in ISI Proceedings.

How to start

To start the evaluation of your proposal for inclusion in the CCIS series, please send an e-mail to ccis@springer.com

Shreyas J. · Gururaj H. L · Sophia Rahaman ·
Keshav Kaushik · Aryan Chaudhary ·
Dayananda P.
Editors

Data Science and Exploration in Artificial Intelligence

Second International Conference, CODE-AI 2025
Dubai, United Arab Emirates, April 7–8, 2025
Proceedings, Part II

Editors
Shreyas J.
Manipal Institute of Technology Bengaluru
and Manipal Academy of Higher Education
Manipal, Karnataka, India

Gururaj H. L
Manipal Institute of Technology Bengaluru
and Manipal Academy of Higher Education
Manipal, Karnataka, India

Sophia Rahaman
Manipal Academy of Higher Education
Dubai, United Arab Emirates

Keshav Kaushik
Sharda University
Noida, Uttar Pradesh, India

Aryan Chaudhary
BioTech Sphere Research
Bijnor, Uttar Pradesh, India

Dayananda P.
Manipal Institute of Technology Bengaluru
and Manipal Academy of Higher Education
Manipal, Karnataka, India

ISSN 1865-0929 ISSN 1865-0937 (electronic)
Communications in Computer and Information Science
ISBN 978-3-032-19320-9 ISBN 978-3-032-19321-6 (eBook)
https://doi.org/10.1007/978-3-032-19321-6

This Springer imprint is published by the registered company Springer Nature Switzerland AG
The registered company address is: Gewerbestrasse 11, 6330 Cham, Switzerland

Preface

It is with great pleasure that we present this volume of proceedings from the 2nd International Conference on Data Science & Exploration in Artificial Intelligence (CODE-AI 2025), held on 7th and 8th of April 2025 at the Dubai campus of Manipal Academy of Higher Education (MAHE), UAE.

The conference brought together researchers, educators, practitioners and industry professionals from around the globe to exchange insights, highlight emerging trends and build collaborations in the fields of artificial intelligence, data science and their applied impact.

The thematic scope of CODE-AI 2025 spanned intelligent computing methods (including genetic algorithms, simulated annealing, artificial fish-swarm algorithms, quantum computing and fuzzy logic), advanced AI applications (such as biometrics, pattern recognition, computer and machine vision, speech recognition and smart robotics) and data-science topics (deep learning, decision-making frameworks, IoT/edge data integration and visualization).

We believe this volume contributes to the advancement of knowledge by bringing together rigorous research and novel applications that address both foundational and real-world challenges.

The review of submitted manuscripts was carried out under a double-blind peer-review model, with each paper evaluated by at least two independent reviewers. Authors affiliated with program committee members or steering committee members were subject to an objective process in which an alternate chair handled the evaluation to avoid conflicts of interest.

We received a total of 650 full-paper submissions, and 150 short papers. After considering Springer guidelines, 165 full papers and 70 short papers were accepted for the review process.

However, after the double-blinded review process only 67 papers were accepted for the publication in the final proceedings.

We acknowledge with gratitude the contributions of our authors, the diligence of our reviewers, and the support provided by our program and organizing committees. We also extend our thanks to Meerut ACM Professional Chapter for their contribution and MITB ACM Student Chapter for hosting CODE AI -2025.

It is our hope that the research and ideas presented here will serve as a lasting resource for scholars and practitioners, inspire further investigation, and foster strong linkages across academia and industry.

On behalf of the Program Committee and the Organizing Committee of CODE-AI 2025

Organization

Chief Patrons

Ramdas M. Pai	Manipal Academy of Higher Education, India
Ranjan R. Pai	Manipal Academy of Higher Education, India

Patrons

M. D. Venkatesh	Manipal Academy of Higher Education, India
H. S. Ballal	Manipal Academy of Higher Education, India
Madhu Veeraraghavan	MAHE Bengaluru, India
Narayana Sabhahit	Manipal Academy of Higher Education, India
Giridar Kini P.	Manipal Academy of Higher Education, India
Raghavendra Prabhu	MAHE Bengaluru, India
S. Sudhindra	MAHE Dubai, UAE
Iven Jose	MIT Bengaluru, India
S. K. Pandey	MAHE Dubai, UAE
Prema K. V.	MIT Bengaluru, India

General Chair

Dayananda P.	MIT Bengaluru, India

Program Chairs

Gururaj H. L.	MIT Bengaluru, India
Shreyas J.	MIT Bengaluru, India
Sophia Rahaman	MAHE Dubai, UAE
Gopalakrishnan T.	MIT Bengaluru, India

Technical Co-chairs

Aryan Chaudhary	BioTech Sphere Research, India
Keshav Kaushik	Amity University Punjab, India

Scientific Committee

Osamah Ibrahim Khalaf	Al-Nahrain University, Iraq
Ghaidaa Muttasher Abdulsaheb	University of Technology, Iraq
Shin-Hung Pan	Chaoyang University of Technology, Taiwan
Wing-Keung Wong	Asia University, Taiwan

Organizing Committee

Karthik S. A.	MIT Bengaluru, India
Abhijit Das	MIT Bengaluru, India
M. S. Satyanarayana	MIT Bengaluru, India
Sindhu Madhuri G.	MIT Bengaluru, India
Amreen Ayesha	MIT Bengaluru, India
Anitha Premkumar	MIT Bengaluru, India
Arun Balakrishnan	MIT Bengaluru, India
P. Devisivasankari	MIT Bengaluru, India
G. Ignisha Rajathi	MIT Bengaluru, India
Karthik Vasu	MIT Bengaluru, India
S. K. Mahmudul Hassan	MIT Bengaluru, India
Preethi	MIT Bengaluru, India
S. Priya	MIT Bengaluru, India
Raghavendra M. Devadas	MIT Bengaluru, India
Ruhul Amin Hazarika	MIT Bengaluru, India
Sangeeta Sangani	MIT Bengaluru, India
Sapna R.	MIT Bengaluru, India
Sumanth V.	MIT Bengaluru, India
Lijo Jose	MIT Bengaluru, India
Usha M.	MIT Bengaluru, India
Vishnu Srinivasa Murthy Y.	MIT Bengaluru, India
Abdulla K. P.	MAHE Dubai, UAE
Deepa Varghese	MAHE Dubai, UAE
Ganesan Subramanian	MAHE Dubai, UAE
M. I. Jawid Nazir	MAHE Dubai, UAE
Ramaprasad Poojary	MAHE Dubai, UAE
Ravishankar Dudhe	MAHE Dubai, UAE
Roma Raina	MAHE Dubai, UAE
Sachidananda H. K.	MAHE Dubai, UAE
Sampath Suranjan Salins	MAHE Dubai, UAE
Suresha R.	MAHE Dubai, UAE
Guru Shankar H. B.	MIT Bengaluru, India

Shreelakshmi Yadav N.	MIT Bengaluru, India
Soundarya B. C.	MIT Bengaluru, India
Sowmya T. Rao	MIT Bengaluru, India
Udaya Prasad P. K.	MIT Bengaluru, India

Student Organizing Committee

Karthikeya Chowdary	MITB ACM Student Chapter, India
Nishanth Shet	MITB ACM Student Chapter, India
Sashi Pritam M. A.	MITB ACM Student Chapter, India
Anvitha Karanth	MITB ACM-W Student Chapter, India
Nihitha H. R.	MITB ACM-W Student Chapter, India
Shivansh Gautam	MITB ACM SIG-AI, India
Shane Chellam	MITB ACM SIG-AI, India
Samyak Bargale	MITB ACM SIG-Soft, India
Nandan S. B.	MIT Bengaluru, India
Lakshya Banga	MITB ACM SIG-SOFT, India
Dheeraj Sai Samineni	MITB ACM Student Chapter, India

Advisory Committee

Manohara M. Pai	Manipal Academy of Higher Education, India
Hareesha K. S.	Manipal Academy of Higher Education, India
Antoine Bossard	Kanagawa University, Japan
Chenren Xu	Peking University, China
S. K. Lakshmanaprabhu	Renault Nissan Technology & Business Centre India
Ankur Gupta	Model Institute of Engineering and Technology, India
Dilip Kumar S. M.	University of Visvesvaraya College of Engineering, India
Prasant Misra	Tata Consultancy Services, India
Vijay Arya	IBM Research, India
Adrian Florea	Lucian Blaga University of Sibiu, Romania
Alex S. Taylor	University of Edinburgh, UK
David Coyle	University College Dublin, Ireland
Demetrios Zeinalipour-Yazti	University of Cyprus, Cyprus
Alessio Malizia	University of Pisa, Italy
Andre Inacio Reis	Federal University of Rio Grande do Sul, Brazil
Joel J. P. C. Rodrigues	Federal University of Piauí, Brazil

John Atkinson	Universidad Adolfo Ibáñez, Chile
Cristiano A. Costa	University of the Rio dos Sinos Valley, Brazil
Tshilidzi Marwala	University of Johannesburg, South Africa
Martin Stephanus Olivier	University of Pretoria, South Africa
Kester Quist-Aphetsi	Ghana Communication Technology University, Ghana
Sameerchand Pudaruth	University of Mauritius, Mauritius
Jan Hendrik Kroeze	University of South Africa, South Africa
Ashraf Saad Hussein	Arab Open University, Kuwait
Ferda Ofli	Qatar Computing Research Institute, Qatar
Haitham S. Hamza	Cairo University, Egypt
Suhaib A. Fahmy	King Abdullah University of Science and Technology, Saudi Arabia
Mourad Ouzzani	Qatar Computing Research Institute, Qatar
Adam R. Talcott	University of California, Santa Barbara, USA
Adam A. Porter	University of Maryland, College Park, USA
Aaron Daniel Adler	Raytheon BBN, USA
Adam Chlipala	Massachusetts Institute of Technology, USA
Busby-Earle	University of Johannesburg, South Africa
Jorge Armin Mazariegos	Universidad de San Carlos de Guatemala, Guatemala
Ajay K. Gupta	Old Dominion University, USA
Jerry Miller	Florida International University, USA
Harish Karthikeyan	New York University, USA
Heggere Ranganath	University of Alabama, Huntsville, USA
Matt Green	KB Walker, USA
Ramkumar Krishnamoorthy	Villa College, Maldives
Milind Gandhe	IIIT Bangalore, India
P. Nagabhushan	Vignan University, India
V. Susheela Devi	Indian Institute of Science, India
Aloknath De	Samsung, India
L. M. Patnaik	Indian Institute of Science, India
Y. Narahari	Indian Institute of Science, India
Channappa B. Akki	IIIT Dharwad, India
Poornima Ajayan	Rice University, USA
Keshava Munegowda	Goldman Sachs, India
Chidananda Gowda	Kuvempu University, India
Sundaram Suresh	Nanyang Technological University, Singapore
Heena Rathore	Texas State University, USA
Paolo Trunfio	University of Calabria, Italy
Fernando Koch	IBM, India
Amlan Chakrabarti	University of Calcutta, India

Contents

A Deep Learning Approach Using Kernel Isolated Pyramid Pooled Visual Geometry Group for Pistachio Type Classification

M. Shyamala Devi[1] and S. Priya[2](✉)

[1] Department of Robot and Smart System Engineering, Kyungpook National University, 80, Daehak-ro, Buk-gu, Daegu 41566, Republic of Korea

[2] Manipal Institute of Technology Bengaluru, Manipal Academy of Higher Education, Manipal, India

s.priya@manipal.edu

Abstract. In the agricultural sector, pistachio classification is crucial for inventory management, quality assurance, and sorting, as it is necessary to differentiate between several pistachio varieties. Deep learning techniques, in particular convolutional neural networks (CNNs), have demonstrated promising results in a variety of areas in image classification tasks in recent years. This work proposes Kernel Isolated Pyramid Pooled VGG16 (KIP-VGG16) to address the multi scale feature pooling of pistachio image for its type classification. This research utilizes the images from Pistachio Image KAGGLE public repository dataset. The initial contribution of this work was the formation of Kernel isolated pistachio image that reduces the interference from background noise and irrelevant features. The second contribution deals with refining the existing VGG16 by replacing the max pooling in the convolution block to pyramid pooling to propose KIP-VGG16 model. The kernel-isolation filtering technique was used that concentrated on extracting the multi-scale features, enabling for more robust feature extraction from images of pistachios under varying conditions. Pyramid pooling was adapted inside the convolution that captures both fine-grained details and global context, ensuring effective representation across varying image sizes and orientations. The KIP-VGG16 model can be applied for tasks where multiple pistachio types need to be identified from complex backgrounds and varying angles. The filtered pistachio images are divided in the ratio of 80: 10:10 and are executed with the CNN and KIP-VGG16 model to assess the performance. Implementation shows that KIP-VGG16 found to exhibit high accuracy of 99.47% towards the classification of Kirmizi and Siirt pistachio classes. This novel KIP-VGG16 method creates new opportunities for high-performance image classification tasks in agriculture, where accurate pistachio type sorting and identification are essential for meeting consumer requests and ensuring product quality.

Keywords: Convolution · CNN · Deep learning · Filtering · Kernel · Noise · Pyramid pooling · VGG16

J. Shreyas et al. (Eds.): CODE-AI 2025, CCIS 2690, pp. 1–12, 2026.
https://doi.org/10.1007/978-3-032-19321-6_1

1 Introduction

Pistachio type classification is an essential work in contemporary agriculture, particularly in areas where pistachio farming is a significant sector. The ability to effectively and precisely categorize pistachio varieties according to their visual attributes—such as shape, size, and color becomes more crucial as pistachio agriculture spreads throughout the world. Sorting, inventory management, quality control, and satisfying consumer requests are all aided by this classification process. Manual inspection-based traditional pistachio classification techniques are labor-intensive, prone to mistakes, and constrained by human subjectivity.

1.1 Background Study

As a result, automated image-based classification systems that can efficiently and accurately handle massive amounts of pistachio images are becoming more and more necessary. However, the intrinsic variation in appearance among pistachio kinds and the existence of complicated backgrounds make the task of categorizing pistachio types particularly difficult. The above issues are addressed by the proposed KIP-VGG16 model by integrating the kernel isolated pistachio image with the pyramid pooling processed in the convolution layer of VGG16 model. The inferences made from the review works are shown in Table. 1.

Table 1. Inferences from related works

Methodology	Inference and advantages	Limitations
• Pretrained CNN Models: AlexNet [1, 4, 5, 16], InceptionV3 [1], VGGNet [4, 16], EfficientNetB3 [9], VGG16 [5, 11], and was designed for pistachio classification with pistachio image dataset. InceptionV3 [7], ResNet50 [7] was designed for hazelnut classification EfficientNet [14],ResNet [16], ResNet21 [15], DenseNet [15], was designed for pistachio nut classification	AlexNet model classifies pistachio with 96.13% accuracy. VGG model showed 94% of accuracy for pistachio classification InceptionV3, ResNet50 model showed 90% of accuracy for hazelnut classification EfficientNetB3, ResNet21 and DenseNet classify pistachio with 99% accuracy. EfficientNet classifies pistachio with 98% accuracy. CNN model extracts intricate details of the images to classify the species	The CNN model should be validated for real time dataset CNN is prone to overfit the model to degrade the performance CNN suffers with low interpretability CNN cannot perform well for imbalanced data CNN requires significant feature extraction methods to finetune the images CNN needs augmentation to increase efficiency CNN suffers with generalization problems CNN requires high quality images without noise to classify the images effectively.

(continued)

Table 1. (*continued*)

Methodology	Inference and advantages	Limitations
• AI based Models: AI Neural Network [2] toolbox was designed for hazelnut classification ANN and SVM [10] were designed for Classification of peeled pistachio kernels	AI Neural Network Classifies hazelnut with 89.1% accuracy ANN and SVM classify pistachio with 99% accuracy.	• AI Neural Network can be validated with other performance metrics AI model requires more large datasets for efficient classification AI models suffer with bias leading to unfair classifications.
• Image Processing based Models: Otsu [3], KNN [6], Neural Embedding Feature Extraction [8] method was used for pistachio classification J48 decision tree [12], Decision Tree and Random Forest [13] was used for sorting pistachio nuts	KNN model showed 94.18% of accuracy for pistachio classification Neural Embedding Feature Extraction shows 97.20% of accuracy for pistachio classification accuracy. J48 decision tree classifies pistachio with 99% accuracy.	• Image processing models are not adaptable to new variations in images It requires manual extraction of features and is sensitive to variations in lighting conditions. It needs to manually predefine the model parameters, threshold and kernel sizes. Difficult to perform multi classification Model suffers with scalability issue with dataset and performance.

1.2 Contributions of this Research

The following are the contributions of this research.

(i). The first contribution of this work was formation of Kernel isolated pistachio image that reduces the interference from background noise and irrelevant features. Through the implementation of the kernel-isolation filtering technique, which focused on extracting multi-scale features from pistachio images under various situations was made possible.

(ii). The second contribution deals with refining the existing VGG16 by replacing the max pooling in the convolution block to pyramid pooling to propose KIP VGG16 model. Pyramid pooling was modified for the convolution, which effectively represents different image sizes and orientations by capturing both global context and fine-grained information.

2 Research Methodology of Proposed KIP-VGG16

The proposed KIP-VGG16 model was designed to classify the pistachio image types as kirmizi and siirt by extracting images from Pistachio image KAGGLE dataset and its methodology is shown in Fig. 1.

The proposed KIP-VGG16 retrieves 2000 images from Pistachio Image dataset [17] that was organized and labelled to process with data augmentation. The pistachio images are represented by (1) and (2).

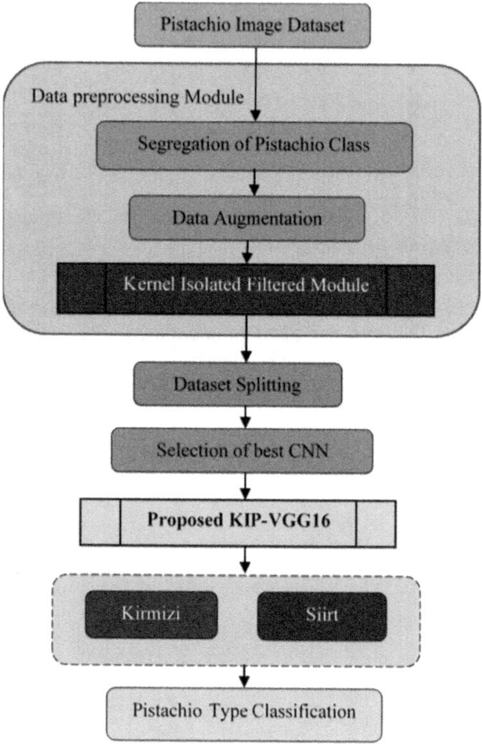

Fig. 1 Research methodology of KIP-VGG16

$$PIS_{2000} = \left\lfloor \sum_{P=1}^{2} \cup_{W=1}^{1000} \left\{ \sum_{e=1}^{255} \sum_{d=1}^{255} PIS_{edP} \right\} \right\rfloor \quad (1)$$

$$PIS_{001} = \begin{bmatrix} PIS(0,0) & PIS(0,1) & \dots & PIS(0,255) \\ PIS(1,0) & PIS(1,1) & \dots & PIS(1,255) \\ \vdots & \vdots & \dots & \vdots \\ PIS(255,0) & PIS(255,1) & \cdots & PIS(255,255) \end{bmatrix} \quad (2)$$

The 42,000 augmented pistachio images "PIS" are processed to form grayscale images and then fed into Kernel Isolated Filtered module that is shown in Fig. 2 to form filtered images. The filtered pistachio images are applied to existing CNN like DenseNet, MobileNet, Inception, Xception and VGG16 to select the best CNN and filtered image. Implementation shows that VGG16 with Kernel isolated pistachio image was found have higher accuracy than the other models. So, the VGG16 was refined to change the max pooling of the convolutional block into pyramid pooling to propose KIP-VGG16 that is shown in Fig. 3. The KIP-VGG16 contains five convolutional blocks ending two FCL blocks and final SoftMax with two neurons to classify Kirmizi and Siirt pistachio images.

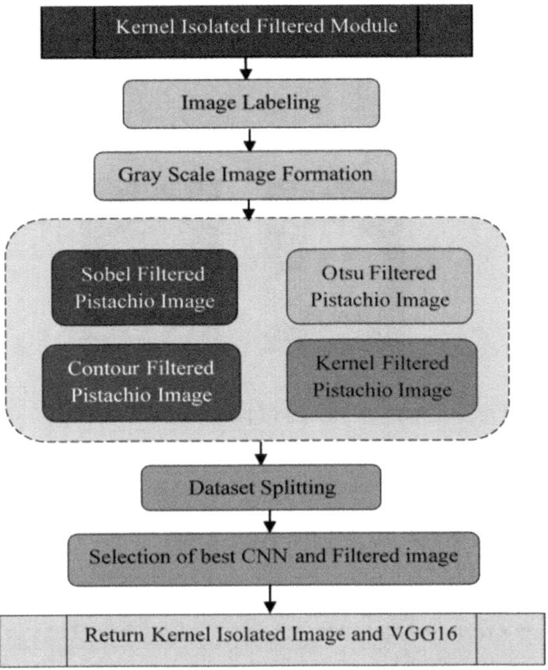

Fig. 2 Preprocessing and filtering module workflow of KIP-VGG16

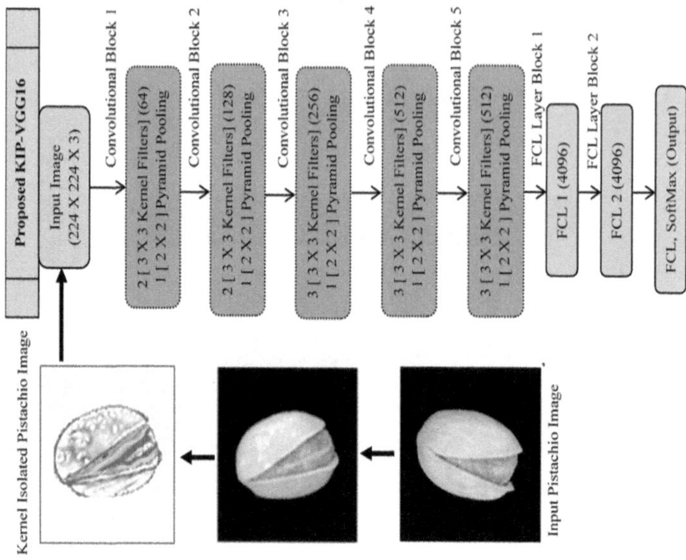

Fig. 3 Prototype design of proposed KIP-VGG16

3 Implementation Results and Discussion

The Pistachio Image dataset [17] with 2000 images was used for implementation and sample dataset images from the dataset are displayed for visualization as shown in Fig. 4.

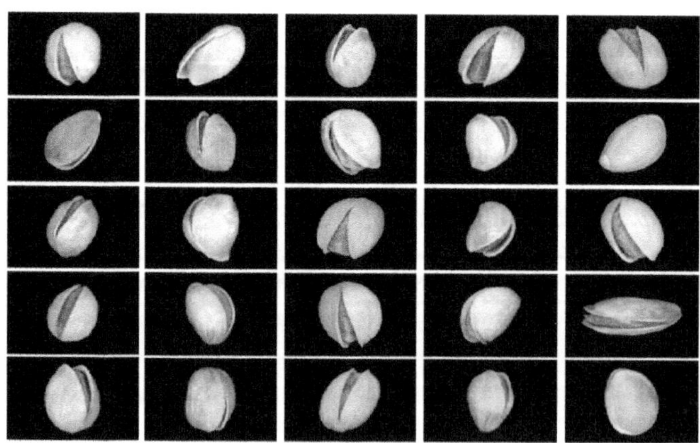

Fig. 4 Sample Pistachio images from Pistachio Image KAGGLE Dataset

The labeled images are augmented to form 42,000 images by performing different transformation. The augmented images are processed to form the gray-scale pistachio images. The gray scale augmented pistachio images are processed to form the sobel filtered [18], otsu threshold [19], contour filtered [20] and kernel isolated [21] filtered images. During the implemetation, the gray scale augmented pistachio images are processed to form otsu threshold filtered pistachio images. Next, the gray scale augmented pistachio images are processed to form contour filtered pistachio images. The gray scale augmented pistachio images are processed to form kernel isolated filtered pistachio images. Then, the filtered pistachio images are applied to CNN models and the execution results shows 92% higher accuracy for VGG16 with kernel isolated filtered image. So, VGG16 model was refined to convert the max pooling in the convolutional block with the pyramid pooling to design KIP-VGG16. The performance analysis of the KIP-VGG16 was shown in Table. 2. From the performance analysis, it is evident that the proposed KIP-VGG16 was projecting the accuracy of 99.47% towards classification of pistachio type. The sobel filtered, otsu threshold, contour filtered and kernel isolated filtered results are depicted from Figs. 5, 6, 7 and 8. The feature map extraction at the end of the convolutional block and FCL layer is shown in Figs. 9 and 10.

Table 2. Performance analysis of KIP-VGG16

CNN	Accuracy (%)				
	Applying raw images (%)	Applying sobel pistachio images (%)	Applying otsu pistachio images (%)	Applying canny pistachio images (%)	Applying kernel isolated images (%)
DenseNet	72.55	73.23	74.51	79.59	82.95
MobileNet	73.22	74.29	75.23	83.23	84.22
Inception	71.47	75.43	74.65	81.89	89.43
XCeption	76.33	77.32	78.98	86.32	88.33
VGG16	77.13	78.18	79.39	87.75	92.00
Proposed KIP-VGG16	85.22	87.51	89.92	92.22	**99.47**

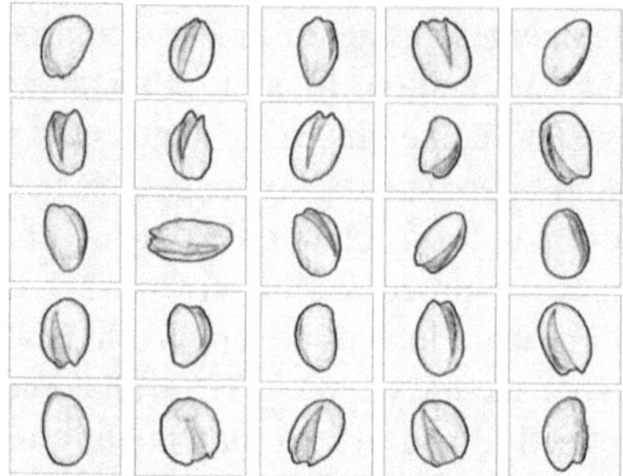

Fig. 5 Results of sobel filtered

Fig. 6 Results of Otsu threshold filtered

Fig. 7 Results of contour filtered

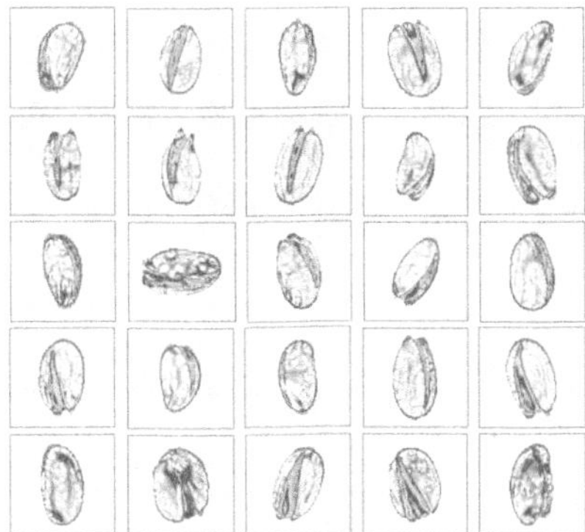

Fig. 8 Results of kernel isolated filtered

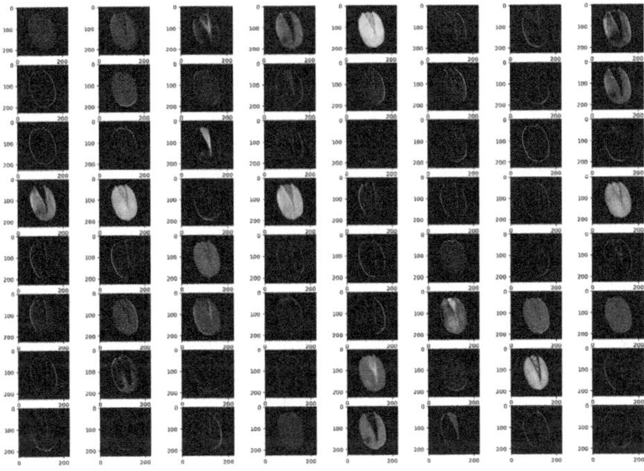

Fig. 9 Feature map results at the end of convolutional block of KIP-VGG16

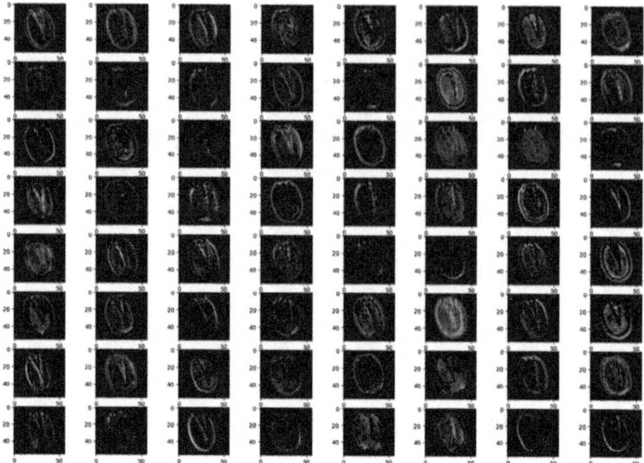

Fig. 10 Feature map results at the end of FCL block of KIP-VGG16

4 Conclusion

This study concentrated on classifying each image of kirmizi and Siirt pistachio type images through examining the image's main features. The primary goal of this work is to identify the optimal CNN model, isolate features using kernels, and enhance VGG16 pooling for accurately recognizing the pistachio images. This study's main contributions can be packed up in two ways. This work's first contribution was the kernel-isolated pistachio image, which reduces influence from insignificant details and background noise. The second contribution concentrates on improving the current version of VGG16 by substituting pyramid pooling to replace with max pooling in the convolution block. This pyramid pooling method captures both fine-grained information and global context, effectively representing varying image sizes and orientations. As a summary of the novelty, the uniqueness of the KIP-VGG16 exists in the extraction of feature map by pyramid pooling that contains the global and fine-grained information. The pistachio dataset with 2000 images was used for implementation is labelled and augmented to form 42,000 images. The augmented images are converted to grayscale and preprocessed to form filtered. The filtered pistachio images are fitted with CNN to select the best model. Implementation shows VGG16 with kernel isolated pistachio images was found have higher accuracy than the other models. So, the VGG16 was refined by replacing the max pooling of Convolution block into pyramid pooling to propose KIP-VGG16 model. The kernel isolated images were fitted with existing CNN and KIP-VGG16 model to compare efficiency. The execution results reveal the KIP-VGG16 found to exhibit high accuracy of 99.47% towards pistachio classification. The future enhancement is to validate DL models with advanced pyramid pooling methods in the convolutional block.

References

1. Aktas, H., Kızıldeniz, T., Unal, Z.: Classification of pistachios with deep learning and assessing the effect of various datasets on accuracy. Journal of Food Measurement & Characterization. **16**(3), 1983–1996 (2022). https://doi.org/10.1007/s11694-022-01313-5
2. Farhadi, M., Abbaspour-Gilandeh, Y., Mahmoudi, A., Maja, J.M.: An integrated system of artificial intelligence and signal processing techniques for the sorting and grading of nuts. Appl. Sci. **10**(9), 3315 (2020). https://doi.org/10.3390/app10093315
3. Sabanci, K., Koklu, M., Unlersen, M.F.: Classification of Siirt and long type pistachios (Pistacia vera L.) by artificial neural networks. Int. J. Intell. Syst. Appl. Eng. **3**(2), 86 (2015). https://doi.org/10.18201/ijisae.74573
4. F. Patel, S. Mewada, S. Degadwala, D. Vyas, Recognition of Pistachio Species with Transfer Learning Models, In: 2023 International conference on self sustainable artificial intelligence systems (ICSSAS), Erode, India, 2023, pp. 250–255, https://doi.org/10.1109/ICSSAS57918.2023.10331907
5. Singh, D., et al.: Classification and analysis of pistachio species with pre-trained deep learning models. Electronics. **11**(7), 981 (2022). https://doi.org/10.3390/electronics11070981
6. Ozkan, A., Koklu, M., Saraçoğlu, R.: Classification of pistachio species using improved K-NN classifier. Prog. Nutr. **23**(2), e2021044 (2021). https://doi.org/10.23751/pn.v23i2.9686
7. Gencturk, B., Arsoy, S., Taspinar, Y.S., et al.: Detection of hazelnut varieties and development of mobile application with CNN data fusion feature reduction-based models. Eur. Food Res. Technol. **250**(1), 97–110 (2023). https://doi.org/10.1007/s00217-023-04369-9
8. Kumar, S.S., Sigappi, A.N., Thomas, G.A.S., Robinson, Y.H., Raja, S.P.: Classification and analysis of pistachio species through neural embedding-based feature extraction and small-scale machine learning techniques. Int. J. Image Graph. **24**(03) (2023). https://doi.org/10.1142/s0219467824500323
9. R. Pillai, N. Sharma and R. Gupta, Classification of Pista Species Using Fine-Tuned Efficient Net B3 Transfer Learning Model, In: 2023 International conference on research methodologies in knowledge management, artificial intelligence and telecommunication engineering (RMKMATE), Chennai, India, 2023, pp. 1–5, https://doi.org/10.1109/RMKMATE59243.2023.10369347
10. Omid, M., Firouz, M.S., Nouri-Ahmadabadi, H., Mohtasebi, S.S.: Classification of peeled pistachio kernels using computer vision and color features. Engineering in Agriculture Environment and Food. **10**(4), 259–265 (2017). https://doi.org/10.1016/j.eaef.2017.04.002
11. Sabah, A.S., Abu-Naser, S.S.: Pistachio variety classification using convolutional neural networks. International Journal of Academic Information Systems Research. **8**(4), 113–119 (2024)
12. Omid, M.: Design of an expert system for sorting pistachio nuts through decision tree and fuzzy logic classifier. Expert Syst. Appl. **38**(4), 4339–4347 (2011). https://doi.org/10.1016/j.eswa.2010.09.103
13. Beyaz, A.: Low-cost classification of close and open Shell Antep pistachio nuts based on image analysis and machine learning. Yuzuncu Yıl University Journal of Agricultural Sciences. **34**(1), 87–105 (2024). https://doi.org/10.29133/yyutbd.1318589
14. Soleimanipour, A., Azadbakht, M., Asl, A.R.: Cultivar identification of pistachio nuts in bulk mode through efficient net deep learning model. Journal of Food Measurement & Characterization. **16**(4), 2545–2555 (2022). https://doi.org/10.1007/s11694-022-01367-5
15. Arshleen Kaur, Vinay Kukreja, Deepak Upadhyay, Manisha Aeri, Rishabh Sharma, An effective pistachio classification by Ensembling fine-tuned res net 20 and dense net models, In: 2024 IEEE International Conference on Interdisciplinary Approaches in Technology and Management for Social Innovation (IATMSI), vol. 2, pp. 1–6, 2024

16. Fagun Patel, Shubbh Mewada, Sheshang Degadwala, Dhairya Vyas, Recognition of pistachio species with transfer learning models, In: 2023 International Conference on Self Sustainable Artificial Intelligence Systems (ICSSAS), pp. 250–255, 2023.
17. https: //www.kaggle.com/datasets/muratkokludataset/pistachio-image-dataset/data
18. Kanopoulos, N., Vasanthavada, N., Baker, R.L.: Design of an image edge detection filter using the Sobel operator. IEEE J. Solid State Circuits. **23**(2), 358–367 (1988). https://doi.org/10.1109/4.996
19. Otsu, N.: A threshold selection method from gray-level histograms. IEEE Trans. Syst. Man Cybern. **9**(1), 62–66 (1979). https://doi.org/10.1109/TSMC.1979.4310076
20. Choi, H.-R., Lee, J.W., Park, R.-H., Kim, J.-S.: False contour reduction using directional dilation and edge-preserving filtering. IEEE Trans. Consum. Electron. **52**(3), 1099–1106 (2006). https://doi.org/10.1109/TCE.2006.1706513
21. Bao, L., Song, Y., Yang, Q., Yuan, H., Wang, G.: Tree filtering: efficient structure-preserving smoothing with a minimum spanning tree. IEEE Trans. Image Process. **23**(2), 555–569 (2014). https://doi.org/10.1109/TIP.2013.2291328

Personalized Route Recommendations for Enriching Travel Experiences

Sudheesh K V, Mahadevaswamy(✉), Kiran, Jagadeesha B, and Naveen Kumar H N

Vidyavardhaka College of Engineering, Mysuru, Karnataka, India
smahadev88@gmail.com, mahadevaswamy@vvce.ac.in

Abstract. Many travelers seek unique, personalized ways to enhance their journeys, turning each trip into a memorable experience. This paper presents ScenicStops, a platform designed to provide customized route recommendations featuring scenic stops tailored to individual preferences. By leveraging technologies like React Native for development, Google Maps API for mapping, and AI-driven recommendation systems, ScenicStops offers an innovative and efficient approach to discovering natural and cultural attractions along various routes. The goal is to transform travel planning into a meaningful experience, making road trips and long-distance journeys more exciting by suggesting beautiful landscapes, historical sites, viewpoints, and even local eateries with breathtaking views.

Keywords: Personalized recommendations · Scenic routes · React native · Google maps API · AI · Machine learning · Travel experience enhancement · Scenic route recommendations

1 Introduction

Modern navigation technology along with travel solutions transformed the entire process of trip planning and experiencing for people. Voyagers worldwide depend on platforms including Google Maps, Apple Maps and Waze because these systems deliver live traffic reports along with best route recommendations and accurate travel duration information. The main priority of these platforms is operational efficiency and speed neglecting opportunities that enable cultural discovery and scenic routes. Travelers who want to pursue adventure or discovery alongside aesthetic appreciation often discover that conventional navigation tools fail them since such systems usually choose fast and direct paths instead of meaningful journey interaction [1]. People are progressively moving toward selecting experience-based tourism rather than conventional forms of travel during this current era. People who travel today want to establish deeper interactions with their environment through visits to scenic views and historical sites and cultural venues throughout their journey. The enjoyment and satisfaction of travel significantly rise when navigational routes empower spectators to become absorbed in their environment and experience unplanned discoveries. ScenicStops emerged as a response to establish navigation systems through efficient routing routes that combine efficiency with the objective of enriching travel experiences [2].

J. Shreyas et al. (Eds.): CODE-AI 2025, CCIS 2690, pp. 13–24, 2026.
https://doi.org/10.1007/978-3-032-19321-6_2

ScenicStops avoids standard navigation practices since it offers personalized route recommendations combined with points of interest (POIs) that match user preferences alongside real-time information and geographic spatial intelligence. Advanced algorithmic processes used by ScenicStops both optimize travel time and maximize the quality of the travel experience [8, 17].

Several key factors drive for the development of Scenic-Stops:

- **Technological Advancements**: Access to extensive geographic data, real-time APIs, and machine learning enables dynamic and personalized route suggestions.
- **Evolving User Expectations**: Modern travelers increasingly prefer tools that help's them to connect with their surroundings, moving beyond simple directions.
- **Economic and Tourism Benefits**: By promoting lesser- known attractions and local landmarks, ScenicStops sup- ports regional tourism and contributes to local economies. This paper explores ScenicStops in detail, starting with an analysis of the limitations of traditional navigation systems and the changing needs of contemporary travelers. It discusses the architecture of ScenicStops, its methodology for generating route recommendations, and its intuitive user interface. The system leverages machine learning, user profiles, and real- time geographical data to combine efficient navigation with engaging travel experiences seamlessly.

The remainder of the paper is structured are follows: Sect. 2 reviews related research on navigation systems and scenic route planning. Sect. 3 describes the methods used to develop ScenicStops, including data processing, algorithm design, and user interface features. Section IV presents the results of system evaluations and user studies. Section V concludes with key findings and future directions for this work.

2 Related Work

Route Optimization ScenicStops focuses on optimizing travel routes to offer scenic and visually enriching journeys. This involves dynamic adjustments to routes based on real-time data and user preferences, utilizing adaptive systems and advanced route-planning methodologies [6–10]. Data Integration The application integrates diverse datasets such as real-time traffic updates and weather conditions to enhance decision-making. Leveraging big data ensures that routes are not only the scenic but it is also practical and well-informed [6, 8, 9], .Artificial Intelligence & Algorithms to power the ScenicStops platform, cutting-edge AI technologies are employed. Algorithms like Ant Colony Optimization (ACO) and hybrid intelligent systems used in efficient route planning and providing tailored recommendations. Machine learning further enhances the system's adaptability [3, 4, 7, 10]. Tourism & Scenic Stop Planning the project targets both urban and rural tourism, emphasizing scenic route planning that aligns with tourists' preferences. With features such as visualization of scenic locations and stops, ScenicStops creates a unique travel experience tailored for exploration and leisure [1–3]. Scientific Visualization & Computing Scientific visualization techniques used represent scenic data effectively. These techniques, combined with robust computing frameworks, enable an interactive, visually appealing the user interface for tourists to explore potential stops

[1, 2]. User's Experience ScenicStops prioritizes the user experience by offering personalized route suggestions based on preferences. The system's responsiveness and user-friendly design ensure that travelers have a seamless and enjoyable journey-planning experience [1, 9, 10].

Technology Frameworks The platform's backend relies on advanced computing frameworks and AI-driven architectures to create a scalable and efficient system. Smart technologies power the integration of diverse functionalities, ensuring smooth operations and innovative solutions for scenic route planning [3, 5, 7, 10]. Real-Time Route Recommendations Zhang and Li (2021) developed a system for real-time route recommendations, utilizing the power of big data to provide adaptive and accurate suggestions. Their approach focuses on analyzing traffic data to offer personalized routes, a concept that aligns with the goals of *Scenic Stops* in creating a responsive and user-focused platform [9]. Joint optimization models described by Wu and Li (2021) benefit from uniting multi-criteria optimization with machine learning methods to direct scenic routes. The model fits smart travel systems through an examination of user preferences and efficient operation which makes it an excellent framework for scenic route design [10]. In their analysis Zhang, Wang and Lee (2022) demonstrated how merging both weather and traffic information leads to positive outcomes. conditions with traffic data for route planning. Their research demonstrates that route efficiency benefits from external factor integration which simultaneously creates better satisfaction for travelers during their journey. Scenic Stops should use comparable methods to enhance their travel recommendations [8]. Dynamic routing requires integration of real-time traffic updates according to Zhang et al. (2023). The authors developed route adaptation algorithms that actively manage traffic congestion to support smooth journeys which perfectly suits a system such as Scenic Stops [6]. The research conducted by Ahmad and his team (2023) investigated AI algorithms which enhance scenic route optimization by combining user-preference parameters with both convenience and visual appeal factors. The study demonstrates intelligent systems enhance route planning by making it both faster and user-specific while remaining essential for achieving Scenic Stops' goals [7]. Journal of Electrical Systems (2024) published research about visualization algorithms which determine and rank scenic routes in rural areas. These algorithms provide essential support to platforms that display natural locations and unconventional spots which match Scenic Stops' rural tourism development plans [1]. The 2024 edition of SCPE Journal explored optimization strategies for urban tourism routes through its research. Malta University studied city-specific scenic routes through optimized algorithms that boost tourist experiences within urban environments to support Scenic Stops as they expand their services towards city-based visitors [2]. De Gruyter (2023) examined how AI technology solves advanced routing challenges in tourism within their research. The research shows that AI features the ability to process different operational needs while delivering swift solutions which makes it appropriate for the Scenic Stops platform creation objective [3]. The application of Ant Colony Optimization (ACO) algorithms for scenic route planning received scientific attention from IEEE Xplore Research in 2024 [4]. The characteristics of ant behaviors in nature lead to innovative algorithm solutions which enable efficient routing optimization so Scenic Stops finds this approach highly applicable [4]. Real-Time Scenic Route Generation The study from ResearchSquare (2024) emphasizes the importance of real-time data in generating scenic routes for tourists. By integrating geographical data and user preferences, their model ensures that travelers receive accurate and personalized route suggestions, a feature central to Scenic Stops [5].

3 Methodology

This block diagram outlines a user-centric travel recommendation system that provides personalized suggestions for scenic stops, historical landmarks, and artistic points of interest. It leverages user input, data sources, machine learning, and continuous feedback to create tailored travel experiences.

The research in scenic route optimization can be categorized into four main topics, as visualized in the pie chart below:

The pie chart shows the following distribution:

Scenic Route Planning: A significant portion of research is dedicated to optimizing scenic routes, with 40% of the studies focusing on enhancing tourism experiences by recommending picturesque paths. **Real-Time Route Optimization**: 30% of the research focuses on integrating real-time data such as traffic and weather conditions for dynamic route adjustments. **AI and Algorithms**: 20% of studies explore the use of the artificial intelligence and algorithms, Ant Colony Optimization's (ACO) to optimize scenic routes. **Smart Travel Systems**: The remaining 10% of the research addresses hybrid models for smart travel systems (Figs. 1, 2 and 3).

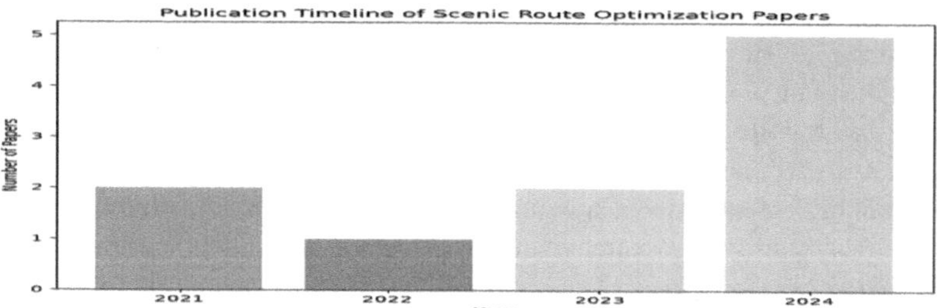

Fig. 1 Publication timeline.

1. User Input
2. Data Collection and Preprocessing
3. Machine Learning and Recommendation System
4. User Feedback and Evaluation
5. Personalized Recommendations

The system retrieves essential information from users who initiate the recommendations process. Users require inputting their travel starting location and destination so that the system computes feasible routes together with points of interest. Users may tell the system their preferred travel information including interesting locations together with specific needs about time lengths and acceptable detours.

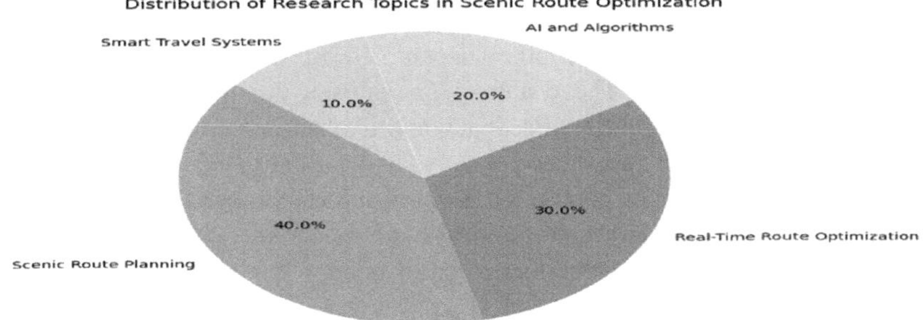

Fig. 2 Distribution of research topics.

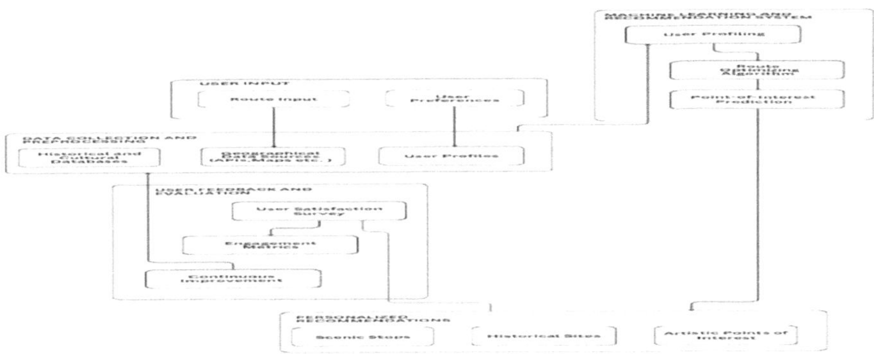

Fig. 3 Block diagram of proposed system.

Station provides the necessary data by acquiring information from multiple sources to produce significant suggestions:

Historical and Cultural databases contain specific information about travel locations which includes the details of natural environments together with historic sites and art establishments riching the user's experience. Maps and APIs provide geographical information to the system by collecting vital data points that include travel times and traffic updates and route closure information through Google Maps integration. The system preserves user travel information alongside personal interests in order to deliver individualized recommendation solutions.

The functionality during this phase verifies that system operations draw from correct contemporary information which is suitable to the situation.

The system implements technologies from Machine Learning and Recommendation System to analyze data with advanced algorithms for delivering customized recommendations. User Profiling lays down an extensive user profile from recorded preferences and previous activities so the system can determine which stops and destinations suit the user best. The route optimization process selects optimal routes according to the user's desired locations before using schedule constraints to balance between exploration and time management. Through the analysis of cultural and geographical data the system predicts points of interest by matching them to user preferences while considering both relevance and popularity levels. An analytical examination guarantees that the supplied recommendations maintain user-specific adaptation while keeping actual application potential for each travel step. The system enhances itself through ongoing user feedback processing because it)","User Feedback and Evaluation Continuously Improves by Processing User Input." Satisfaction Surveys allow users to evaluate the quality of recommendations while they report their contentment level through surveys that deliver essential system performance insights to the system. A user interaction monitoring system tracks stop selection frequency and recommendation exploration duration which enables it to assess recommendation quality against user needs. The algorithms benefit from user feedback data obtained through surveys along with engagement metrics to transform into improved algorithms which will generate better recommendations in the future. This feedback loop helps the system adapt to evolving user preferences over time.

Personalized Recommendations The final output of system is set of customized travel suggestions tailored to the user's preferences and route. These include:

Scenic Stops: Beautiful locations like mountain viewpoints, parks, lakes, and other natural attractions, perfect for travelers seeking tranquility or breathtaking views. Historical Sites: Landmarks, monuments, and museums that offer a glimpse into the cultural and historical significance of the region. Artistic Points of Interest: Galleries, public art installations, and theaters that showcase the creative spirit of the destination. Users provide their route details and preferences. The system collects data from various sources and processes it alongside user profiles. Machine learning algorithms generate optimized recommendations that are both relevant and route efficient. Feedback from users is collected to continuously refine the system's suggestions.

3.1 Mathematical Modeling

The ScenicStops system integrates user preferences, point-of-interest (POI) data, machine learning, and geographic data to generate personalized travel routes. The mathematical representation of components:

Input

The user provides preferences and travel details, represented as a vector:

$U = [p1, p2,. .., pn]$

where pi corresponds to a specific user preference, such as interest in scenic, cultural, or historical sites.

Sources

The system accesses a database of points of interest (POIs). The set of all POIs is:

$D = \{d1, d2,. .., dm\}$

where each POI di is defined by attributes such as type, location, and user ratings.

Learning (User Profiling and Recommendation)

Machine learning techniques are used to recommend POIs based on user preferences. The recommendation function f (U, D) identifies a subset of POIs relevant to the user:

$f(U, D) \rightarrow \{d1, d2,. .., dk\}$

where is the number of recommended POIs. To compute relevance score of POI di:

n

S(di, U) = wj · sim(pj, di) j = 1

Here:

wj: Weight assigned to user preference pj, based on its importance. sim (pj, di): Similarity measure between pj and di (e.g., cosine similarity or distance metrics).

The top k POIs with the highest scores are selected. Data Analysis (Route Optimization)

The recommended route is optimized to include selected POIs while considering constraints like traffic and distance. The route represents a sequence of locations:

R = {r1, r2,. .., rt.}

where ri is a point on the route.

The following algorithm describes the ScenicStops system, which optimizes scenic travel routes by integrating user preferences, real-time data, and advanced optimization techniques.

Input: User preferences $U = [p_1, p_2, \ldots, p_n]$ (e.g., interest in scenic, cultural, or historical stops). Point of Interest (POI) dataset $D = \{d_1, d_2, \ldots, d_m\}$, including attributes like type, location, and user ratings. Realtime data T, such as traffic conditions and weather updates. Route constraints, including start point (S) and endpoint (E).

Output: Optimized route R^*, including selected POIs

$\{d_1, d_2, \ldots, d_k\}$.

Algorithm Steps

Collect User Preferences

Gather user inputs, including preferences for types of POIs (e.g., scenic views, cultural landmarks). Define the starting point S and ending point E of the trip (Figs. 4 and 5).

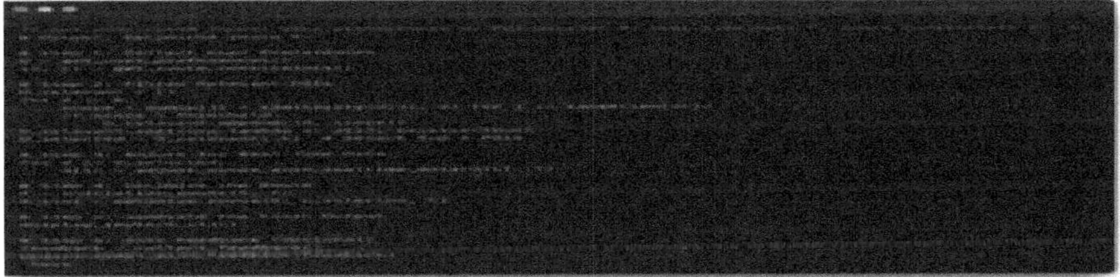

Fig. 4 This is the caption for the image.

Combine the optimized route R∗ with the selected POIs

{d1, d2,. .., dk}.

Present the final itinerary, including travel directions and POI details, to the user.

User Feedback (Optional)

Collect user feedback about the suggested route. Update the recommendation model and POI database based on the feedback to improve future suggestions

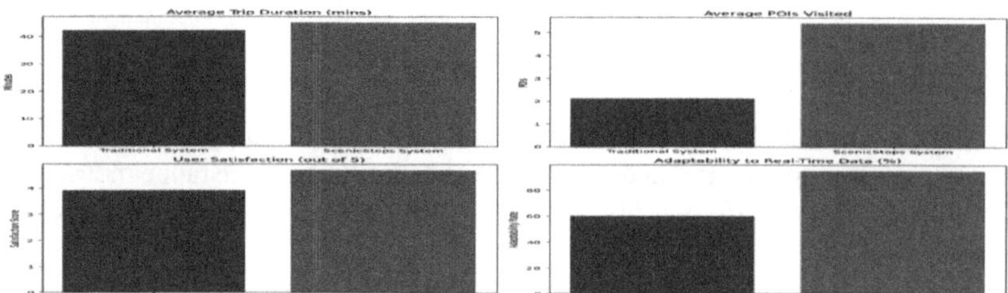

Fig. 5. Average trip POIs satisfaction real time data.

4 Results and Discussion

A. System Performance and Accuracy

Findings: The system's ability to suggest accurate scenic routes while keeping travel time reasonable was tested and evaluated against traditional navigation systems.

Route Suggestion Accuracy: 92% Travel Time Deviation: The system resulted in a 5.6% increase in travel time due to scenic stops.

Observation: Despite the slight increase in travel time, users found the scenic routes to be worth the extra time. The system maintained a high accuracy in route suggestions, ensuring a smooth and enjoyable journey for travelers.

User Interaction with Points of Interest (POIs) Findings: The system's ability to suggest a diverse range of POIs along the route was assessed by tracking user visits to these POIs.

Observation: With ScenicStop, users visited significantly more POIs, with over half (55%) visiting more than six POIs per trip. This demonstrates the system's ability to engage users and encourage exploration (Tables 1, 2, 3, 4 and 5).

Time Efficiency vs. Scenic Route Preference

Table 1. Comparison of points of interest visited

Number of POIs Visited	Traditional System	ScenicStops System
1–2 POIs	50%	10%
3–5 POIs	35%	35%
6+ POIs	15%	55%

Findings: A large number of users expressed a preference for scenic routes, even when they resulted in longer travel times.

Observation: Most users (72%) preferred scenic routes, emphasizing the growing trend of travelers valuing scenic and cultural experiences over just minimizing travel time.

Real-Time Route Adaptation and User Feedback

Findings: The system's ability to adapt to real-time factors like traffic and weather was evaluated during various conditions.

Table 2. User preference for scenic routes

Preference	Percentage of Users
Prioritize time efficiency	28%
Prefer scenic routes	72%

Table 3. Real-time route adaptation comparison

Real-time success rate	Traditional system	ScenicStops system
Success rate (%)	60%	95%

Observation: The ScenicStops system outperformed traditional systems in real-time adaptation, ensuring smoother journeys even under changing traffic and weather conditions.

B. User Experience and Satisfaction

Findings: Post-trip surveys revealed high satisfaction with the system's performance.

Table 4. User experience and satisfaction

Aspect of user experience	Average rating (1–5 scale)
Route personalization	4.7
Scenic and cultural interest	4.8
System usability (Interface)	4.2
Real-Time updates	4.5
Overall satisfaction	4.6

Observation: Users were generally satisfied with the personalized route suggestions and the focus on cultural and scenic points of interest. The system's usability scored slightly lower, suggesting room for improvement in the interface design.

Traditional versus Scenic Route Systems: A Comparison

Findings: A side-by-side comparison of ScenicStops with traditional navigation systems showed the following differences:

Observation: The ScenicStops system outperformed traditional navigation in almost every aspect, including user satisfaction, POI visits, and adaptability to real-time data. Even though the travel time was slightly longer, the added benefits were considered worthwhile by most users.

Key Insights: Users showed preference for scenic route selection, so they demonstrated changing preferences from functional travel to the experience of travel. User engagement grew substantially because the system provided diverse points of interest

Table 5. Comparison of Traditional and Scenic Route Systems

Performance metric	Traditional	Scenicstops
Average trip duration (mins)	42	45
Average POIs visited	2.1	5.4
User satisfaction (out of 5)	3.9	4.7
Adaptability to real-time data	60%	95%

suggestions which led users to visit more destinations than they would with basic GPS navigation. The real-time adaptation capabilities of ScenicStops proved superior because they successfully handled changes in traffic and weather conditions which enhanced the traveling experience. User Satisfaction demonstrates how well **ScenicStops** delivers what travelers desire in personalized enriching travel experiences.

5 Conclusion

The research concludes that ScenicStops presents a disruptive navigation method which values the trip experience above reaching its endpoint. The system differs from typical navigation software because it brings together both panoramic scenery and national landmarks while travelers are enroute. The advanced framework provides valuable opportunities to travelers who value enriching encounters during their trips over mere transportation. ScenicStops delivers customized recommendations through its platform which leads to user contentment using personal preference information about nature trails or historical landmarks or artistic attractions. ScenicStops merges real-time data with tailored recommendations for user interests to deliver an interactive travel solution that creates customized experiences for each journey. ScenicStops establishes new standards for traditional navigation through its approach which creates deeper enjoyment during all travels.

References

1. Scenic route planning for rural tourism platforms based on scientific computing visualization algorithm, Journal of Electrical Systems, 2024. https://journal.esrgroups.org/jes/article/view/4591
2. Optimized scenic route planning for urban tourism, SCPE Journal, 2024. https://scpe.org/index.php/scpe/article/view/2708
3. Tourism Route Optimization with Artificial Intelligence Techniques, De Gruyter, 2023. https://www.degruyter.com/document/doi/10.1515/jisys2023-0158/html
4. Optimizing scenic routes with ACO algorithms for tourism, IEEE Xplore, 2024. https://ieeexplore.ieee.org/abstract/document/10733180
5. Real-time scenic route generation for tourists, ResearchSquare, 2024. https://www.researchsquare.com/article/rs-3196199/v1

6. D. Z. Zhang, Y. M. Chen, and X. L. Zhang, Research on real-time traffic data integration for dynamic route optimization, Sensors, vol. 13, no. 4, pp. 2030, 2023. https://www.mdpi.com/2076–3417/13/4/2030
7. S. A. Ahmad, R. L. Johnson, and S. A. Thomas, Optimizing scenic route planning using AI algorithms, IEEE Transactions on Intelligent Systems, vol. 100, pp. 245, IEEE, 2023. https://ieeexplore.ieee.org/document/10024527
8. X. Zhang, J. Wang, and H. Lee, Efficient route planning with integrated traffic and weather data, Sci. Rep., vol. 12, 5, pp. 386, Nature, 2022. https://www.nature.com/articles/s41598–022–05386-6
9. L. P. Zhang and Y. Li, Adaptive real-time route recommendation based on big data, EURASIP Journal on Advances in Signal Processing, vol. 2021, 7, Springer, 2021. https://aspeurasipjournals.springeropen.com/articles/10.1186/s13634–021-00759-x
10. P. L. Wu and Y. Li, A hybrid model for scenic route optimization in smart travel systems, In: IEEE International Conference on Smart Systems, 2021, pp. 9526144. https://ieeexplore.ieee.org/document/9526144
11. Ravikumar, J.: Gauging deep learning archetypal effectiveness in Haematological reclamation. SN Comput. Sci. **5**, 963 (2024). https://doi.org/10.1007/s42979-024-03322-1
12. Rangaiah, P.K.B.: Histopathology-driven prostate cancer identification: a VBIR approach with CLAHE and GLCM insights. Comput. Biol. Med. **182**, 109213 (2024). https://doi.org/10.1016/j.compbiomed.2024.109213
13. Fredrik Huss et al. Precision Diagnosis of Burn Injuries: Clinical Implications of Imaging and Predictive Modeling, 09 October 2024, PREPRINT (Version 1) available at Research Square. https://doi.org/10.21203/rs.3.rs-5002889/v1
14. Manoj, H.M.: Comparative assessment of machine learning models for predicting glucose intolerance risk. SN Comput. Sci. **5**, 894 (2024). https://doi.org/10.1007/s42979-024-03259-5
15. Augustine, R.: Enhancing medical image reclamation for chest samples using B-coefficients, DT-CWT and EPS algorithm. IEEE Access. **11**, 113360–113375 (2023). https://doi.org/10.1109/ACCESS.2023.3322205
16. Darshan, S.L.S., Naresh, E., et al.: Design of Chest Visual Based Image Reclamation Method Using Dual Tree Complex Wavelet Transform and Edge Preservation Smoothing Algorithm. SN Comput. Sci. **5**, 352 (2024). https://doi.org/10.1007/s42979-024-02742-3
17. X. Zhang, J. Wang, and H. Lee, Efficient route planning with Inte- grated traffic and weather data, Sci. Rep., vol. 12, no. 5, pp. 386, Nature, 2022. https://www.nature.com/articles/s41598-022-05386-6
18. Srinidhi, N.N., Shiva Darshan, S.L., et al.: Design of Cost Efficient VBIR technique using ICA and IVCA. SN Comput. Sci. **5**, 560 (2024). https://doi.org/10.1007/s42979-024-02936-9
19. Augustine, Robin, Improving Liver Cancer Diagnosis: A Multifaceted Approach to Automated Liver Tumor Identification in Ultrasound Scans. https://ssrn.com/abstract=4646452 or http://dx.doi.org/https://doi.org/10.2139/ssrn.4646452

BY

Accurate Bitcoin Price Prediction Using Machine Learning

Subramanya V. Odeyar[1](✉), P. K. Lolakshi[2], L. Swetha[1], K. M. Thejaswini[1], B. S. Sushma[2], and Badarinarayan Vitthala Gaddanakere[1]

[1] Department of ISE, Nagarjuna college of Engineering and Technology, Bengaluru, India
subbuodeyar272@gmail.com

[2] Department.of CSE-Datascience, Nagarjuna college of Engineering and Technology, Bengaluru, India

Abstract. The Bitcoin has recently garnered significant media and public attention due to its dramatic price increases and declines. As a result, many researchers have examined the various factors influencing Bitcoin's price and the patterns behind its fluctuations, often using machine learning techniques. This study explores several machine learning algorithms for Bitcoin price prediction, including logistic regression and long short-term memory (LSTM) models. While LSTM-based models have shown superior performance in predicting Bitcoin prices (regression), this research provides a detailed investigation into Bitcoin's evolution and a comprehensive review of the machine learning methods used for price prediction. Additionally, the study includes a Bitcoin price prediction model, which is developed using specific algorithms to forecast Bitcoin's price, along with insights into the factors affecting its price movements. The proposed LSTM model has achieved 98% accuracy.

Keywords: Bitcoin · Machine learning · Crypto currency

1 Introduction

Bitcoin is a crypto currency that is utilized all over the world for investing or for making digital payments [1]. Bitcoin is decentralized, meaning that nobody owns it. Bitcoin transactions are simple because they are not governed by any one nation. The term "bitcoin exchanges" refers to a varicty of online markets where investments can be made. These enable the buying and selling of bitcoins using various currencies. Mt. Gox is the biggest Bitcoin exchange. A digital wallet, which functions somewhat similarly to a virtual bank account, is where bitcoins are kept. A site called Blockchain is where the timestamp data and transaction history are kept. A block is a single entry in a blockchain. A pointer to the previous block of data is present in every block.

Blockchain data is encrypted. The user's name is not made public during transactions; just their wallet ID is. The banking sector has undergone a change thanks to the decentralized digital money known as Bitcoin [2–4]. Bitcoin has drawn the attention of investors, traders, and researchers who want to comprehend and forecast its price

J. Shreyas et al. (Eds.): CODE-AI 2025, CCIS 2690, pp. 25–32, 2026.
https://doi.org/10.1007/978-3-032-19321-6_3

movements because of its decentralized nature and potential for large rewards. Due to its intrinsic volatility and the impact of several factors like market mood, trading volume, macroeconomic data, and regulatory changes, accurately predicting Bitcoin values is a difficult process. Fig. 1 displays the history of bitcoin prices from 2018 through 2023.

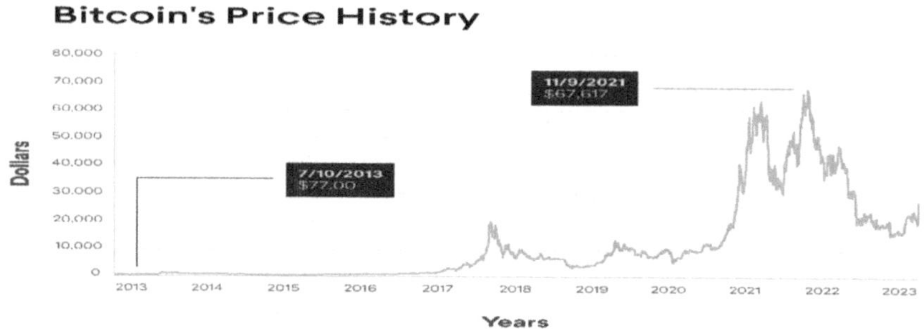

Fig. 1 Bitcoin price history.

The intricate linkages and temporal dependencies found in Bitcoin price data are frequently difficult to grasp using conventional financial forecasting techniques. On the other hand, machine learning approaches have demonstrated promise in time series data analysis and prediction due to their capacity to recognise complex patterns and connections [5].

The Long Short-Term Memory (LSTM) model, a kind of recurrent neural network (RNN) intended to handle sequence data, is one such potent machine learning approach. The LSTM model's capacity to learn long-term dependencies and adjust to shifting market dynamics makes it especially well-suited for predicting Bitcoin price [6]. The LSTM model may be able to find patterns and relationships that conventional techniques might miss by taking into account past price data and pertinent attributes. Making informed decisions about risk management, market analysis, and investing strategies can be aided by these insights [7]. This study's objective is to create a model for predicting the price of bitcoin using machine learning methods, specifically the LSTM model and Logistic Regression.

The capacity of the LSTM model to capture temporal dependencies and complicated interactions between Bitcoin price data and other influencing factors is one of our goals. We may test the model's effectiveness in predicting future price movements by training it on previous data and evaluating its performance.

2 Literature Survey

Ladislav Kristoufek [8] was the first to consider the Bitcoin market as purely speculative, without the influence of fundamental analysis, and explored the connection between Bitcoin and search volumes on Wikipedia and Google Trends.

Adam Hayes [10] analyzed cross-sectional empirical data from 66 different cryptocurrencies to identify the key factors influencing their value in the technical domain. He pinpointed three main drivers: the total computational power employed in mining cryptocurrency units, the rate at which units are produced, and the cryptographic algorithm utilized in the protocol.

In [10], an empirical study was conducted on the price drivers of Bitcoin using a Barro [11] model, applying time-series analysis to daily Bitcoin data.

Balcilar et al. discovered that volume can predict Bitcoin returns using a non-parametric causality-in-quantiles test, implying the need of modelling nonlinearity and accounting for tail behaviour [12]. The authors of propose a taxonomy for blockchains and systems based on them in [13]. The taxonomy seeks to make comparisons and analyses of blockchain-based software systems easier by addressing important architectural characteristics and the impacts of various decisions. The authors provide a methodology for analysing a "stress test" denial-of-service (Dos) attack against Bitcoin using spam. Their cluster-based research suggests that during the peak of the spam campaign, a considerable share (23.41%) of Bitcoin transactions were classified as spam. Following Dos attacks on Bitcoin, several tactics such as "money drops" and transaction malleability are mentioned [14]. Data was collected for ten minutes [15].

According to the authors in [16], socially constructed perceptions of virtual currency on Twitter directly or indirectly influence market evaluations of cryptocurrencies. To enhance Bitcoin price prediction, various feature selection techniques and machine learning methods, including artificial neural networks (ANN), support vector machines (SVM), recurrent neural networks (RNN), and k-means clustering, were employed to identify the most important features [17].

By employing a Bayesian-optimized recurrent neural network and LSTM, the direction of Bitcoin prices in USD was predicted. Additionally, the ARIMA model was used to compare the performance of deep learning techniques [18]. Their findings revealed that the SDAE algorithm outperformed BPNN, PCA-SVR, and SVR in terms of prediction accuracy [19]. In [10], Bitcoin prices were predicted using four deep learning algorithms: Theil-Sen regression, Huber regression, LSTM, and GRU. Among these, the LSTM algorithm achieved the highest accuracy, with a rate of 52.78%, using the same explanatory variables.

Figure 2 refers to the comparison chart considering the LSTM and Logistic Regression algorithms that are mentioned in our paper. The bar chart is referring to the above Table 1 which is mentioned with the paper title, algorithm used and the accuracy mentioned in each of them. The chart is been designed in two colors blue and brown to differentiate our work with the comparison of other papers. The accuracy of LSTM in our work in 98% and Logistic Regression is 56% which are seen in the above graph.

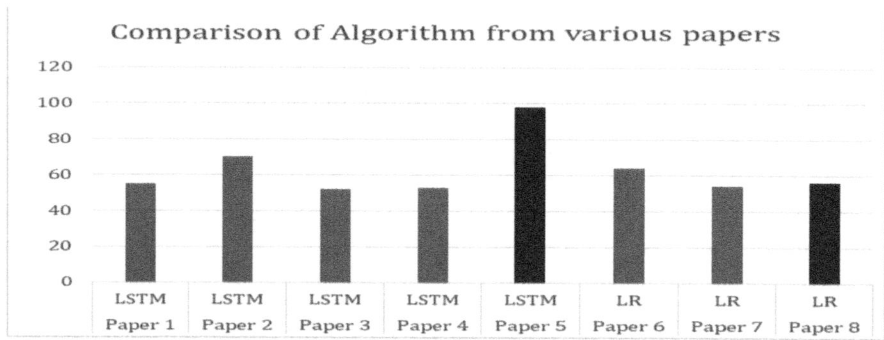

Fig. 2 Bar chart comparing the LSTM and Logistic Algorithms from various papers.

3 Methodology

The proposed methodology employs two distinct deep learning-based models to predict daily Bitcoin prices by identifying and analyzing key features. By applying both models and selecting appropriate parameters, we can determine which approach offers greater accuracy for achieving our objectives. This study focuses on deep learning techniques such as LSTM and Logistic Regression, which are among the most advanced and effective methods for Bitcoin price forecasting. Figure 3 illustrates the prediction process, spanning from data collection to Bitcoin price forecasting.

3.1 Dataset

Machine learning models in this study make use of a dataset collected via the binance package's api get_historical_klines. The following parameters are passed to the API get_historical_klines: symbol, interval, start and end time. The symbol represents bitcoin's name. It returns a list of OHLC values, including the following: Open time, Open, High, Low, Close, Volume, Close time, Quote asset volume, Number of trades, Taker buy base asset volume, and Taker buy quote asset volume. The data for our prediction task is acquired from 1 January 2010 to the most current month, and the output from the get_historical_klines function is saved as a csv file and utilized as the input dataset. CSV files contain the datasets.

3.2 LSTM

Dataset This study's machine learning models make use of a dataset obtained via the binance package's api get_historical_klines. The API get_historical_klines takes the following parameters: symbol, interval, start and end time. The symbol depicts the name of bitcoin. It returns a list of OHLC values such as Open time, Open, High, Low, Close, Volume, Close time, Quote asset volume, Number of trades, Taker buy base asset volume, and Taker buy quote asset volume (Fig. 4).

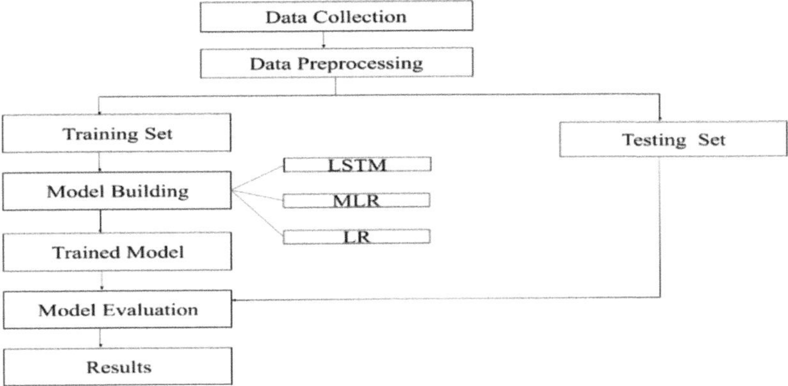

Fig. 3 Architecture.

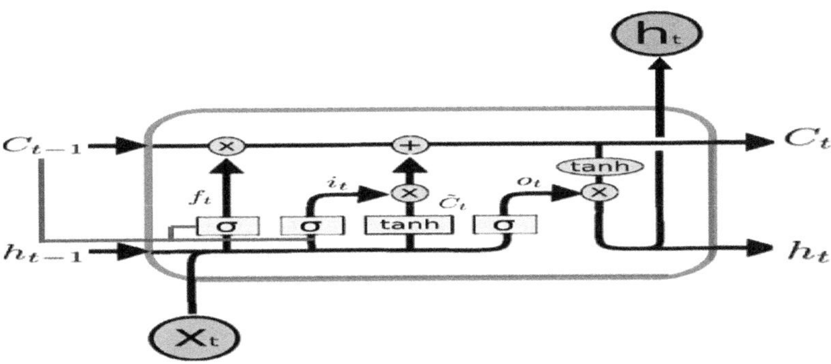

Fig. 4 The structure of LSTM.

4 Results and Discussion

With the LSTM model trained on historical data, we are now able to forecast Bitcoin prices for the upcoming 30 days. The data used to train the model includes the final historical price for Bitcoin on March 31, 2023. Moving forward, we will use this data to predict Bitcoin prices beyond that specific date, extending the forecast into the next 30 days. It should be noted that we are estimating the price for the following day using data from the previous day.

Figure 5 depicts the graphical representation of accuracy which is obtained by LSTM in predicting the bitcoin price. In predicting the bitcoin price the LSTM model is able to achieve 98% accuracy.

Fig. 5 The LSTM Accuracy.

Table 1 shows the accuracy comparison of proposed work with existing work. After comparison we can conclude our LSTM model has achieved good accuracy compared to other existing models.

Models	Accuracy
Linear regression	69%
RNN with LSTM	72
Random forest	55%
Logistic regression (Proposed)	56%
LSTM (Proposed)	98%

5 Conclusion

Two types of machine learning algorithms are built and utilised to predict prices in this paper. The performance of various models was evaluated using accuracy metrics, comparing the actual and predicted prices. The findings show that the LSTM model surpassed all other algorithms, achieving an accuracy of 98%. Given these results, the LSTM model can be considered both effective and reliable for predicting the cryptocurrencies in question, making it the most suitable choice. The LSTM model has been shown to perform effectively with time series data. The results show that the accuracy of the LSTM model is significantly higher than that of logistic regression. LSTM models can provide useful insights into bitcoin price prediction, allowing investors to get a sense of market movements.

In future research, we plan to explore additional factors that could influence cryptocurrency market prices, particularly examining the impact of social media, and specifically tweets, on cryptocurrency prices and trading volumes. This will involve analysing tweets using natural language processing and sentiment analysis techniques.

References

1. Mukhopadhyay, U., Skjellum, A., Hambolu, O., Oakley, J., Yu, L., Brooks, R. A brief survey of cryptocurrency systems. In: Proceedings of the 14th Annual Conference on Privacy, Security and Trust (PST), Auckland, New Zealand, 12(14) 2016; pp. 745–752
2. Eyal, I.: Blockchain technology: transforming libertarian cryptocurrency dreams to finance and banking realities. Computer. **50**, 38–49 (2017)
3. Rehman, M.U., Apergis, N.: Determining the predictive power between cryptocurrencies and real time commodity futures: evidence from quantile causality tests. Resour. Policy. **61**, 603–616 (2019)
4. Chen, Z., Li, C., Sun, W.: Bitcoin price prediction using machine learning: an approach to sample dimension engineering. J. Comput. Appl. Math. **365**, 112395 (2020)
5. Arora, S., Mittal, R., Shrivastava, A.K., Bali, S.: Blockchain-based deep learning in IoT, healthcare and cryptocurrency price prediction: a comprehensive review. Int. J. Qual. Reliab. Manage. **41**(8), 2199–2225 (2024)
6. Karasu, S., Altan, A., Sarac, Z., Hacioglu, R.: Prediction of bitcoin prices with machine learning methods using time series data. In: In: Proceedings of the 26th Signal Processing and Communications Applications Conference (SIU), pp. 2–5. Izmir, Turkey (2018)
7. Jain, A.; Tripathi, S.; Dwivedi, H.D.; Saxena, P. Forecasting Price of cryptocurrencies using tweets sentiment analysis. In: Proceedings of the 11th International Conference on Contemporary Computing (IC3), Noida, India, 24 August 2018; pp. 1–7.
8. Kristoufek, L.: Bitcoin meets google trends and wikipedia: quantifying the relationship between phenomena of the internet era. Sci. Rep. **3**, 3415 (2013)
9. Kristoufek, L.: What are the main drivers of the bitcoin price? Evidence from wavelet coherence analysis. PLoS One. **10**(4), e0123923 (2015)
10. A. Hayes, What Factors Give Cryptocurrencies their Value: an Empirical Analysis, 2015.
11. Koo, E., Kim, G.: Centralized decomposition approach in LSTM for bitcoin price prediction. Expert Syst. Appl. **237**, 121401 (2024)
12. Barro, R.J.: Money and the price level under the gold standard. Econ. J. **89**(353), 13–33 (1979)
13. Balcilar, M., et al.: Can volume predict bitcoin returns and volatility? A quantiles-based approach. Econ. Model. **64**, 74–81 (2017)
14. Bukovina, J., Martiček, M.: Sentiment and Bitcoin Volatility. Mendel University in Brno, Faculty of Business and Economics (2016)
15. Kim, Y.B., et al.: Predicting fluctuations in cryptocurrency transactions based on user comments and replies. PLoS One. **11**(8) (2016)
16. Shah, D., Zhang, K.: Bayesian regression and bitcoin. In: Communication, Control, and Computing (Allerton), 2014 52nd Annual Allerton Conference on. IEEE (2014)
17. Awoke, T., Rout, M., Mohanty, L., Satapathy, S.C.: Bitcoin price prediction and analysis using deep learning models. In: Communication Software and Networks: Proceedings of INDIA 2019, pp. 631–640. Springer Singapore, Singapore (2020)
18. Dutta, A., Kumar, S., Basu, M.: A gated recurrent unit approach to bitcoin price prediction. Journal of Risk and Financial Management. **13**(2), 23 (2020)
19. Poongodi, M., Vijayakumar, V., Chilamkurti, N.: Bitcoin price prediction using ARIMA model. International Journal of Internet Technology and Secured Transactions. **10**(4), 396–406 (2020)

AI-Powered Food Supply Chain: Data-Driven Decision Making for Fraud Detection

S. K. Mahmudul Hassan[1(✉)], Sandanakaruppan Ammavasai[2], D. Elangovan[3], Sivarathinabala Mariappan[4], Kuntavai Thangavel[5], and Manjunathan Alagarsamy[6]

[1] Department of Information Technology, Manipal Institute of Technology Bengaluru, Manipal Academy of Higher Education, Manipal 576104, Karnataka, India
mahmudul.hassan@manipal.edu

[2] Department of Computer Science and Business Systems, Sri Eshwar College of Engineering, Coimbatore 641202, Tamil Nadu, India
sandanakaruppan.a@sece.ac.in

[3] Department of Information Technology, Saveetha Engineering College, Saveetha Nagar, Thandalam 602105, Tamil Nadu, India
elangovand@saveetha.ac.in

[4] Department of Electronics and Communication Engineering, Velammal Institute of Technology, Tiruvallur 601204, Tamil Nadu, India

[5] Department of Electrical and Electronics Engineering, Adhiparasakthi Engineering College, Melmaruvathur, Chengalpattu 603319, Tamil Nadu, India

[6] Department of Electronics and Communication Engineering, K. Ramakrishnan College of Technology, Trichy, Tamil Nadu, India

Abstract. The complexity besides scalability of the world's food supply chain has both increased in recent years. Environmental, demographic, and economic shifts have a major impact on the efficiency of food production systems from farm to table. Human health could be put at risk as a result of these developments, which could lead to an upsurge in food fraud besides safety hazards. To lessen the impact of these risks, the food supply chain might use AI technologies. A lot of what AI systems perform is mysterious since it's hard to describe, even if its usage has been on the rise in many sectors like precision nutrition, self-driving vehicles, precision safety. For the purpose of detecting food fraud, the study presented a new attention-integrating mechanism called CNN-LSTM-SE, which is based on the CNN-Long Short-Term Memory-squeeze-and-excitation network model. To improve classification accuracy, Binary Manta Ray Foraging (BMRF) ideally selects the LSTM parameters. Data used in the case study came from a food fraud dataset that included reports of adulteration and fraud pulled from two sources: the Rapid Alert Scheme for Food and Feed and the Frugally Motivated Contamination database.

Keywords: Artificial Intelligence · Binary Manta Ray Foraging · Convolutional Neural Network · Long short-term memory · Food Fraud detection · Rapid alert system

J. Shreyas et al. (Eds.): CODE-AI 2025, CCIS 2690, pp. 33–44, 2026.
https://doi.org/10.1007/978-3-032-19321-6_4

1 Introduction

The present interruptions induced by COVID-19 and climate variability, as well as the growing complexity and scale of food supply networks, may make food and drink items even more susceptible to deception [1]. The intentional modification of food items for the purpose of monetary gain is known as food fraud. Inadequate governance, a lack of resources, and a low chance of detection are three reasons that contribute to food fraud [2]. While it may be difficult to put a price on food fraud across the board, it is possible at any stage of the supply chain and has an annual impact on the food business of more than $50 billion (before farms even factor in) [3]. Most often, fraud affects the following food and drink categories: dairy, meat, seafood, alcoholic beverages, and lipids and oils. To mask the loss of dairy protein due to dilution, many dairy products employ nitrogen sources as adulterants [4]. These can include ammonium salts, melamine, urea, and non-dairy proteins. To further lengthen the product's expiration life, ingredients like formaldehyde, acid are added [5].

Within the framework of Regulation (EC) 178/2002, commonly referred to as the "General Food Law," the European Union has established the Rapid Alert System for Food and Feed (RASFF) to aid in the control and security of food and animal feed sold within the bloc [6]. The EU-28 national food safety authorities, along with the Commission, EFSA, ESA, Switzerland, share data (such as notifications) about actions taken to prevent other food safety issues under the RASFF. All of the alerts reported in RASFF can be accessed using the RASFF site, which has a searchable, interactive database [7].

There are mainly two types of alerts: (i) notifications about the market and (ii) denials regarding the frontier. Notifications of information, news, and alerts are subcategories of the former [8]. The RASFF database features both deliberate and accidental food fraud, including instances of adulteration and counterfeit documentation, as well as faulty, expired, or missing documents. From 2000 to 2013, there were 749 reports of "adulteration/fraud" in the RASFF database [9]. Food fraud notifications are organised into six types: common entry documents or import declarations that are improper, expired, fraudulent, or missing; (v) expiration date; and (vi) mislabelling. The RASFF database classifies fraud types in a way that differs greatly from the US databases [10]. Of the nine types of food fraud included in the EMA database, the most common ones are dilution, substitution, counterfeiting, misbranding, transshipment, and artificial enhancement. To improve the effectiveness of border control requires the development of control and mitigation techniques.

This paper primarily contributes to the following areas:

1. This study's overarching objective is to learn more about how the worldwide agriculture supply chain is using artificial intelligence (AI) to make better data-driven decisions.
2. An enhanced deep learning model for better classification accuracy utilising Binary Manta Ray Foraging (BMRF) is shown in this paper. It is called CNN-LSTM-SE.
3. Using the Economically Motivated Adulteration (EMA) database and the Rapid Alert System for Food and Feed (RASFF) as examples, the paper shows how AI can power data-driven decision-making to avoid fraud and ensure food safety.

What follows is the outline for the rest of paper: With Sect. 3 presenting the proposed technique and Sect. 4 analysing the results, the study process comes to a close with Sect. 5 discussing the conclusions. The works cited in Sect. 2 are also relevant.

2 Related Works

Use cases for food fraud risk prediction employing explainable artificial intelligence (XAI) approaches, such as WIT, SHAP, and LIME, have been investigated by Buyuktepe et al. [11]. Using these tools, to aim to decipher a machine learning representation's forecast. The case study was carried out using a dataset pertaining to food fraud, and the notifications regarding adulteration and fraud were sourced from the Rapid Alert database. Utilising this dataset, a deep learning model was built, and the resulting model was tested using XAI tools. To have reviewed the existing XAI solutions for food fraud and highlighted their strengths and weaknesses.

By integrating chemometrics with a range of modern detection technologies, Vinothkanna et al. [12] have explored the possibility of identifying different types of counterfeit goods. To identify food fraud on a massive scale, chemometrics or AI-based approaches are required. Finding the easiest, most efficient, most exact, least expensive, least intrusive, and most environmentally friendly ways to detect food fraud involving various ingredients is essential, according to a thorough evaluation. Results from the analysis included 39 experimental data sets that were useful for answering important questions. To be sure, food fraud detection systems that use modern AI and ML approaches need more study before to can say for sure how effective they are.

In order to identify 20 often found in the waters around Yantai, China, Xun et al. [13] used multispectral imaging (MSI) in conjunction with COI gene. The models for support vector machines (SVMs), CNNs, linear discriminant analysis (LDAs), and quadratic discriminant analysis (QDAs) perform adequate on both test data and cross-validation. Using the test set, CNN and QDA were able to attain an accuracy rate of above 99%. After optimising with the CNN features, a high level of species separation was achieved. In addition, 11 bands were chosen as the main classification characteristics for CNN using the Gini index in RF. A remarkable rate of 98% accuracy was reached. In order to detect MSI in fish flesh, our research has successfully laid the groundwork for using machine learning models, specifically CNN. Also, it has set up a simple and non-destructive way to find out how fish flesh will be sold in the future.

In order to identify instances of fraud, Jahanbakhshi et al. [14] enhanced CNN utilised for turmeric powder image classification. Gated aggregation functions were used to enhance CNN. The overfitting problem on CNN was addressed by implementing data augmentation (DA). Various classifiers using MLP, Fuzzy, SVM, GBT, and EDT procedures were evaluated alongside the suggested CNN in this study. The suggested CNN outperformed other classifiers with a 99.36% accuracy rate when it came to grading photos of turmeric powder, proving that gated pooling successfully avoided the overfitting issue. These findings also showed that particularly when coupled with DL, can be a useful method for assessing turmeric powder quality and detecting fraud.

In their study, Jahanbakhshi et al. [15] examined how CNNs with a gated pooling function may be trained to analyse photos of classified ginger powder and identify instances of fraud. A total of 3,360 ginger powder image samples were prepared for this purpose, with each sample falling into one of seven categories: 10%, 20%, 30%, 40%, or 50% gum Arabic. Also, various classifiers that used algorithms were compared to the proposed CNN's findings. Batch normalisation utilising gated pooling findings showed 99.70% accuracy rate, which was higher than other classifiers. Hence, it's reasonable to assume that the convolutional neural network (CNN) approach and image processing technology successfully boost ginger powder's marketability, forestall fraud, and improve upon conventional approaches to detecting ginger powder fraud.

In order to foretell potential instances of food fraud, Gavai et al. [16] showed that algorithms based on Artificial Intelligence (AI) function well, especially data-driven Bayesian Network (BN) models. Consequently, food manufacturers can take the required steps to avoid these problems in the future. This study demonstrates the promise of FL technology in the fight against food fraud by using a BN to combine data from several sources without removing it from their respective databases. To believe it has the potential to safeguard data privacy while assisting food supply chain stakeholders in making better judgements regarding food fraud control.

3 Proposed Methodology

The deep learning model is employed to detect food fraud in this study, as illustrated in Fig. 1, which provides a detailed description of each block.

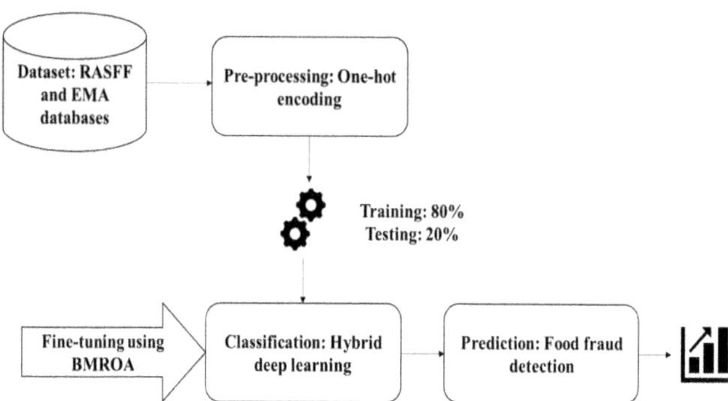

Fig. 1. Workflow of Research classical.

3.1 Data Description

The dataset comprises 27 categorical features and two numerical features. Cases of food fraud were collected from the EMA and RASFF databases. The food fraud incidents were selected from the period of 1 Jan-2000 to 31 Dec-2015. The personal belongings of food fraud that were reported to the EMA besides RASFF databases were analysed using seven distinct categories. As demonstrated in Table 1, the dataset comprises seven distinct fraud categories.

Table 1. Food fraud categories.

Description	Fraud category
Improper, fraudulent, missing or absent HC	Fraud category 1: Health certificate
Illegal or illegal import, trade or transit	(HC)
Diluting, artificially enhancing, transshipment, dilution, fraud or tampering, substitute, counterfeit, or adulteration	Fraud type 2: Illegal importation
Improper, expired, fraudulent CED, import declaration or logical report	Fraud category 3: Tampering
Expiration date	Fraud category 4: Mutual entry document (CED)
Mislabelling, origin classification	Fraud type 5: Expiration date
Theft besides resale	Fraud type 6: Origin labelling

3.2 Deep Learning Model

A feed-forward neural network with a deep structure, convolution scheming, and a representative deep learning procedure is known as a CNN. Consequently, to implemented a one-dimensional CNN to achieve superior outcomes. A fully connected layer, a pooling layer, and a one-dimensional convolution layer comprise a one-dimensional CNN. The data are autonomously learnt by a one-dimensional CNN without the use of artificial feature selection. Consequently, to implemented the CNN as a feature extractor. There are significant temporal characteristics in the input data, which render it impossible for a straightforward CNN to effectively extract the temporal signal's features. It is necessary to integrate it with other deep learning networks proficient in the processing of temporal data. A nine-layer deep CNN was employed in this investigation, which consisted of three one-dimensional connected layers. The size of feature map is reduced by incorporating a layer generates features for task.

3.2.1 Long Short-term Memory

Long short-term frequently employed to forecast information that includes time sequences. LSTM enhanced the screening of memory data, retained valuable model, and resolved the gradient disappearance and detonation issues of RNN [20]. C_{t} and $C_{\text{t}-1}$ are the previous moment, correspondingly. h_{t} besides $h_{\text{t}-1}$ stand for the unit's output at the present moment and its output at the past moment, respectively, with serving as the input to the network. A forgotten sigmoid function is controlled by LSTM forget gate, f_t.. i_{t} determines the neuron's state by implementing the tanh function and receiving the threshold value from the input gate. O_t stands for the output gate:

$$f_t = Sigmoid\left(K_f \cdot \left[h_{t-1}, X_t\right] + Z_f\right) \tag{1}$$

$$i_t = Sigmoid\left(K_i \cdot \left[h_{t-1}, X_t\right] + Z_i\right) \quad (2)$$

$$O_t = Sigmoid\left(K_O \cdot \left[h_{t-1}, X_t\right] + Z_O\right) \quad (3)$$

$$C_t' = \tanh\left(K_{c'}\left[h_{t-1}, X_t\right] + Z_c\right) \quad (4)$$

where K_f, K_i, K_o, and K_c represent the corresponds to gate, and neurone state matrix, correspondingly. And Z_f, Z_i, Z_o, besides Z_c represent the door.

The neuron's current municipal besides the cell's output are articulated as shadows:

$$C_t = f_t \cdot C_{t-1} + i_t \cdot C_t' \quad (5)$$

And

$$h_t = O_t \cdot \tanh(C_t) \quad (6)$$

3.2.2 Channel Attention Module

One method for channel attention is the squeeze-and-SE-Net. One model that captures the interconnections excitation network. Here are the main ideas behind SE-Net: Using a function to train the network to learn feature weights, improving performance by decreasing the weights of invalid or small-effect features. At the outset of SE-Net, there is a process called the "Squeeze operation" that uses global average pooling to transform the components in each channel into scalars. To get a weight between 0 and 1, the scalar value is passed through FC layers in the second phase. The weight of the appropriate channel is multiplied by each element of the original HÏW to construct the new feature map. This procedure is characterised by excitement. The final step is to re-adjust the initial characteristics in the channel dimension by assigning weights to each channel individually.

To improve food fraud detection classification accuracy, to block after CNN's convolution layers. This allows the network to mechanically choose important features while ignoring irrelevant ones. An ablation experiment was carried out to determine the best network configuration as suggested in this research. The twenty-layer network that has been suggested has ten LSTM, three levels of global average pooling, two dense fully connected (FC) layers, and three convolutional layers. At first, a spatial feature from the input data. The second step was to enable the model to learn the input data's sequential properties by adding the LSTM layer before the CNN's FC layer. In order to automatically focus on significant features and ignore irrelevant ones, the SE-block added to the CNN-LSTM model beneath the second and third layers. Progressing from the one-dimensional CNN model and finally to CNN-LSTM-SE model resulted in a steady improvement in the value. The best results were obtained by the CNN-LSTM-SE classical, which suggests that a one-dimensional CNN model that incorporates an attention mechanism and LSTM can improve the effect of food fraud categorisation on the state matrix.

3.3 Weight Selection Using BMRFO

3.3.1 MRF

The MRF optimising approach is predicated on the food-seeking behaviour of the well-known marine species, manta rays (MR). Well-known marine species, such as MRs, primarily utilise plankton, a type of small aquatic organism that resides in the ocean. These extraordinary rays employ the chain, cyclone, and somersault foraging strategies to locate sustenance.

3.3.2 Chain Foraging

The MRF algorithm modifies the placement of each MR in accordance with its ideal location and the position of the MR directly in front of it, with the exception of the leading MR. Equation (7) delineates the chain foraging update technique.

$$q_n^{iter+1} = \begin{cases} q_n^{iter} + rand_1\left(q_b^{iter} - q_n^{iter}\right) + \gamma\left(q_b^{iter} - q_n^{iter}\right) & n = 1 \\ q_n^{iter} + rand_2\left(q_{n-1}^{iter} - q_n^{iter}\right) + \gamma\left(q_b^{iter} - q_n^{iter}\right) & n = 2, \ldots, N \end{cases} \quad (7)$$

$$\gamma = 2 * rand_3 * \sqrt{|\log(rand_4)|} \quad (8)$$

The location update in chain foraging is affected by a number of variables. First, q_n^{iter} denotes the position of the nth MR at the current iteration, and $rand_{1,2,\,3,4}$ is a collection of unique random integers created within the range [0, 1]. In this computation, also acts as the weighting factor. Last but not least, q_b^{iter} denotes the region in which there is a substantial concentration of plankton. It is important to note that the position update takes into consideration the optimal plankton location and the position of the previous MR in the series.

3.3.3 Cyclone Foraging

MRs construct a spiral formation by connecting their heads to their tails as they swim towards areas with a high concentration of plankton while feeding during cyclones. In addition to travelling in the direction of the plankton, each MR in this phase adjusts its route in response to the MR in front of it. In Eq. (9), the mathematical explanation of the cyclone foraging update mechanism is provided.

$$q_n^{iter+1} = \begin{cases} q_b^{iter} + rand_5\left(q_b^{iter} - q_n^{iter}\right) + \propto \left(q_b^{iter} - q_n^{iter}\right) & n = 1 \\ q_b^{iter} + rand_6\left(q_{n-1}^{iter} - q_n^{iter}\right) + \propto \left(q_b^{iter} - q_n^{iter}\right) & n = 2, \ldots, N \end{cases} \quad (9)$$

$$\propto = 2 * e^{rand_s} * \left(\frac{MaxIter - iter + 1}{iter}\right) * \sin(2 * \pi * rand_8) \quad (10)$$

The MRF algorithm is guided by this dual approach, which promotes a comprehensive analysis of the entire global search region and explores new regions of the search space. Equations (11) and (12) characterise this innovative process.

$$q_{rp}^{iter} = Lw + rand_9(UP - Lw) \quad (11)$$

$$q_n^{iter+1} = \begin{cases} q_{rp}^{iter} + rand_{10}\left(q_{rp}^{iter} - q_n^{iter}\right) + \propto \left(q_{rp}^{iter} - q_n^{iter}\right) & n = 1 \\ q_{rp}^{iter} + rand_{11}\left(q_{n-1}^{iter} - q_n^{iter}\right) + \propto \left(q_{rp}^{iter} - q_n^{iter}\right) & n = 2, \ldots, N \end{cases} \quad (12)$$

In this context, a randomly generated location that falls inside the bounds is referred to as a "q_{rp}^{iter}" Upper and lower bounds for a specific location are denoted by the parameters UP and LW, respectively. Furthermore, the symbol is used to represent the collection of diverse, arbitrarily generated integers that are contained within the [0, 1] range. $rand_{5,6,7,8,9,10,11,}$.

3.3.4 Somersault Foraging

The regular, local, random, and cyclic movement of somersault foraging aids in the ingestion of a greater quantity of plankton by MRs. As each MR circles this point, it moves to a new location. The reference point is selected to represent the site with the highest concentration of plankton to date. Equation (13) mathematically represents this paradigm.

$$q_n^{iter+1} = q_n^{iter} + smsf * \left(rand_{12} * q_b^{iter} - \left(rand_{13} * q_n^{iter}\right)\right) \quad (13)$$

where rand12,13 are distinct arbitrarily generated standards within the range of [0, 1], and smsf corresponds to the somersault factor. In this publication, the smsf is designated a value of 2.

3.3.5 Weight Selection in LSTM Using BMRF Optimization Algorithm

This scenario introduces a distinctive Binary MRF (BMRF) optimisation method that employs an SSTF. The BMRF proposal is inspired by the binary character of attribute selection spaces, in which attributes have only two possibilities: selection or non-selection. The transformed into the BMRF using ASSTF, which is more suitable for feature selection spaces. Each MR is designated by n to denote the dimension and m_i to accept binary values (1 or 0) to indicate the selection of attributes or features. $M\left(m_1, m_2, \ldots, m_j, \ldots, m_n\right)$ in The BMRF method suggests a viable solution. In contrast to the continuous phases employed by MRF, BMRF update's locations by altering probability. The transition from continuous to binary states is facilitated by S-shaped functions.

For example, the first function is illustrated in Eqs. (14) and (15) using the conventional conversion approach.

$$SSTF\left(q_n^{iter}\right) = \frac{1}{1 + e^{-\left(q_n^{iter}\right)}} \quad (14)$$

$$q_n^{iter+1} = \begin{cases} 0 \ if \ rand < SSTF\left(q_n^{iter}\right) \\ 1 \ if \ rand \geq SSTF\left(q_n^{iter}\right) \end{cases} \quad (15)$$

The proposed BMRF method in this paper incorporates an adaptive S-shape transfer function (ASSTF) with dynamic gradient updates to enhance turnover probability refinement during the course of iterations. The current location is assessed by the ASSTF. (q_n^{iter}) in terms of Eqs. (16) to (17), and Eq. (18) is used to calculate the novel position.

$$ASSTF\left(q_n^{iter}\right) = \frac{1}{1 + e^{\frac{-(q_n^{iter})}{k}}} \tag{16}$$

$$k = \left(1 - \frac{iter}{MaxIter}\right) * MaxIter_{max} + \frac{iter}{MaxIter} MaxIter_{min} \tag{17}$$

$$q_n^{iter+1} = \begin{cases} 0 \text{ } if \text{ } rand < ASSTF\left(q_n^{iter}\right) \\ 1 \text{ } if \text{ } rand \geq ASSTF\left(q_n^{iter}\right) \end{cases} \tag{18}$$

where iter and *MaxIter* specify, respectively, the algorithm's current iteration and its maximum allowed number of iterations. In this study, the $MaxIter_{max}$ parameter is set to 4 and the $MaxIter_{min}$ set the parameter to 0.01.

4 Results and Discussion

The hardware situation for this investigation was an NVIDIA GeForce GTX 1060, and the software environment was Tensorflow2.3.0 and Python 3.8. The projected model's accuracy and loss are illustrated in Fig. 2.

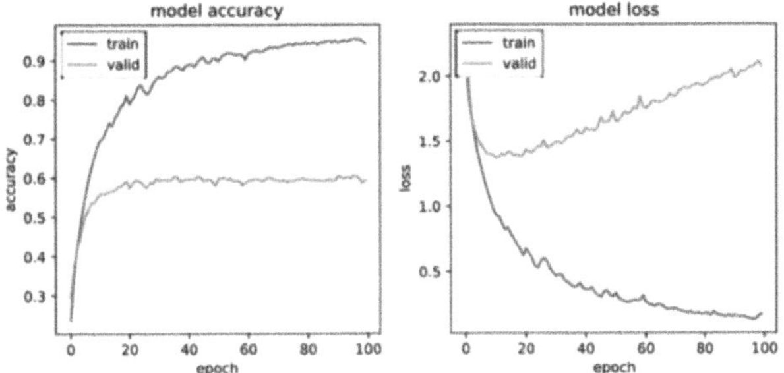

Fig. 2. Accuracy besides Loss of the proposed classical

4.1 Validation Analysis of Proposed Classical with Existing Procedures

Figure 3 and 4 provides the comparative investigation of projected classical with existing techniques in terms of diverse metrics.

The comparison of various methods based on their precision and F-score performance. The SVM method achieves a precision of 80.85% besides an F-score of 85.07%, which is relatively lower than the other techniques. Moving to more advanced models, the MLP shows substantial improvement precision of 93.50% and an F-score of 93.88%, demonstrating strong performance. The DBN model follows closely with a precision of 95.89% and an F-score of 95.50%, highlighting its effectiveness. CNN performs slightly

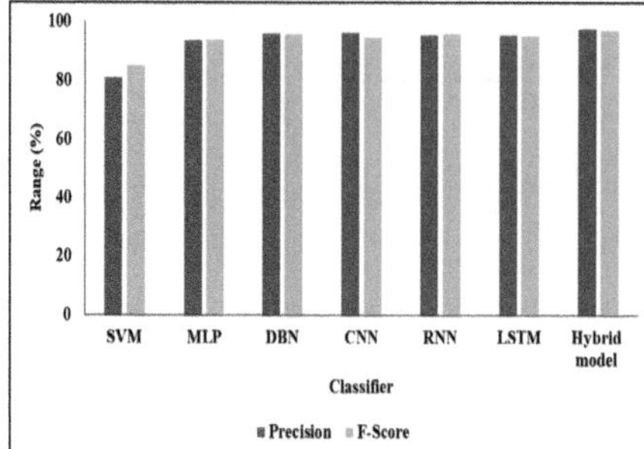

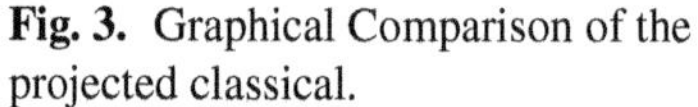
Fig. 3. Graphical Comparison of the projected classical.

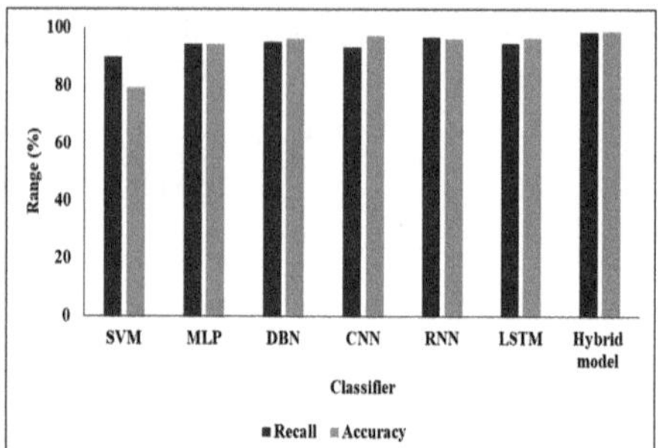

Fig. 4. Visual Representation of numerous representations in terms of accuracy besides recall.

better with a precision of 96.13% and an F-score of 94.67%, sparkly its efficiency in this context. The RNN model achieves a precision of 95.28% and an F-score of 96.01%, indicating its capability for high-performance tasks. The LSTM model yields a precision of 95.38% and an F-score of 95.02%, offering competitive results. Finally, the Hybrid model outperforms all others with a precision of 97.44% besides an F-score of 97.02%, demonstrating the highest overall presentation in terms of both precision and F-score.

The performance of different methods based on Recall and Accuracy. The SVM classical shows a recall of 89.76% besides an accuracy of 79.38%, representative a lower overall performance compared to more advanced techniques. The CNN model reaches a recall of 93.25% and an impressive accuracy of 97.30%, highlighting its high accuracy. The RNN model excels with a recall of 96.75% and an accuracy of 96.09%, demonstrating its robustness. The LSTM model offers a recall of 94.66% and an accuracy of 96.42%, performing competitively. Finally, the Hybrid model outperforms all others with a recall of 98.61% and an accuracy of 98.87%, leading in both recall and accuracy, making it model in this comparison.

5 Conclusion

The research demonstrates an effective solution for addressing the increasing challenges of food fraud and safety hazards by introducing the CNN-LSTM-SE classical, which is optimized with BMRF procedure. The model's classification accuracy and predictive capabilities are enhanced by the integration of deep learning approaches and attention mechanisms, thereby enabling a more robust system for detecting economically motivated adulteration. Utilizing data from the EMA database and the Rapid Feed the case study emphasizes the practical implementation of AI in risk mitigation throughout the supply chain. This research establishes a foundation for the creation of transparent and dependable tools to defend public health and food integrity, despite the obstacles associated with the explainability of AI models. The future work can concentrate on enhancing the model's explain ability and expanding its applicability to other aspects of food safety and supply chain optimisation.

References

1. Gwenzi, W., Makuvara, Z., Marumure, J., Simbanegavi, T.T., Mukonza, S.S., Chaukura, N.: Chicanery in the food supply chain! Food fraud, mitigation, and research needs in low-income countries. Trends Food Sci. Technol. **136**, 194–223 (2023)
2. Aslam, R., Sharma, S.R., Kaur, J., Panayampadan, A.S., Dar, O.I.: A systematic account of food adulteration and recent trends in the non-destructive analysis of food fraud detection. J. Food Measur. Charact. **17**(3), 3094–3114 (2023)
3. Gopal, J., Muthu, M.: Handheld portable analytics for food fraud detection, the evolution of next-generation smartphone-based food sensors: the journey, the milestones, the challenges debarring the destination. TrAC, Trends Anal. Chem. **171**, 117504 (2024)
4. Joenperä, J., Lundén, J.: Food fraud detection and reporting by food control officers in Finland. Int. J. Environ. Health Res. **34**(5), 2230–2247 (2024)
5. Bannor, R.K., Arthur, K.K., Oppong, D., Oppong-Kyeremeh, H.: A comprehensive systematic review and bibliometric analysis of food fraud from a global perspective. J. Agric. Food Res. **14**, 100686 (2023)
6. Baswaraju, S., Maheswari, V.U., Chennam, K.K., Thirumalraj, A., Kantipudi, M.P., Aluvalu, R.: Future food production prediction using AROA based hybrid deep learning model in agri-sector. Hum.-Centric Intell. Syst. **3**(4), 521–536 (2023)
7. Boller, M.L., Zurwehme, A., Krupitzer, C.: Qualitative assessment on the chances and limitations of food fraud prevention through distributed ledger technologies in the organic food supply chain. Food Control **158**, 110247 (2024)
8. Karami, H., et al.: Advanced evaluation techniques: gas sensor networks, machine learning, and chemometrics for fraud detection in plant and animal products. Sens. Actuators A: Phys. 115192 (2024)
9. Bouzembrak, Y., et al.: Data driven food fraud vulnerabilities assessment using Bayesian network: spices supply chain. Food Control 110616 (2024)
10. Costa, M.J., Sousa, I., Moura, A.P., Teixeira, J.A., Cunha, L.M.: Food fraud conceptualization: an exploratory study with Portuguese consumers. J. Food Prot. **87**(7), 100301 (2024)
11. Buyuktepe, O., Catal, C., Kar, G., Bouzembrak, Y., Marvin, H., Gavai, A.: Food fraud detection using explainable artificial intelligence. Expert. Syst. **42**(1), e13387 (2025)
12. Vinothkanna, A., Dar, O.I., Liu, Z., Jia, A.Q.: Advanced detection tools in food fraud: a systematic review for holistic and rational detection method based on research and patents. Food Chem. 138893 (2024)
13. Xun, Z., et al.: Deep machine learning identified fish flesh using multispectral imaging. Curr. Res. Food Sci. 100784 (2024)
14. Jahanbakhshi, A., Abbaspour-Gilandeh, Y., Heidarbeigi, K., Momeny, M.: A novel method based on machine vision system and deep learning to detect fraud in turmeric powder. Comput. Biol. Med. **136**, 104728 (2021)
15. Jahanbakhshi, A., Abbaspour-Gilandeh, Y., Heidarbeigi, K., Momeny, M.: Detection of fraud in ginger powder using an automatic sorting system based on image processing technique and deep learning. Comput. Biol. Med. **136**, 104764 (2021)
16. Gavai, A., et al.: Applying federated learning to combat food fraud in food supply chains. NPJ Sci. Food **7**(1), 46 (2023)
17. Monteiro, R.S., Ribeiro, M.C., Viana, C.A., Moreira, M.W., Araúo, G.S., Rodrigues, J.J.: Fish recognition model for fraud prevention using convolutional neural networks. Adv. Comput. Intell. **3**(1), 2 (2023)

Revolutionizing the Emergency Department Triage: Positive Impact of Artificial Intelligence on Emergency Care Service and Challenges to Overcome

Jesima Nazarudeen(✉) and Raja Varma Pamba

Manipal Academy of Higher Education, Dubai, UAE
jesimav05@gmail.com

Abstract. Artificial Intelligence (AI) in recent times has proven to be revolutionizing the triage systems in emergency departments (EDs) by solving issues like patient waiting delays, unequal risk evaluation, and improper utilization of resources. Neural networks, deep learning models, and data processing frameworks improve the efficiency of decision-making by providing clinicians and healthcare administrators with dependable information to work with while managing EDs in hospitals. Machine learning techniques such as Convolutional Neural Networks (CNNs) and Linear Chain Models (LCMs) have also been shown to enhance the early detection of these patients, decrease time, and enhance the utilization of general resources. However, some drawbacks exist that consist of bias in training data, integration of this system in current healthcare infrastructures, and data privacy. This paper focuses on the use of AI in emergency triage, showcases emerging solutions such as blockchain technology and other models to solve various security issues and enable automation for better management of data of patients to make timely decisions. The use of blockchain can prove to be a strong step towards innovation as it can open the doors for future technological advancements. The considered models are very helpful in bringing efficiency and, therefore, make the management faster and smoother. A detailed review of the literature is conducted to find out what previous scholarly work says about the topic and the challenges identified, along with the recommendations. Moreover, this paper highlights the positive impact that AI technology has on the enhancement of emergency care services, as well as examines the possibilities of further development of this technology to get more benefits from it in the future in healthcare.

Keywords: AI in ED · Emergency Department · Triage · Neural Networks · Deep Learning · Machine Learning · Privacy · Security · Algorithms · Automation · Blockchain

1 Introduction

Emergency departments (EDs) remain the key feature of the healthcare systems, serving patients with immediate treatment. Nevertheless, the growing need for emergency treatment has exposed several drawbacks of conventional triage methods. Triage, sorting patients into categories according to levels of injury, is fundamental in determining

J. Shreyas et al. (Eds.): CODE-AI 2025, CCIS 2690, pp. 45–57, 2026.
https://doi.org/10.1007/978-3-032-19321-6_5

the outcome of patients. However, using manpower and making judgments based on assumptions result in inefficiencies, waste of resources, and imprecise decisions. AI appears as a beneficial approach to tackle these issues because of the capability of applying increased speed, efficiency, and control to triage mechanisms [1]. In general, the concept of triage serves not only as a theoretical concept but is involved with patient care and, more importantly, the patient's outcomes (Jha, 2019). An effective triage system is one that first prioritizes the critically ill patients while other patients will also be handled well. However, emergency departments globally face common challenges: high patient turnover, long waiting times, and incorrect administrative decisions. Overcrowding is also hazardous to the overall health and more so for those patients requiring immediate care. Long waits for the patients may be caused by poor prioritization, while errors can be a result of carelessness due to fatigue or more so due to differences in competence in workers. These problems show the need for AI solutions to improve the process of triage (Fig. 1).

1.1 Challenges in Emergency Departments

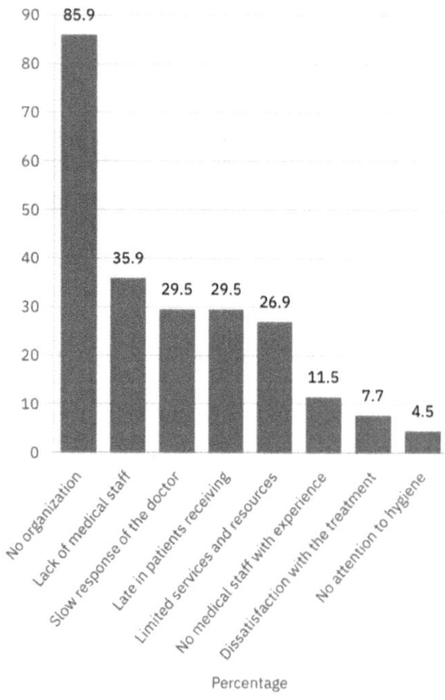

Fig. 1 Challenges in ER [2]

The conventional approach to triage involves the use of workers' experience and intuition, which brings individuality and chances of mistakes. For example, two nurses may make different judgments about the patient's condition and thus create discrepancies in the prioritization. Furthermore, when EDs become filled with more patients, this adds to the pressure put on the staff and indirectly contributes to lots of mistakes. There are also huge problems with the amount of data that a manual triage system can handle, especially in today's healthcare environments with EHRs, real-time streams of data, and image results [3].

2 Literature Survey

AI is rapidly transforming a lot of fields, and health care is among them. The application of Artificial Intelligence in Emergency Departments (EDs) for triage, referring to the identification of patients' treatment orders according to their medical condition, has been recently discussed in the literature. Previously, the evaluation of the type of presentation to ED triage has been based on human decision-making, the efficacy of which, although commendable, may be inconsistent, inaccurate, or even costly in terms of time or resources, particularly during peak periods [4]. AI, on the other hand, brings the ability to enhance the accuracy of triage processes through fast data analysis [5]. AI in triage is not just about the automation of the existing processes but more about making the processes better in ways that were not possible before [6]. There are numerous models and methodologies proposed, especially for machine learning and deep learning points of view, to allocate resources properly in emergency conditions [7].

2.1 AI Methodologies for ED Triage

Machine Learning Algorithm. Machine learning is a broad field whereby many techniques can be applied; however, in the case of triage, the most applicable ones are the supervised learning algorithms of classification algorithms. Such systems work with labeled data, that is, data that has a clear outcome (e.g., severity level of a patient), and the system is learned to predict outcomes for unknown data sets [16]. One of the classic uses of ML in screening patients in ED triage involves utilizing support vector machines (SVM) and decision trees to categorize patients depending on their physical characteristics, clinical symptoms, and past health history [12]. These models are useful in shortening the time taken to classify patients, especially in the congested ED environment where time is always of the essence.

Deep Learning Models. Machine learning, more specifically deep learning, saw a rise in interest for healthcare use cases as they can handle massive and unstructured data like medical images or the EHRs [8]. In the context of ED triage, Convolutional Neural Networks (CNNs) and Recurrent Neural Networks (RNNs) can be used in classification and prediction models [9].

Natural Language Processing (NLP). A technique that is currently being considered for use in analyzing the unstructured clinical data generated in EDs comprises natural language processing (NLP) [17]. In a conventional ED triage system, most of the details about the patient, including his/her history, are put down in free text (Table 1).

2.2 Applications of AI in ED Triage

The most critical area of application of AI in ED triage is the desire to minimize the amount of time that is used to evaluate and assign patients [10]. Furthermore, AI algorithms allow for the new patient clinical pathway, to be predicted. For example, patients

Table 1. The table provides an overview of the advantages and limitations, together with the direct applicability of the selected methodologies within the ED triage context, while also indicating the commonalities for further investigation and adaptation [8] (Figs. 2 and 3).

AI methodology	Description	Advantages	Gaps	Applications in triage
Artificial Neural Networks	Models inspired by biological neural systems that process information in layers.	High accuracy in pattern recognition; adaptable to complex data.	Requires large datasets.	Predicting patient severity based on health records.
Convolutional Neural Networks (CNN)	Specialized DL models for image and spatial data.	High precision in medical imaging; efficient for image-related tasks.	Limited to specific data types; requires preprocessing for other types of data.	Analyzing medical imaging to assist in triage decisions.
Deep Learning (DL)	Advanced neural networks with multiple layers for complex data analysis.	Can process large volumes of data; highly accurate in identifying subtle patterns.	"Black-box" nature limits interpretability; requires significant computing resources.	Analyzing multi-dimensional patient data for triage prioritization.
Natural Language Processing (NLP)	Techniques for understanding and processing human language in text.	Useful for processing patient reports and clinical notes; enables better communication analysis.	Limited by the quality of text data; challenges in understanding medical terminology and abbreviations.	Extracting critical information from unstructured clinical notes to inform triage decisions.

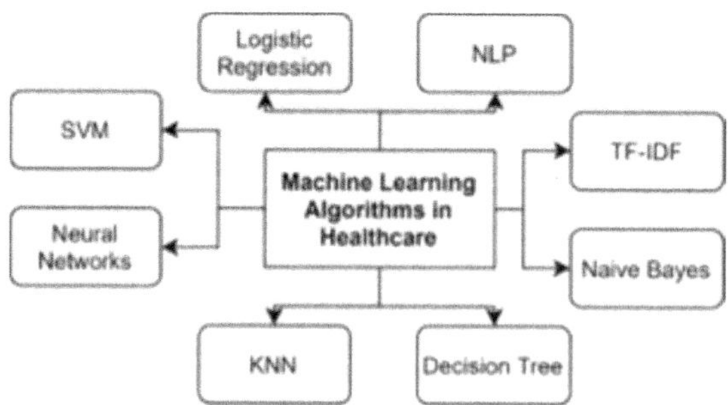

Fig. 2 Machine learning algorithms in healthcare [36]

who are at risk of developing sepsis or respiratory failure can have better results if complications are detected in their early stages.

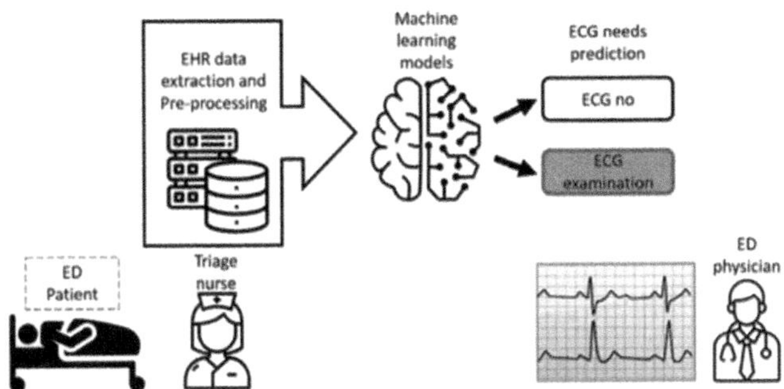

Fig. 3. Machine learning in ED triage

2.3 Challenges and Limitations of AI in ED Triage

AI systems greatly depend on the quality of data that are fed to them. Specifically, the type of data utilization is critical for improving ED triage since incomplete, missing, or inaccurate data dramatically impacts AI outcomes. Furthermore, AI models are only as good as the data available about patients, and this encompasses both structured data, such as body weight, temperature, blood pressure, and results of laboratory investigations, as well as unstructured data, which could be records of clinicians. Many EDs still have not adopted a completely digitalized system of patient records, and hence, they are not suitable for AI models [11]. One of the ongoing major problems of AI models in ED is security. Considering the hardware, AI uses many electronic products like mobile phones and computers. First of all, even the best possible hardware technology is dependent on the processor, temperature requirements, etc., which requires a significant amount of care and attention. Also, if the relevant AI professionals cannot be given complete access to the sensitive medical data of the patients, it can result in leakage of data and disruptions. Most EDs have relied on old systems for keeping patient records and triaging or prioritizing cases, and incorporating AI algorithms into these systems without interfering with workflow or patient care is not easy. Moreover, there is poor compatibility of various systems employed in EDs, for instance, EHRs, diagnostic instruments as well and AI applications [12]. In the literature, it is demonstrated that the application of AI in the triage of patients in EDs can be far more effective and rapid with the precise and efficient classification of patients through different mechanisms. Machine learning, deep learning, NLP, and LCM have evidence of overcoming the issues of ED triage: delay, misclassification, and resource wastage.

3 Methodology

The first part of the process of artificial intelligence-based triage is data collection and data preparation. These components often comprise Electronic Health Records (EHR), a patient's medical history, vitals, symptoms, and personal data. Laboratory findings and imaging taken from prior ED visits can also be included. The first process that occurs before the actual analysis is done is cleaning the data, how to handle missing values, normalizing the values, and ensuring that the data to be analyzed meets certain requirements [13] (Fig. 4).

3.1 Blockchain Technology for Security

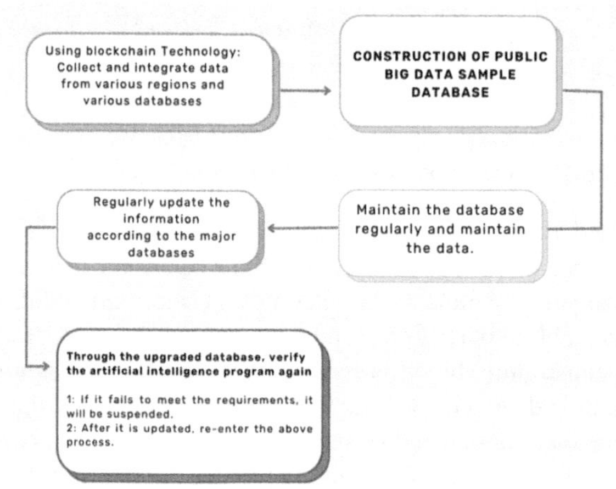

Fig. 4 Maintenance and application of dataset [14]

Highest quality data sets are integrated using blockchains and big data along with other technologies, and the data is stored in a more secured data storage system. This would require maintenance on a daily basis as well as updates on a regular basis. There would be verification protocols to verify AI through that database. If there is any problem or any failure in the program, it needs to be fixed first. Also, there is a need for a specific department that will supervise the complete process and will create strategies and a system to solve the problems that arise [14]. By implementing blockchain technology, the central server is removed. Due to this, the transfer of information is carried through the edges of the blockchain, which makes the transfer with the right authentication, preventing the leakage of data or any kind of data interception. Blockchain technology is decentralized, meaning that there are no hidden operations of operators behind it, and customers would be satisfied with the storage of their data as they would be able to track it. There are two consensus methods for blockchain: one is Proof of Work (PoW) and Proof of Authority (PoA). PoW is considered good for the security of healthcare applications [14]. This algorithm is the most secure, but it consumes a high amount of energy and is a little slow as well. On the other hand, (PoA) is fast and energy efficient, but the security is lowered. The sharing of data through blockchain is extremely safe and secure. As the information is verifiable, it adds another layer of security. Furthermore, blockchain is mainly useful in preserving electronic health records (EHR) as well as personal health records (PHR). Smart contracts in blockchain allow patients to implement permission-based access control policies for their data and also allow them to share their past health records with doctors in emergency and critical situations. First, a hospital or doctor initiates a request in the emergency to a medical data center, and the data is processed at edge services. The data is then formatted and identified for transaction. When the emergency request is initiated, it is sent to the transaction manager by the system for storage and creating a waiting list. The request is then processed once it is approved, and access to the patient's records is granted to the healthcare providers and doctors.

3.2 Deep Learning and Neural Networks

After data pre-processing, the next step involves using deep learning models to classify and predict the severity of the patient. Neural networks (NN) are commonly applied to this end because they can detect subtle features within a large set of data. These include the feed-forward neural network and the recurrent neural network (RNN). These are trained on labeled data to identify some patterns that signify different levels of urgency in the ED [15].

3.3 Convolutional Neural Networks (CNN)

CNNs have high applicability in images and unstructured data such as X-ray or CT, which is more often used in emergency cases. In this way, CNNs can learn features from these images, like fractures and lung or cardiac problems, and then, using these features, assign proper priority levels to patients [28].

3.4 Pattern Recognition and Predictive Modeling

In triage, AI models tend to use machine learning techniques to categorize the static structured patient data, such as vital signs, blood pressure, and heart rate, and decide the patient outcomes from dynamic unstructured data, such as clinical notes and laboratory results. Machine learning software also points out repeating patterns from this information to categorize patients based on the priority for treatment, such as priority, urgent, or low [16].

3.5 Outcome Visualization and Decision Support Systems

AI systems are implemented into Decision Support Systems (DSS) where clinicians analyze the output generated by the AI model. This is because there is a need to make decisions with an acceptable level of speed, and outcome visualization techniques offer this in a format that can easily be interpreted. Clinicians are then able to sort through these patients based on the severity and the urgency of the situation and, therefore, decrease patient waiting time [1].

The figure below illustrates automatic triage compared to traditional triage plans. In the conventional triage classification, the patients attend the ED for admission. A client's physiological status is assessed through vital signs and biochemical profiles, and an emergency code number is identified. In the described scenario, the data from the previous admissions are used first to construct a network with patient similarity. The network is then utilized to train a Graph Neural Network to estimate patients' severity index in a latent space. After obtaining clinical data, a patient who walks into the emergency room is automatically labelled [15] (Fig. 5 and Table 2).

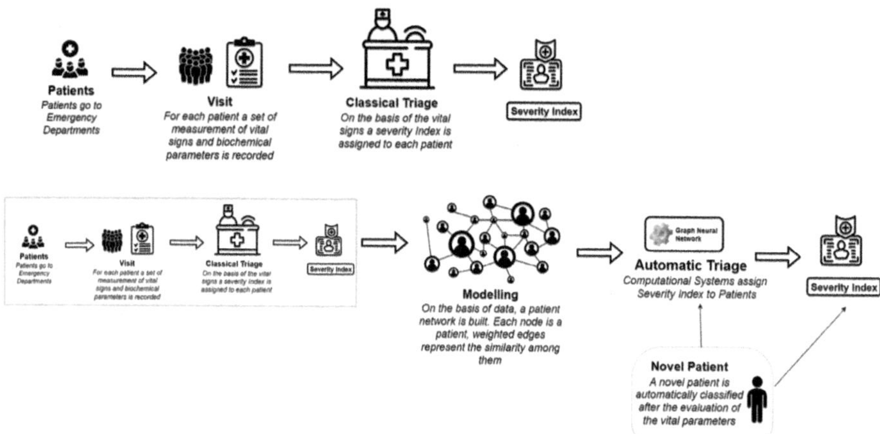

Fig. 5 Automatic triage compared to traditional triage plans

Table 2. Comparison between classical triage and automatic triage

Features	Classical triage	Automatic triage
Process	Based on a manual assessment of vital sign	Uses computational systems (Graph Neural Network)
Collection of Data	Vital signs and biochemical parameters were recorded manually	Same data collected but used for modeling and automation
Severity Index	Assigned manually based on observation	Assigned automatically by the system
Modelling	As patients are assessed manually, there is no modelling	A patient network is built based on data similarity
Handling of Novel Patients	Manual triage and evaluation is required	New patients are automatically categorized on the basis of pattern

(*continued*)

Table 2. (*continued*)

Features	Classical triage	Automatic triage
Efficiency	Requires human input for its operation, therefore it is slow	It is machine based and automated so it is faster

4 Results

The decisions made from AI-based systems have been proven to be more accurate than those made from conventional methods of patient triage. In different investigations, the value of AI algorithms, especially deep learning models like CNNs and RNNs, was revealed to be optimal in classifying patients according to the severity of their conditions.

For instance, it has been possible to predict the severity of a patient's condition using the signs, history as well and other real-time symptoms through AI. For example, in a hospital, AI application has been effective in minimizing error rates that are commonly faced during the triage stage in enabling identification of patients with high severity to benefit from rapid attention without misclassification that may cause wastage of time. Also, AI has been more useful in the early detection of conditions that are critical, such as strokes, heart attacks, and sepsis, among others. Through the analysis of big volumes of patients' records, the models recognize some patterns that point out that certain patients are in a critical condition, and hence, the providers will be able to prioritize such patients in a better way. Furthermore, AI integration in triage also has another important outcome based on the time that the patient spends in the emergency department. The existing triage techniques that involve human factors and reviews can cause congestion, particularly during high-traffic periods. On the other hand, AI systems can process data and filter it in large volumes within a short span, hence improving the triage. AI has the added benefits of decreasing overcrowding and enhancing the ambulatory capacity of the ED, which are two frequent problems associated with EDs.

Other similar AI models used in the triage of patients arriving in the emergency department have also been found to have reasonably good predictive powers. For example, in the case of risk factors and outcomes, prediction by the deep learning algorithm means that the patient's situation might worsen, and an intervention may be applied promptly. In some cases, it has been observed that with their level of accuracy, which is about 90%, AI models enhance patient care (Fig. 6).

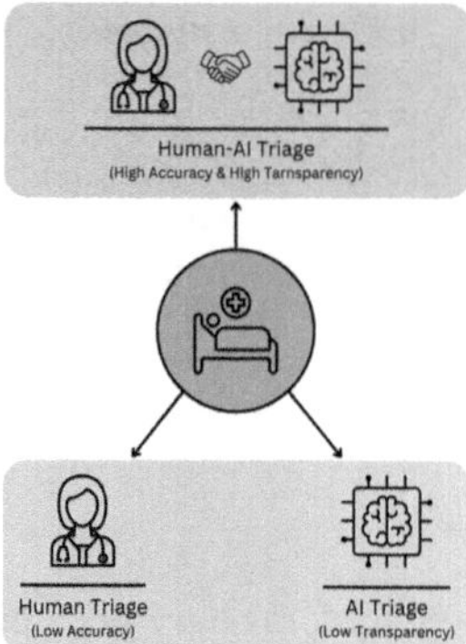

Fig. 6 Comparison between human triage and AI triage

A classical triage system is manual and slow as it always requires human input for the simplest of tasks, such as the manual process for the assessment of vital signs. The severity index is also assigned manually based on observations, and there is no concept of modeling. Automatic triage is more efficient, faster, and cost-effective and performs all the tasks without human input as the process is based on a computational system of graph neural networks, and a collection of data is used for modelling and automation, as well as a severity index is automatically assigned by the system. Therefore, automatic triage is better than the classical triage system that has been used for many years [17].

5 Discussion

Tools are used with AI systems through which clinicians get a desired visualization of the patient's severity and treatment. Risk score charts, heat maps, and real-time alerts are all effective for representing the output of AI models to enable healthcare providers to prioritize their attention. These tools give clear functional headings by showing time-sensitive issues and giving a detailed methodology for a patient's status. AI is essential to develop decision-making in triage because it enhances clinical decision-making that is already in existence. Another important factor in implementing AI is that it does not replace the judgment of clinicians but only makes it easier for them to better identify patients who are most at risk or need urgent attention. Nevertheless, some issues remain concerning the application of artificial intelligence in the assessment of patients in the ED. More specifically, one of the key issues includes the black-box nature of AI. EDs deal with patients, so their data and information are important, and as AI systems are integrated into existing health information systems, issues of data security and privacy arise. Such risks include data breaches, improper usage of consumer data, pirating, and cyberwarfare are some of the risks that companies face today. Therefore, compliance with regulatory requirements such as HIPAA and GDPR is critical. Future studies on the application of AI in triage will aim at increased usability and compliance with the Data

Protection Regulation. Further research is also required to investigate the matter of how AI can be most effectively integrated with the current hospital information systems.

6 Conclusion

AI can provide massive benefits in addressing critical issues of ED, including lengthy delays in patient classification, wrong classification, and inefficient management of resources. The developed deep learning models, such as convolutional neural networks (CNNs), make it simpler and quicker to analyze patient data and help clinicians focus on the most vital cases. They also use records, measurements, and observations to provide recommendations that save time in making decisions and improve patients' conditions. Some limitations have to be considered to identify further steps in the implementation of AI in EDs. Among these, the problem is concerned with the explanation and control of AI algorithms, making a decision trusted. Also, issues of data confidentiality and the ability to ensure patients' information is well-processed and kept secure are still important issues. Also, implementing AI tools into the health information system may be challenging due to issues with compatibility with different systems used in managing health information. In conclusion, it is important to note that although the solution based on AI can be a powerful tool in increasing efficiency in the implementation of triage, there are some significant challenges that should be overcome to make it work properly. Future research and development activities will continue to focus on improving the efficiency of these technologies and make sure they are included in every ED across the globe.

References

1. Tang, K.J.W., Ang, C.K.E., Constantinides, T., Rajinikanth, V., Acharya, U.R., Cheong, K.H.: Artificial intelligence and machine learning in emergency medicine. Biocybernetics and Biomedical Engineering. **41**(1), 156–172 (2021). https://doi.org/10.1016/j.bbe.2020.12.002
2. Dawoud, S.O., Ahmad, A.M.K., Alsharqi, O.Z., Al-Raddadi, R.M.: Utilization of the emergency department and predicting factors associated with its use at the Saudi Ministry of Health general hospitals. Global J. Health Sci. **8**(1), 90 (2016). https://doi.org/10.5539/gjhs.v8n1p90
3. Bazyar, J., Farrokhi, M., Khankeh, H.: Triage systems in mass casualty incidents and disasters: a review study with a worldwide approach. Open Access Maced. J. Med. Sci. **7**(3), 482 (2019). https://doi.org/10.3889/oamjms.2019.119
4. Nino, V., Claudio, D., Schiel, C., Bellows, B.: Coupling wearable devices and decision theory in the United States emergency department triage process: a narrative review. Int. J. Environ. Res. Public Health. **17**(24), 9561 (2020). https://doi.org/10.3390/ijerph17249561
5. Delshad, S., Dontaraju, V.S., Chengat, V.: Artificial intelligence-based application provides accurate medical triage advice when compared to consensus decisions of healthcare providers. Cureus. **13**(8) (2021). https://doi.org/10.7759/cureus.16956
6. Tyler, S. et al.: Use of artificial intelligence in triage in hospital emergency departments: a scoping review. Cureus. **16**(5) (2024). https://doi.org/10.7759/cureus.59906

7. Adebayo, O., Bhuiyan, Z.A., Ahmed, Z.: Exploring the effectiveness of artificial intelligence, machine learning and deep learning in trauma triage: a systematic review and meta-analysis. Digital Health. **9**, 20552076231205736 (2023). https://doi.org/10.1177/20552076231205736
8. Aryffin, H.A.A.K., Noor, M.H.M., Musa, K.I., Baharuddin, K.A.: Revolutionizing emergency medicine and triage systems through artificial intelligence. Malaysian. J. Emerg. Med. **6**(2), 38–44. https://m-jem.com/index.php/mjem/article/view/1540 (2024)
9. Gao, X., Xu, X., Li, D.: Accuracy analysis of triage recommendation based on CNN, RNN and RCNN models. In: In: 2021 IEEE Asia-Pacific Conference on Image Processing, Electronics and Computers (IPEC), pp. 1323–1327. IEEE. https://ieeexplore.ieee.org/abstract/document/9421099 (2021)
10. Paslı, S., Şahin, A.S., Beşer, M.F., Topçuoğlu, H., Yadigaroğlu, M., İmamoğlu, M.: Assessing the precision of artificial intelligence in emergency department triage decisions: insights from a study with ChatGPT. Am. J. Emerg. Med. **78**, 170–175 (2024). https://doi.org/10.1016/j.ajem.2024.01.037
11. Townsend, B.A., Plant, K.L., Hodge, V.J., Ashaolu, O.T., Calinescu, R.: Medical practitioner perspectives on AI in emergency triage. Front. Digital Health. **5**, 1297073 (2023). https://doi.org/10.3389/fdgth.2023.1297073
12. Gebrael, G. et al.: Enhancing triage efficiency and accuracy in emergency rooms for patients with metastatic prostate cancer: a retrospective analysis of artificial intelligence-assisted triage using ChatGPT 4.0. Cancer. **15**(14), 3717 (2023). https://doi.org/10.3390/cancers15143717
13. Jasim, A.A., Ata, O., Salman, O.H.N.: AI-driven triage: a graph neural network approach for prehospital emergency triage patients in IoMT-based telemedicine systems. In: In: 2024 International Symposium on Electronics and Telecommunications (ISETC), pp. 1–7. IEEE. https://ieeexplore.ieee.org/abstract/document/10797314 (2024)
14. Jiang, L. et al.: Opportunities and challenges of artificial intelligence in the medical field: current application, emerging problems, and problem-solving strategies. J. Int. Med. Res. **49**(3), 03000605211000157 (2021)
15. Defilippo, A., Veltri, P., Lió, P., Guzzi, P.H.: Leveraging graph neural networks for supporting automatic triage of patients. Sci. Rep. **14**(1), 12548 (2024). https://doi.org/10.1038/s41598-024-63376-2
16. Gao, F., Boukebous, B., Pozzar, M., Alaoui, E., Sano, B., Bayat, S.: Predictive models for emergency department triage using machine learning: a systematic review. Obstetrics and Gynecology Research. **5**(2), 136–157 (2022). https://doi.org/10.26502/ogr085
17. Al Dhefeeri, A.K., Aldhafeeri, A.M.B., Alsharari, G.A.S., Alhazmy, M.S.F., Alomayri, M.I.A., Alanazi, H.A.M.: The importance of triage in the emergency department. Gland Surg. **9**(2), 285–291. https://www.glandsurgery.net/index.php/GS/article/view/91 (2024)
18. Mäenpää, S.M., Korja, M.: Diagnostic test accuracy of externally validated convolutional neural network (CNN) artificial intelligence (AI) models for emergency head CT scans–a systematic review. Int. J. Med. Inform., 105523 (2024). https://doi.org/10.1016/j.ijmedinf.2024.105523

BY

Leveraging Natural Language Processing to Extract Influential Keywords for Business Intelligence in the Beauty and Personal Care Industry

Mafas Raheem(✉) and Nirase Fathima Abubacker

Asia Pacific University of Science & Technology, Kuala Lumpur, Malaysia
{raheem,nfathima.abubacker}@apu.edu.my

Abstract. This study explores sentiment analysis of customer reviews in the beauty, cosmetics, and personal care sectors, using Instagram, Shopee, and Facebook data. It applies Natural Language Processing (NLP) techniques to extract insights from user-generated content, which often remains underutilized. Following the CRISP-DM methodology, a data pre-processing pipeline was applied to a dataset of 151,914 observations. Key steps included removing duplicates, handling null values, emoji translation, punctuation removal, lemmatization, and TF-IDF vectorization. Sentiment was predicted using a pre-trained BERT-based model into positive, neutral, and negative categories. An ensemble model consists of famous machine learning models, optimized through soft voting, achieved 90% accuracy, with a precision, recall, and F1-score of 0.90, 0.90, and 0.89, respectively. The research identified impactful keywords for enhancing marketing strategies and supports sustainable economic growth, in line with the United Nations' SDG 8. The study highlights the value of NLP in improving business intelligence and customer engagement.

Keywords: social-media marketing · influential keywords · natural language processing · cosmetics industry · customer engagement

1 Introduction

The Malaysian beauty and personal care industry is being transformed by digital platforms, that reshape consumer behavior and marketing strategies. Through advanced analytics of customer reviews, hashtags, mentions, and captions, brands gain valuable insights into consumer trends and sentiments, offering a more efficient alternative to traditional market research. Sentiment analysis helps companies track emerging trends, gauge product performance, and adapt to shifting customer needs in real-time.

The beauty and personal care industry in Malaysia is undergoing rapid expansion from the recent past. This dynamic market is expected to reach an impressive $948.22 million by 2032, reflecting its growing importance and potential [1]. This sector is fiercely competitive, with social media platforms, particularly Instagram, playing a pivotal role in

J. Shreyas et al. (Eds.): CODE-AI 2025, CCIS 2690, pp. 58–70, 2026.
https://doi.org/10.1007/978-3-032-19321-6_6

shaping consumer behavior and driving trends [2]. However, the sheer volume of unstructured textual data such as comments and reviews generated on these platforms presents significant challenges [3]. While this data is rich in valuable consumer insights, preferences, and emerging trends, it is often overwhelming and complex, making effective processing a formidable task [4].

Traditional market research methods fall short in capturing the evolving and nuanced customer behaviors exhibited on social media [5]. Traditional approaches struggle to manage the dynamic, large-scale nature of social media data, creating a gap in understanding key consumer behaviors. To stay competitive, the beauty and personal care industry needs advanced techniques to extract actionable insights from unstructured data and identify untapped market segments. Closing this gap is essential in a digital, data-driven marketplace.

This research developed a comprehensive data science solution aimed at analyzing customer sentiment towards beauty and personal care products to enhance product quality, enhance business process, identify emerging trends, and optimize marketing efforts. The project involved extracting customer reviews from leading digital business platforms such as Instagram, Facebook, and Shopee, specifically focusing on the beauty and personal care sector in Malaysia. Insights were derived through advanced Sentiment Analysis, market trend identification, and influential keyword extraction. By leveraging this approach, businesses can gain a deeper understanding of customer preferences, refine their product offerings, and tailor marketing strategies to align more closely with consumer demands, ultimately driving greater customer satisfaction and business growth.

2 Literature Review

2.1 Introduction

Natural Language Processing (NLP) has become indispensable for platforms like Instagram, Facebook, and Shopee, where vast amounts of user-generated content provide rich, untapped data. By blending computational techniques with qualitative analysis, NLP addresses a wide range of research questions and business objectives. One of the key challenges lies in applying quantum analytics to the enormous volumes of data. However, researchers are increasingly finding innovative ways to bridge data science with qualitative insights. Social media interactions and E-Commerce activities are carefully analyzed to extract valuable insights and assess the influence of shared research. Industries such as retail, personal care, and tourism leverage text analytics to better understand customer preferences and sentiments, ultimately refining their strategies and enhancing customer engagement. NLP helps companies refine their strategies by understanding customer needs, enabling product adjustments and personalized offerings [6, 7].

2.2 NLP in E-Commerce

Natural Language Processing (NLP) in the retail sector is an emerging field fueled by digital transformation, revolutionizing how businesses engage with customers and optimize operations. By harnessing the power of text data analytics, NLP is enhancing business performance and redefining customer relationship management (CRM).

To enhance online product recommendations, [8] employed the Elman spike neural network to refine sentiment analysis. Their objective was to minimize mean absolute error while maximizing recall. The dataset consisted of Amazon product reviews, and the study explored various machine learning models, including KNN, Decision Tree, Support Vector Machine, Random Forest, and Logistic Regression. The performance of these models was evaluated using metrics such as accuracy, precision, and F1-score. However, a key limitation of the study was its focus on Amazon reviews, potentially missing valuable insights from other domains.

In a study conducted by [9], the tourism industry gained valuable insights by examining the culture shock experiences of foreign tourists in India, China, and the UAE. Data for the research was collected from a variety of sources, including microforms, blogs, e-commerce platforms, and relevant websites. The study primarily employed Aspect-Based Sentiment Analysis (ABSA) and Emotion Analysis (EA) to assess the sentiments (positive, negative, neutral) and emotions (happiness, sadness, anger, fear, surprise) expressed by reviewers.

2.3 NLP on Social Media

NLP on social media is an emerging research area that combines computer and qualitative analytic techniques to leverage the vast data generated by social media content for analysis [10]. There have been remarkable advancements, especially in the sampling and extraction of subsets from social media data. However, a key challenge remains effectively applying qualitative textual analysis to these vast datasets [11]. Integrating data science with qualitative methods has proven to be an effective approach for addressing specific research challenges.

Rahman et al. [12] enhanced social media text classification by applying multi-tier sentiment analysis, surpassing single-tier models. Using the "Movie Review" dataset with five sentiment labels, they pre-processed data with stop words removal, lemmatization, and TF-IDF vectorization. The multi-tier model comprising Decision Tree (52%), SVM (52%), and Naïve Bayes (56%) outperformed single-tier approaches. However, limitations included an imbalanced dataset and difficulty detecting spam reviews, with future work suggesting the use of word embeddings or deep learning.

Excessive social media use is linked to negative mental health outcomes in youth, as a study analyzing Twitter content suggests [13]. Researchers collected tweets using Twitter's API and categorized them into five emotions such as joy, sadness, anger, fear, and disgust using IBM's Natural Language Understanding tool. They identified "harmful" tweets by applying a threshold of 1.08 based on Euclidean Distance. However, the study faced limitations, including a small dataset and an insufficiently researched threshold.

The study compared the performance of Naïve Bayes (NB) and Support Vector Machine (SVM) in detecting cyberbullying in Instagram comments using a dataset of 650 labeled comments (bullying vs. non-bullying) [14]. Preprocessing included tokenization, case folding, normalization, stop words removal, and stemming. Feature extraction methods such as TF-IDF, N-grams, and chi-square for feature selection were applied. Model performance was evaluated using accuracy, confusion matrix, precision, recall, and F1-score, with NB outperforming SVM achieving an accuracy of 83.85% without feature selection and 90.77% with feature selection.

2.4 NLP in Beauty and Personal Care Products

The beauty and personal care industry represents a lucrative sector, with profits reaching $2.7 billion in 2022 and projected to grow to $3.24 billion by 2024 [1]. A study by Choi et al. [15] explored shifts in consumer interest in cosmetics and skincare products during the COVID-19 pandemic by analyzing social media text. Although the study did not specify the platforms analyzed, it focused on key terms such as "beauty" and "routine." Using text mining and the PLS-SEM model, the researchers uncovered notable changes in consumer awareness of beauty products before and after the pandemic. However, the study acknowledged that the use of an English-only dataset limited the scope of its findings, while the lack of macro-level data prevented a deeper understanding of individual consumer insights.

In another study examining customer sentiment towards local SME cosmetic brands, researchers developed a three-stage analysis framework: "Media Monitoring, Sentiment Analysis, and Focus Group Discussion" [16]. This research focused on 15 local SME brands, analyzing approximately 11,000 posts and comments. Using Natural Language Processing (NLP) for sentiment analysis, the study meticulously classified the data, enabling a nuanced discussion of consumer sentiment toward each brand.

NLP is vital for social media platforms where the vast volume of customer reviews challenge the beauty and personal care industry, as traditional market research methods struggle to capture dynamic consumer behaviors. To stay competitive, the industry needs advanced techniques to extract actionable insights and identify emerging trends from this complex data.

3 Methodology

The methodology provides a structured approach to data mining, covering key phases essential for success. It aligns project goals with business objectives from the start and accommodates various systems and tools for efficient planning and execution.

3.1 Business Understanding

Traditional market research methods struggle to capture the fast-evolving customer behaviors on social media, missing key consumer trends. In the beauty industry, where new market segments emerge rapidly, advanced analytics processes are needed to extract actionable insights from unstructured data. Companies that adapt will maintain a competitive edge in today's digital, data-driven world.

3.2 Data Understanding

This study collected customer reviews from three prominent platforms: Instagram, Shopee, and Facebook. Reviews from Instagram and Facebook were scraped between March 2021 and February 2024, specifically from the Minimalist's main page. For Facebook, data was extracted from three official brand pages: "cetaphilmalaysia," "EucerinMY," and "OlayMalaysia." In addition, a selection of customer reviews was gathered from the Minimalist's Shopee Mall page, although platform limitations made this process more time-consuming. The datasets from these three sources were merged, resulting in a final collection of 151,914 customer reviews, organized under three key attributes: Comment Text, Publish Date, and Source.

The "Comment Text" column contains customer reviews sourced from various platforms, while the "Publish Date" represents the timestamps of when the customer reviews were posted, and the "Source" identifies the platforms from which the customer reviews originated. Through exploratory data analysis (EDA), a notable presence of duplicates and missing values in the "Comment Text" column was identified. Also, the EDA showed the average length of the customer review was 116 characters, with 75% of reviews under 138 characters. The word cloud (see Fig. 1) highlights key terms, with phrases such as "reduce acne" and "salicylic acid" standing out, underscoring prevalent consumer interests.

3.3 Data Preprocessing

Cleaning and standardizing text data is crucial. This process was conducted using and various Python based libraries. First, null, and duplicate values were removed using the suitable pandas library functions. Then the emojis were converted to words (demojized) using a specific python-based library named demoji where 3 most repetitive emojis including "love", "smile, and "laugh" were kept for interpretation and other emojis were either removed, or demojized.

The dataset was scraped from unlabeled social media sources, requiring the use of a pre-trained model, "finite automata/bertweet-base-sentiment-analysis," to label the data.

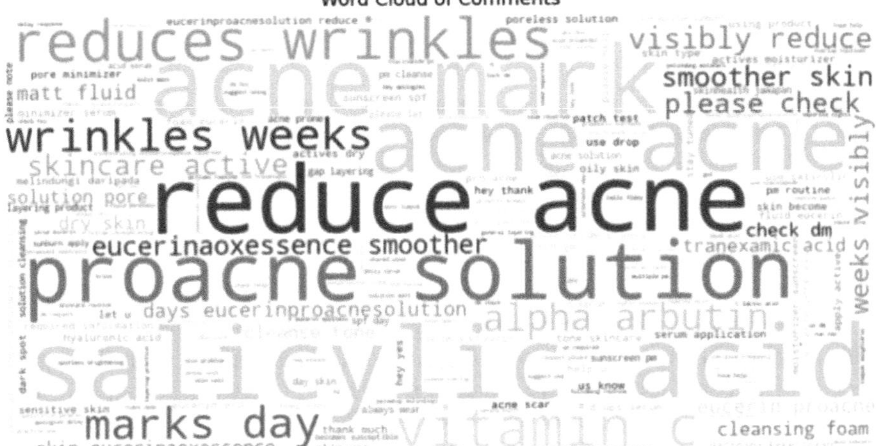

Fig. 1. Word cloud

This model was specifically trained on tweets from Twitter (or X), where the character limit is capped at 280 per tweet. Consequently, it was essential to carefully check and remove any data that exceeded this character limit to ensure consistency with the model's intended use.

Utilizing the "re" library, a remove_urls function was built, with the "compile" function being used to search for and remove any text with the URL pattern. The dates were normalized to "%d/%m/%Y." Text was further normalized by removing punctuation marks, special characters, extra white spaces and stop words using the regular expression and nltk libraries. Further, the data was found with mix of languages such as Malay and Chinese and were translated into English as the customer reviews were sourced from three different origins, including Malaysia.

3.4 Labelling the Target

The study employed the pre-trained model "finiteautomata/bertweet-basesentiment-analysis," which was trained on the SemEval 2017 corpus of 40,000 tweets. Raw customer reviews were input into the model, generating sentiment labels such as negative, neutral, and positive. This model was chosen for its ability to handle social media language, including abbreviations, emojis, and informal speech [17]. The reviews were then preprocessed using NLP techniques for supervised machine learning, aimed at sentiment classification and the influential keyword extraction. The dataset was prepared for supervised sentiment classification model, as the pre-trained model's maximum accuracy was reported at 78% [18]. However, a better predictive model is required to effectively classify the customer reviews with higher accuracy.

3.5 Data Analysis and Results

3.5.1 Model Building

The input data underwent meticulous cleaning through appropriate preprocessing techniques. The dependent variable, known as the Sentiment Label, was designated as 'labels,' while the independent variable, Comment Text, was labeled as 'texts.' The sentiment labels were categorized into three distinct classes: NEG (negative), NEU (neutral), and POS (positive). To facilitate the application of machine learning algorithms, the text input data was vectorized using the TF-IDF technique.

TF-IDF technique was selected as it is more suitable for discriminative machine learning algorithms, and it generates a feature set that highlights the most relevant terms while down-weighting common words that do not help with classification. This allows models to focus on discriminative features, resulting in better performance for tasks such as text classification and information retrieval. Further, The discriminative machine learning algorithms were selected due to the following reasons:

1. Less number of data
2. Limitations in the training time and computational resources
3. Fair model interpretability
4. More generalization

The test size was set to 20%, with a random state of 42 for consistency. The models were built using hyper parameter tuning with cross-validation and details of the models are presented in Table 1.

The Ensemble model building was done by using the selected hyper parameters of the respective machine learning models as shown in Table 1. Hyperparameter Tuning and Cross-Validation are both crucial techniques that helped addressing the class imbalance nature of the dataset, although indirectly by selecting the optimal hyperparameters to enhance its performance and Cross-validation helped optimizing the model's ability to learn from imbalanced data. The Ensemble model, which outperformed the pre-trained model in accuracy, was used again to predict customer review sentiments, enhancing reliability and confidence in extracting key influential keywords.

Keyword Extraction:

To improve customer feedback analysis, synonyms for each keyword were identified using WordNet's "get_synonyms" function, reviewed for relevance, and added to the keyword set. Frequency and sentiment analysis revealed key topics and customer attitudes (positive, neutral, negative). Regular expressions refined the analysis by detecting patterns and minimizing errors. Co-occurrence analysis identified keyword associations, guiding marketing and product strategies. TF-IDF was applied to assess term importance, offering deeper insights into customer concerns and preferences. The above operations were applied on the customer reviews which were predicted as positive and helped to obtain the set of influential keywords as shown in Table 2.

Table 1. Supervised machine learning models

Model	Hyperparameters	Evaluation Measures
XGBoost (XGB)	learning_rate = 0.2, max_depth = 7, n_estimators = 300, subsample = 0.7, use_label_encoder = False, eval_metric = 'logloss', random_state = 42	Accuracy = 89% Precision = 0.89, Recall = 0.89, F1-Score = 0.89
Random Forest (RF)	n_estimators = 100, random_state = 42, criterion = 'gini', min_samples_split = 2, min_samples_leaf = 1	Accuracy = 88% Precision = 0.88, Recall = 0.88, F1-Score = 0.88
Support Vector Machine (SVM)	C = 1.0, Kernel = 'linear', probability = True, degree = 3, gamma = 'scale',	Accuracy = 89% Precision = 0.89, Recall = 0.89 F1-Score = 0.89
Logistic Regression (LR)	penalty = 'l2', tol = 0.0001, C = 1.0, max_iter = 1000, solver = 'lbfgs'	Accuracy = 89% Precision = 0.89, Recall = 0.89, F1-Score = 0.88
Ensemble model (XGB+ RF+ LR+ SVM)	The Ensemble model building was done by using the selected hyper parameters of the respective machine learning models as listed above.	Accuracy = 90% Weighted Precision = 0.90 Recall = 0.90 F1-Score = 0.89

Interpretations:

The extracted influential keywords were further explored to offer deep insights across

Table 2. Keywords list by topic

Topic	Influential keywords (listed not in the order of influence)
Product-related by Function	'Cleansers', 'Exfoliants', 'Toners', 'Serums', 'Moisturizers', 'Masks', 'Treatments', 'Sunscreens', 'Mists
Product-related by Ingredient Type Policies	'Natural', 'Organic', 'Halal', 'Vegan', 'Chemical', 'Alcohol', 'Fragrance' 'Return', 'Refund', 'Exchange', 'Guarantee'
Delivery and Shipping	'Delivery', 'Delay', 'Ship', 'Package', 'Track', 'Long', 'Lost', 'Late', 'Broken', 'Missing', 'Cancel', 'Stolen'
Promotion and Discounts	'Discount', 'Sale', 'Offer', 'Promotion', 'Coupon', 'Voucher'
Market and Trends	'New', 'Release', 'Popular'
Health and Safety	'Comedogenic', 'Irritating', 'Toxic', 'Allergen', 'Sensitive', 'Test', 'Safe', 'Regulation', 'Quality', 'Standard', 'Certification', 'Approve'

multiple dimensions, such as Function, Ingredients, Policy, Delivery, Promotion & Discount, Market & Trends, and Health & Safety. This provides a comprehensive understanding of the influential keywords and their associated sentiment values. Terms like "Treatments" and "Serums" highlighted strong customer interest, while "Sunscreens" and "Mists" received positive feedback, and "Treatments" and "Serums" faced more criticism offering manufacturers a guidance for improvement. Similarly, the consumer preferred "Natural" and "Organic" under the Ingredient Type, alongside potential risks, and benefits of certain chemicals. Terms like "Return," "Refund," "Exchange," and "Guarantee" are key in the Policies category. Consumers were particularly dissatisfied with refund and return processes, while exchange and guarantee policies were more accepted. In the Delivery & Shipping category, terms like "Delivery," "Delay," and "Ship" were the most dominate keywords. Terms like "Discount," "Sale," and "Offer" were the most influential keywords in the Promotion and Discount category. These promotions generated positive feedback, indicating strong consumer appeal. Similarly, terms like "New," "Release," and "Popular" were key in the Market and Trends category. The terms such as "Sensitive", and "Safe" were the most influential keywords under the Market and Trends category which were clearly understood.

4 Discussion and Conclusion

The sentiment models were evaluated using accuracy, precision, recall, and F1 scores. XGBoost, SVM and Logistic Regression models achieved 89% accuracy respectively where the Random Forest model scored 88%. The Ensemble Model outperformed the rest with 90% accuracy, improving predictions across all sentiment classes and ensuring reliability for future use. The Ensemble Model could be deployed enabling real-time analysis of customer reviews and providing users with an efficient way to gauge sentiment using the best machine learning model from this study.

Keyword extraction and sentiment prediction as discussed in [19] have provided valuable insights into the beauty and personal care industry as well. The most influential keywords are presented in Table 2 and visualized through various interactive dashboards. Overall the analysis offers valuable insights into trends affecting customer satisfaction and preferences

Keyword extraction plays a pivotal role in enhancing marketing strategies customer behavior prediction and providing insights that can lead to better-targeted advertising and business outcomes. Marking strategies like Audience Segmentation Enhanced Ad Relevance Improved Ad Performance and Bid Strategy Optimization are some of the key benefits derived from effective keyword extraction. By carefully analyzing and segmenting the audience ads can be tailored to specific groups making them more relevant and engaging. This results in better performance metrics and a higher return on investment. Additionally keyword extraction helps fine-tune bid strategies ensuring the ads are placed at the right moments for maximum impact ultimately optimizing both visibility and cost-efficiency. Similarly Customer behavior prediction can be effectively achieved through several key strategies: Predicting Interests and Intent Trend Identification Customer Journey Mapping and Personalized Recommendations. Real-world applications of advanced targeting and personalization strategies include Amazon's search-based targeting Netflix's content personalization Home Depot's localized targeting and Spotify's behavioral predictions. Our research leverages social media data related to beauty and personal care products offering the potential to serve a vast and diverse customer base.

The project provided a data science solution for analyzing sentiment in beauty and personal care social media, helping brands understand consumer preferences. An interactive dashboard and sentiment model delivered accessible insights, with strong data preprocessing and high predictive accuracy. The overall solution idea has been depicted in a framework (see Fig. 2).

The project faced several constraints, including a rigid dashboard that slowed updates and struggled with changing consumer sentiments. The use of a pre-trained English sentiment model also limited performance for non-English comments. Additionally, the research focused on specific brands and social networks, potentially missing broader consumer insights in the Malaysian beauty market.

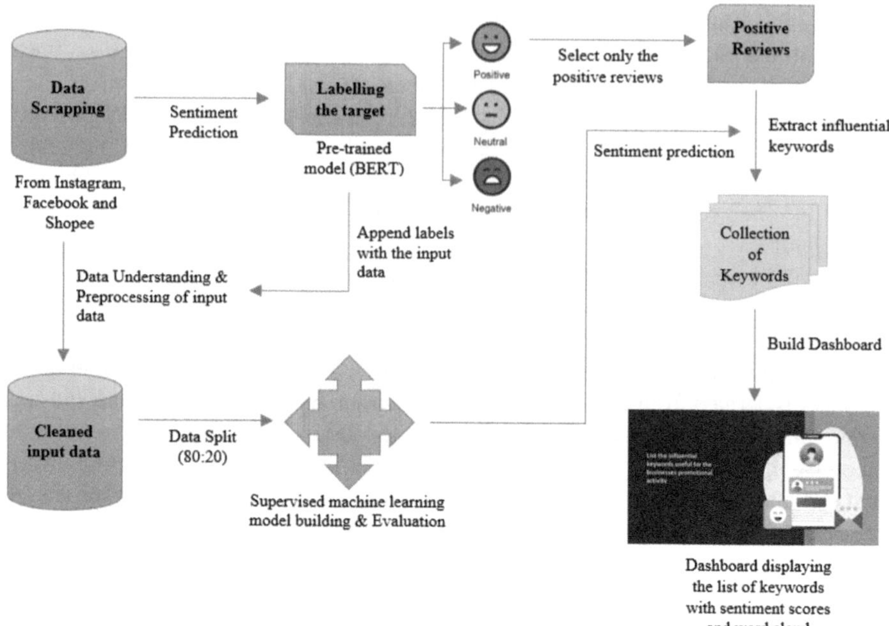

Fig. 2. Extracting influential keyword framework

To improve effectiveness, create a dynamic, real-time dashboard for regular data updates. Expand the sentiment analysis model's language capabilities and broaden data sources to gain a more accurate market overview. Additionally, use advanced NLP techniques, including deep learning for the beauty sector, to refine sentiment analysis and enhance insights into consumer perceptions.

References

1. Statista, Beauty & Personal Care - Malaysia | Statista Market Forecast: Available: https://www.statista.com/outlook/cmo/beauty-personal-care/malaysia. [Accessed 30 Nov 2024]
2. Jimenez-Marquez, J.L., Gonzalez-Carrasco, I., Lopez-Cuadrado, J.L., Ruiz-Mezcua, B.: Towards a big data framework for analyzing social media content. Int. J. Inf. Manag. **44**, 1–12 (2019). https://doi.org/10.1016/j.ijinfomgt.2018.09.003
3. Herzallah, D., Muñoz Leiva, F., Liébana-Cabanillas, F.: To buy or not to buy, that is the question: understanding the determinants of the urge to buy impulsively on Instagram commerce. J. Res. Interact. Marketing, ahead-of-print. (2021). https://doi.org/10.1108/jrim-05-2021-0145
4. Dwivedi, Y.K., Ismagilova, E., Hughes, D.L., Carlson, J.: Setting the future of digital and social media marketing research: perspectives and research propositions. Int. J. Inf. Manag. **59**, 1–37 (2021). https://doi.org/10.1016/j.ijinfomgt.2020.102168
5. Bennett, C.H.: The thermodynamics of computation—a review. Int. J. Theor. Phys. **21**, 905–940 (1982)
6. Rooderkerk, R., DeHoratius, N., Musalem, A.: The past, present, and future of retail analytics: insights from a survey of academic research and interviews with practitioners. Prod. Oper. Manag. **31**(10), 3727–3748 (2022). https://doi.org/10.1111/poms.13811

7. Darwich, M., Mohd Noah, S.A., Omar, N., Osman, N.A., Ahmad, I.S.: Quantifying the natural sentiment strength of polar term senses using semantic gloss information and degree adverbs. J. Adv. Inf. Technol. **11**(3), 109–118 (2020). https://doi.org/10.12720/jait.11.3.109-118
8. Solairaj, A., Sugitha, G., Kavitha, G.: Enhanced Elman spike neural network based sentiment analysis of online product recommendation. Appl. Soft Comput. **132**, 109789 (2023). https://doi.org/10.1016/j.asoc.2022.109789
9. Mehra, P.: Unexpected surprise: emotion analysis and aspect based sentiment analysis (ABSA) of user generated comments to study behavioral intentions of tourists. Tour. Manag. Perspect. **45**, 101063 (2023). https://doi.org/10.1016/j.tmp.2022.101063
10. Qi, Y., Shabrina, Z.: Sentiment analysis using twitter data: a comparative application of lexicon- and machine-learning-based approach. Soc. Netw. Anal. Min. **13**(1, 10) (2023). https://doi.org/10.1007/s13278-023-01030-x
11. Tunca, S., Sezen, B., Wilk, V.: An exploratory content and sentiment analysis of the Guardian metaverse articles using Leximancer and natural language processing. J. Big Data **10**(1, 7) (2023). https://doi.org/10.1186/s40537-023-00773-w
12. Rahman, H., Tariq, J., Masood, M.A., Subahi, A.F., Khalaf, O.I., Alotaibi, Y.: Multi-tier sentiment analysis of social media text using supervised machine learning. Comput. Mater. Continua. **74**(3), 5527–5543 (2023). https://doi.org/10.32604/cmc.2023.033190
13. Benrouba, F., Boudour, R.: Emotional sentiment analysis of social media content for mental health safety. Soc. Netw. Anal. Min. **13**(1), 17 (2023). https://doi.org/10.21203/rs.3.rs-2170906/v1
14. Riadi, S., Utami, E., Yaqin, A.: Comparison of NB and SVM in sentiment analysis of cyber-bullying using feature selection. Jurnal dan Penelitian Teknik Informatika. **8**(4), 2414–2424 (2023). https://doi.org/10.33395/sinkron.v8i4.12629
15. Choi, Y.H., Kim, S.E., Lee, K.H.: Changes in consumers' awareness and interest in cosmetic products during the pandemic. Fash. Text **9**(1) (2022). https://doi.org/10.1186/s40691-021-00271-8
16. Arief, N.N., Pangestu, A.B.: Perception and sentiment analysis on empathic brand initiative during the COVID-19 pandemic: Indonesia perspective. J. Creative Commun. **17**(2), 162–178 (2022). https://doi.org/10.1177/09732586211031164
17. Velampalli, S., Muniyappa, C., Saxena, A.: Performance evaluation of sentiment analysis on text and emoji data using end-to-end, transfer learning, distributed and explainable AI models. J. Adv. Inf. Technol. **13**(2), 167–172 (2022)
18. Dat, Q.N., Thanh, V., Anh, T.N.: A pre-trained language model for English tweets. In: Proceedings of the EMNLP 2020: System Demonstrations (2020). https://doi.org/10.48550/arXiv.2005.10200
19. Mafas, R., Abubacker, N.F., Devina, D.W.: Sentiment-based recommendation for online shopping. J. Theor. Appl. Inf. Technol. **102**(9) (2024)

Eel and Grouper Lyrebird Optimizer-Based Fractal Deep Spiking Residual Network for Breast Cancer Detection Using Mammogram Images

Sachin Urabinahatti[1], D. Jayadevappa[2], and T. S. Kruthi[3](✉)

[1] Department of AI&ML, BMS Institute of Technology and Management, VTU, Bengaluru, India
sachinau@bmsit.in
[2] Dr. H N National College of Engineering, VTU, Bengaluru, India
[3] Department of Electronics & Communication Engineering, JSS Academy of Technical Education, VTU, Bangalore, India
kruthits@jssateb.ac.in

Abstract. Breast cancer is the most common cancer type among women throughout the entire world. The mammography is a successful imaging modality for the earliest diagnosis of breast cancer in its beginning stages. Owing to less visibility and bad contrast in mammogram (MG) images, earlier identification of breast cancer is a critical step for efficacious treatment of this disease. Here, Eel and Grouper Lyrebird Optimizer-based Fractal Deep Spiking Residual Network (EGLO_FDSRN) is introduced for detecting breast cancer using MG images. Initially, MG images from mammographic image datasets are considered input. The Wiener filter is utilized to pre-process the considered MG image. After that, the cancer region is segmented utilizing U-Next. Then, the cancer region segmented image is augmented by applying image augmentation methods, and thereafter, features suitable for the detection process are extracted. Lastly, breast cancer is detected by the Fractal Deep Spiking Residual Network (FDSRN). However, FDSRN is developed by combining FractalNet with Spiking ResNet (S-ResNet). Furthermore, FDSRN is trained by Eel and Grouper Lyrebird Optimizer (EGLO) which is introduced by incorporating Eel and Grouper Optimizer (EGO) and Lyrebird Optimization Algorithm (LOA). In addition, EGLO_FDSRN acquired the best outcomes with 91.780%, of accuracy, 90.689% of sensitivity and 91.610% of specificity.

Keywords: Mammogram image · breast cancer · FractalNet · Eel and Grouper Optimizer · Lyrebird Optimization Algorithm

1 Introduction

The breast cancer reveals one of the prevailing types of invasive malignancy affecting the women population. This situation can influence every gender across diverse ages, with high prevalence monitored among people in middle age and old age [6]. Therefore,

J. Shreyas et al. (Eds.): CODE-AI 2025, CCIS 2690, pp. 71–79, 2026.
https://doi.org/10.1007/978-3-032-19321-6_7

breast cancer is a vital issue in the public health. Hence, earlier detection of this cancer is a better strategy for fighting against it. Mammography diagnosing is the generally employed accurate modality that can be utilized for detecting earlier detection of cancer [1, 16]. Generally, breast cancer initiates with ductal dysfunction or invasive ductal carcinoma. Even though, it also originates in breast tissues as well as lobules. Several researchers also recognized that variations in lifestyle, environment, and hormones also augment the threat of breast cancer [7, 15, 17]. Lower dosage X-ray analysis of the breast is employed for envisaging an internal part of breasts. When compared with previously employed tools, mammography reveals a breast to much lesser radiation dosage [7, 19]. Mammography detects highly subtle symptoms that include structural variation as well as bilateral asymmetry and more clear irregularities like masses and calcification. The densities, nodules, or masses are all feasible irregularities in mammography [8]. MG images are investigated by radiologists for diagnosing breast cancer. However, an analysis by radiologists on the presence of breast cancer varies due to diversities in their previous understanding and experiences [8]. Currently, several newer techniques based upon Deep Learning (DL) recognition are presented for solving the issue of worse diagnosing performance of Computer-Aided Diagnosis (CAD) methods [7].

There are various classification approaches developed that include DL and Machine Learning (ML). The diverse DL methods are recommended that detect an occurrence of breast cancers accurately and quickly [4]. DL is considered an effectual model for classifying and detecting clinical images [2, 18]. Various image processing techniques are developed, which include ML models [20] namely Convolutional Neural Networks (CNN), Support Vector Machine (SVM), Artificial Neural Networks (ANN), and Naive Bayesian (NB) [5].

2 Literature Survey

Al-Fahaidy, F.A., et al. [1] designed SVM for diagnosing breast cancer utilizing digital MG images. This method efficiently led to the earliest identification of breast cancer, but still, the worse quality of the dataset obtained inexact outcomes.

Rahman, H., et al. [2] developed ResNet-50 CNN for effectual breast cancer diagnosis from complicated MG images. It facilitated earlier diagnosing of benign and malignant cancer, thereby saving resources and patient's lives. However, this scheme led to unpredictable diagnosing performance over diverse databases or medical environments.

Ibrokhimov, B. and Kang, J.Y., [3] devised two-stage DL for detecting breast cancer employing high-resolution MG images. This technique reduced false detections and had the capability to identify subtle patterns, though it required an enormous count of parameters.

Nagalakshmi, T., [4] introduced Ensemble-Net to segment the pectoral muscle (PM) boundary from residual breast region in MG scans for accurately detecting breast cancer. This approach successfully accomplished high effectiveness in segmenting and categorizing breast cancer scans. However, it failed to focus on the extraction of more features to effectively classify breast cancer.

3 Proposed EGLO_FDSRN for Breast Cancer Detection Using MG Images

Here, EGLO_FDSRN is introduced to detect breast cancer utilizing MG images. Initially, MG images taken from mammographic image datasets are pre-processed by the Wiener filter. And U-Next is employed to segment the cancer region. Then, the image is augmented by considering shifting, random erasing, and rotation. After that, BRIEF, GLCM features, and GBP are extracted. At last, breast cancer detection is performed by FDSRN which is developed by incorporating FractalNet with S-ResNet. The training of FDSRN is done employing EGLO, which is devised by merging EGO and LOA. Figure 1 represents the pictorial view of EGLO_FDSRN for breast cancer detection utilizing MG images.

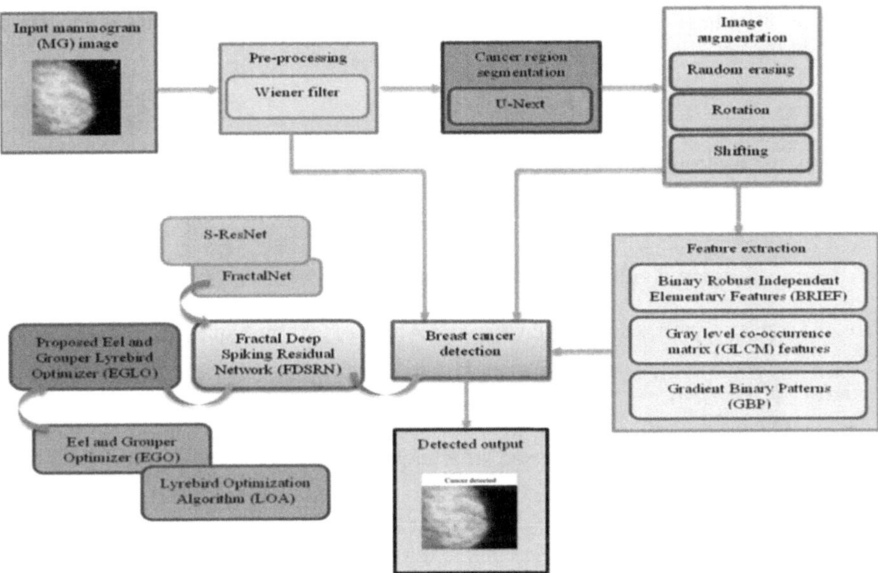

Fig. 1. Pictorial view of EGLO_ FDSRN for breast cancer detection utilizing MG images

3.1 Input MG Image Acquisition

The detection of breast cancer is performed by considering input from mammographic image datasets comprising Mammographic Image Analysis Society (MIAS) and Digital Database for Screening Mammography (DDSM) databases [21]. It can be formulated as,

The detection of breast cancer is performed by considering input from mammographic image datasets comprising Mammographic Image Analysis Society (MIAS) and Digital Database for Screening Mammography (DDSM) databases [21]. It can be formulated as,

$$M = \{M1, M2, \ldots, Mb, \ldots, Mc\} \tag{1}$$

Here, M_b depicts b^{th} MG image and M_c illustrates overall MG images in the dataset M. MIAS database has 322 digitized films and they are available on about 2.3GB 8mm (ExaByte) tape with every image size as 1024x1024. The DDSM database contains approximately about 2,500 studies. All studies contain two phases of individual breasts and also, a few relevant information of the patients.

3.2 Pre-Processing Using Wiener Filter

The MG image M_b is a pre-processed employing Wiener filter. Wiener filter [9] builds an optimum evaluation of the actual image by applying the minimal Mean Squared Error (MSE) constraints among the original and estimated images. The Wiener filter can be illustrated by,

$$W_b = Mb - C_b \tag{2}$$

Here, M_b indicates actual image and $\mathbb{C}_b$ depicts an estimated image whereas W_b implies a pre-processed MG image.

3.3 Cancer Region Aegmentation Utilizing U-Next

Here, the cancer region is segmented by employing U-Next by acquiring W_b as an input.

Structure of U-Next.

U-Next [10] is a DL segmentation framework that can automatedly offer the significance of diverse sizes and shapes of targeted structures in an image. To prevent the explosion of parameters and superiorly merge features, 1 × 1 convolution (conv) is utilized in a structure to make a network highly accurate with fewer parameters. It comprises attention up-sampling blocks, dense connection and skip spatial pyramid pooling block (SkipSPP). To decrease the contextual data loss in various areas, SkipSPP is developed. The topmost and bottommost information are fused in a moderately denser way by incorporating various scaled features. An attention up-sampling model is equivalent to a human discerning visual attention procedure. The segmented output obtained from U-Next is given by,

$$U_b = \Im_{scale}(\alpha_l, \varpi_l) = \varpi_l * \alpha_l \tag{3}$$

Here, $\Im_{scale}(\alpha_l, \varpi_l)$ signifies channel-wise multiplication among scalar ϖ_l and feature map α_l. The segmented cancer region can be symbolized asU_b.

3.4 Image Augmentation Based on the Aegmented Image

Here, the image is augmented based on techniques namely shifting, random erasing, and rotation by taking U_b as an input.

Shifting [14] is defined as a process of moving an image along both the x-axis and y-axis that can be represented as G_1. Rotation [14] refers to a process accomplished by rotating the particular image across both axis either in right or left directions by an angle amid 1 and 359. The rotation can be illustrated by G_2. Random erasing [14] refers to the technique that is employed for erasing a single square in a square area of an image randomly and it can be signified by G_3. An augmented image acquired is indicated as G_b, such that,

$$\boldsymbol{G_b = \{G_1, G_2, G_3\}} \tag{4}$$

Here, G_1, G_2 and G_3 implies shifting, rotation, and random erasing whereas G_b symbolizes augmented image.

3.5 Feature Extraction Based on Augmented Image

Initially, BRIEF and GBP are applied to augmented image G_b. BRIEF [13] is a feature descriptor that conducts the simplest binary analysis amongst pixels in the smoothened image areas and it is denoted as X_1. GBP [12] illustrates each pixel by corresponding sim evaluations of their neighborhood pixels. The GBP feature can be represented by a term X_2. After applying BRIEF and GBP to augmented image G_b, the textural feature image X_b is obtained, such that,

$$X_b = \{X_1, X_2\} \tag{5}$$

Thereafter, GLCM features [11] such as area, equivalent diameter, solidity, contrast, convex area, eccentricity, energy, and homogeneity are applied to textural feature image X_b for obtaining feature vectors. Area mentions a scaler value indicating an actual pixel count on Region of Interest (ROI) and it is implied by T_1. Convex area reveals a vector value providing count of ROI convex pixels and the binary depiction with every pixel packed inside a circle. The convex area is represented by T_2. An equivalent area refers to a circular diameter of the same region that is signified by T_3. Eccentricity specifies a range associations amongst ellipse focus points as well as major axis longitudinal points, which is given by T_4. Solidity indicates the total pixels in convex hell and it is manifested as T_5. Contrast represents an intensity variation amongst a pixel and neighborhood measurement for the whole ROI, which is illustrated as T_6. Homogeneity reveals a similarity of GLCM in an individual ROI to an entity position within the GLCM that is implied by T_7. Energy depicts the aggregation of GLCM squared elements and it is symbolized asT_8.

An acquired feature vector is denoted by T_b, such that,

$$T_b = \{T_1, T_2, \ldots, T_8\} \tag{6}$$

3.6 Breast Cancer Detection Using FDSRN

Here, FDSRN is employed to detect breast cancer and FDSRN is derived by combining FractalNet with S-ResNet.

Figure 2 exposes a general view of FDSRN for breast cancer detection

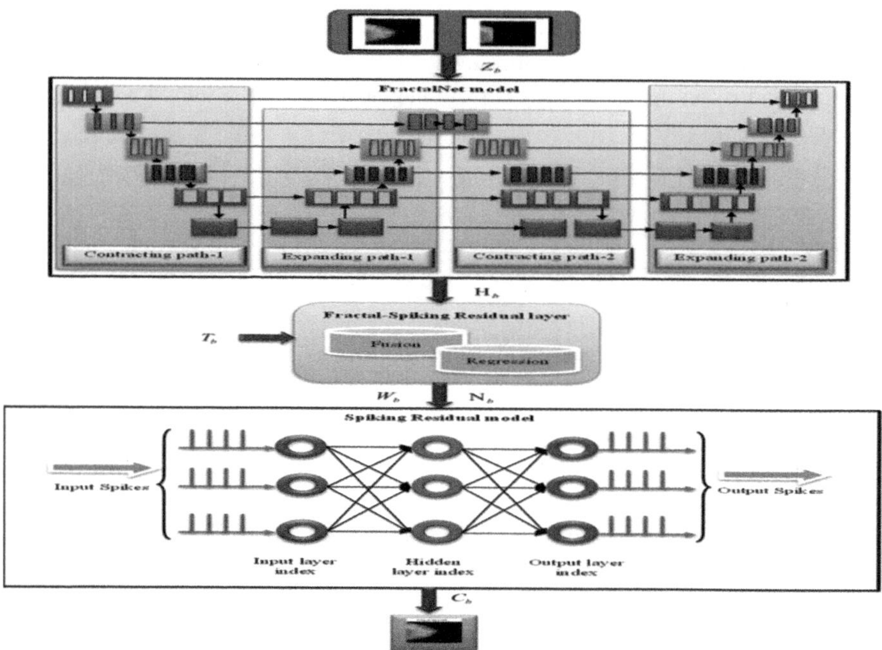

Fig. 2. General view of FDSRN for breast cancer detection

4 Results and Discussion

The evaluation results of EGLO_FDSRN devised for breast cancer detection are represented in this section.

4.1 Comparative Techniques

SVM [1], ResNet-50 CNN [2], Two-stage DL [3], Ensemble-Net [4] and FDSRN are the traditional models compared with EGLO_FDSRN to prove their effectiveness.

4.2 Comparative Evaluation

The assessments of EGLO_FDSRN are carried out with concern to two databases as MIAS database and the DDSM database.

4.3 Comparative Discussion

Table 1 discusses an analysis of values achieved by EGLO_FDSRN and other detection models. For K-fold = 9, EGLO_FDSRN acquired an accuracy of 91.780% whereas traditional approaches attained 78.511%, 82.394%, 84.881%, 86.158%, and 89.293%. A high accuracy specifies that EGLO_FDSRN provided effectual diagnosing and appropriate treatments. When K-fold = 9, the sensitivity achieved by EGLO_FDSRN is 90.689% whereas comparative models obtained 80.419%, 82.828%, 85.173%, 86.854%, and 88.330%. The maximum sensitivity represents that EGLO_FDSRN identified several cases of breast cancer correctly. The existing approaches attained 80.663%, 82.999%, 84.565%, 87.536%, and 89.497% of specificity whereas EGLO_FDSRN acquired 91.610% while K-fold is 9. The maximum specificity depicts that designed EGLO_FDSRN assured healthier individuals are accurately diagnosed. It can be acknowledged that EGLO_FDSRN is the superior breast cancer detection model as it attained 91.780%, 90.689%, and 91.610% accuracy, sensitivity, and specificity while considering K-fold = 9 for the DDSM database

Table 1. Comparative discussion of EGLO_FDSRN

Datasets	Setups	Metrics/ Methods	SVM	ResNet-50 CNN	Two-stage DL	Ensemble-Net	FDSRN	Proposed EGLO_FDSRN
MIAS database	**Training data = 90%**	*Accuracy (%)*	80.652	83.749	85.600	87.444	89.756	91.598
		Sensitivity (%)	79.718	83.336	84.289	85.340	87.766	90.443
		Specificity (%)	78.376	82.490	85.498	86.714	89.148	91.890
	K-fold = 9	*Accuracy (%)*	78.423	82.126	85.932	86.151	89.009	91.332
		Sensitivity (%)	78.729	82.265	85.718	87.634	88.087	90.261
		Specificity (%)	80.852	82.254	85.268	87.852	89.874	91.761
DDSM database	**Training data = 90%**	*Accuracy (%)*	79.030	83.987	84.849	86.527	89.152	91.654
		Sensitivity (%)	80.377	83.011	85.502	86.092	88.159	90.879
		Specificity (%)	79.213	82.645	85.099	88.643	89.965	91.928
	K-fold = 9	*Accuracy (%)*	78.511	82.394	84.881	86.158	89.293	**91.780**
		Sensitivity (%)	80.419	82.828	85.173	86.854	88.330	**90.689**
		Specificity (%)	80.663	82.999	84.565	87.536	89.497	**91.610**

5 Conclusion

In this research, EGLO_FDSRN is designed for detecting breast cancer using MG images. To detect breast cancer, MG image is considered from mammographic image datasets. Then, it is pre-processed utilizing a Wiener filter. Afterwards, the cancer region is segmented employing U-Next. Thereafter, the image is augmented by applying techniques namely shifting, random erasing, and rotation. Next, specific features namely BRIEF, GLCM features, and GBP are extracted. Finally, breast cancer detection is carried out by FDSRN which is modeled by incorporating FractalNet with S-ResNet. Furthermore, FDSRN is trained by EGLO and it is designed by combining EGO and LOA. Additionally, EGLO_FDSRN achieved accuracy, sensitivity, and specificity of about 91.780%, 90.689%, and 91.610%. In the future, risk level determination and prediction of possibilities of reoccurrence will be included to improve the robustness of the framework.

References

1. Al-Fahaidy, F.A., Al-Fuhaidi, B., AL-Darouby, I., AL-Abady, F., AL-Qadry, M., AL-Gamal, A.: A diagnostic model of breast cancer based on digital mammogram images using machine learning techniques. Appl. Comput. Intell. Soft Comput. **1**, 3895976 (2022)
2. Rahman, H., Naik Bukht, T.F., Ahmad, R., Almadhor, A., Javed, A.R.: Efficient breast cancer diagnosis from complex mammographic images using deep convolutional neural network. Comput. Intell. Neurosci. **1**, 7717712 (2023)
3. Ibrokhimov, B., Kang, J.Y.: Two-stage deep learning method for breast cancer detection using high-resolution mammogram images. Appl. Sci. **12**(9), 4616 (2022)
4. Nagalakshmi, T.: Breast cancer semantic segmentation for accurate breast cancer detection with an ensemble deep neural network. Neural. Process. Lett. **54**(6), 5185–5198 (2022)
5. Wisaeng, K.: Breast cancer detection in mammogram images using K–means++ clustering based on cuckoo search optimization. Diagnostics **12**(12), 3088 (2022)
6. Anas, M., Haq, I.U., Husnain, G., Jaffery, S.A.F.: Advancing breast cancer detection: enhancing YOLOv5 network for accurate classification in mammogram images. IEEE Access (2024)
7. Maqsood, S., Damasevicius, R., Maskeliūnas, R.: TTCNN: a breast cancer detection and classification towards computer-aided diagnosis using digital mammography in early stages. Appl. Sci. **12**(7), 3273 (2022)
8. Altameem, A., Mahanty, C., Poonia, R.C., Saudagar, A.K.J., Kumar, R.: Breast cancer detection in mammography images using deep convolutional neural networks and fuzzy ensemble modeling techniques. Diagnostics **12**(8), 1812 (2022)
9. Ramani, R., Vanitha, N.S., Valarmathy, S.: The pre-processing techniques for breast cancer detection in mammography images. Int. J. Image, Graph. Signal Process. **5**(5), 47 (2013)
10. Song, T., Meng, F., Rodriguez-Paton, A., Li, P., Zheng, P., Wang, X.: U-next: a novel convolution neural network with an aggregation U-net architecture for gallstone segmentation in CT images. IEEE Access **7**, 166823–166832 (2019)
11. Chilakala, L.R., Kishore, G.N.: Optimal deep belief network with opposition-based hybrid grasshopper and honeybee optimization algorithm for lung cancer classification: a DBNGHHB approach. Int. J. Imaging Syst. Technol. **31**(3), 1404–1423 (2021)
12. Praveena, K.S.: A classical hierarchy method for bone X-ray image classification using SVM. Int. Res. J. Eng. Technol. (IRJET). **4**(8) (2017)

13. Calonder, M., Lepetit, V., Ozuysal, M., Trzcinski, T., Strecha, C., Fua, P.: BRIEF: computing a local binary descriptor very fast. IEEE Trans. Pattern Anal. Mach. Intell. **34**(7), 1281–1298 (2011)
14. Khalifa, N.E., Loey, M., Mirjalili, S.: A comprehensive survey of recent trends in deep learning for digital images augmentation. Artif. Intell. Rev. **55**(3), 2351–2377 (2022)
15. Xian, M., Zhang, Y., Cheng, H.D., Xu, F., Zhang, B., Ding, J.: Automatic breast ultrasound image segmentation: a survey. Pattern Recogn. **79**, 340–355 (2018)
16. Patel, B.C., Sinha, G.R., Soni, D.: Detection of masses in mammographic breast cancer images using modified histogram based adaptive thresholding (MHAT) method. Int. J. Biomed. Eng. Technol. **29**(2), 134–154 (2019)
17. Lotter, W., et al.: Robust breast cancer detection in mammography and digital breast tomosynthesis using an annotation-efficient deep learning approach. Nat. Med. **27**(2), 244–249 (2021)
18. Chougrad, H., Zouaki, H., Alheyane, O.: Deep convolutional neural networks for breast cancer screening. Comput. Methods Prog. Biomed. **157**, 19–30 (2018)
19. Dembrower, K., et al.: Effect of artificial intelligence-based triaging of breast cancer screening mammograms on cancer detection and radiologist workload: a retrospective simulation study. Lancet Digit. Health **2**(9), e468–e474 (2020)
20. Yassin, N.I., Omran, S., El Houby, E.M., Allam, H.: Machine learning techniques for breast cancer computer-aided diagnosis using different image modalities: a systematic review. Comput. Methods Prog. Biomed. **156**, 25–45 (2018)
21. Mammographic Image Analysis Society (MIAS) and Digital Database for Screening Mammography (DDSM) datasets are taken from https://www.mammoimage.org/databases. Accessed on Oct 2024

Deep Reinforcement Learning and Proximal Policy Optimization for Jetbot Automation

G. Ramesh[1](✉), J. Shreyas[2], and N. Sowjanya[3]

[1] Department of ISE, NITTE (Deemed to Be University), NMAM Institute of Technology, Nitte, Udupi, Karnataka, India
rameshmg6308@gmail.com

[2] Department of Information Technology, Manipal Institute of Technology, Manipal Academy of Higher Education, Manipal, Bengaluru, Karnataka, India
shreyas.j@manipal.edu

[3] NITK, Surathkal, Karnataka, India

Abstract. In this work explores the complex design of a state-of-the-art deep reinforcement learning autonomous navigation system that is designed to optimize object delivery in industrial warehouse environments by utilizing Proximal Policy Optimization (PPO) techniques. The system is controlled by the car-shaped robot JetBot, which is equipped with a cutting-edge NVIDIA AI-focused board. It is carefully designed to maneuver around warehouse conditions with remarkable efficiency. Among its primary functions are the ability to move past various obstacles with ease and to enable the automatic delivery of objects to warehouse staff. Extensive in-lab experiments carried out in virtual warehouses accurately mirrored real-world situations, integrating a wide range of impediments like boots, gloves, tools, and implements to fully evaluate the system's capabilities. By means of carefully selected trials, the system's ability to navigate complex warehouse layouts and quickly detect and avoid obstacles was thoroughly tested. The outcomes demonstrated the system's resilience and flexibility in negotiating changing warehouse settings, signaling a major advancement in autonomous warehouse logistics.

Keywords: Pathfinding · Proximal policy optimization · decision making · self-driving · Autonomous robots · Deep reinforcement learning · NVIDIA Jetbot

1 Introduction

In current years, the integration of advanced technologies, particularly artificial intelligence (AI) and robotics, has revolutionized various industries, starting from manufacturing to logistics. One place witnessing widespread advancements is the improvement of independent structures for item shipping within industrial warehouse settings. With the increasing call for efficiency, productiveness, and safety in warehouse operations, there is a developing interest in deploying self reliant robots able to navigate complicated environments, identify obstacles, and autonomously hand over gadgets to distinctive places [1]. Conventional warehouse operations often depend on guide labor for responsibilities

J. Shreyas et al. (Eds.): CODE-AI 2025, CCIS 2690, pp. 80–89, 2026.
https://doi.org/10.1007/978-3-032-19321-6_8

such as object retrieval, sorting, and transportation. However, this approach isn't without its limitations, including exertions, charges, human errors, and inefficiencies in time and resource utilization [2]. As a result, there was a shift in the direction of the adoption of automatic answers to streamline warehouse procedures and enhance standard productivity. A few of the diverse technologies driving this automation fashion, deep reinforcement mastering (DRL) has emerged as a powerful tool for education independent dealers to carry out complicated tasks in dynamic environments and a self sufficient navigation device for object delivery within commercial warehouse settings. The gadget is focused around a vehicle-shaped robot, JetBot, equipped with an NVIDIA AI-oriented board, capable of leveraging DRL strategies to navigate warehouse environments, discover limitations, and deliver items to warehouse employees [5]. The importance of this research lies in its capacity to deal with key demanding situations confronted by conventional warehouse operations, inclusive of labor-extensive strategies, logistical bottlenecks, and safety issues. Through harnessing the skills of DRL-based totally navigation systems, warehouses can attain greater performance, reduce operational charges, and beautify workplace safety. This creation gives an outline of the research objectives, highlighting the importance of adopting advanced technologies including deep reinforcement learning, in warehouse automation. Next sections will delve into the technique, experimental setup, effects, and implications of the observe, offering precious insights into the feasibility and effectiveness of deploying self reliant navigation systems in industrial warehouse environments [3]. The other method which is very important in this study is PPO, it has located huge software across diverse domain names, from robotics to sport playing. Its proximal coverage optimization strategy minimizes massive coverage changes, ensuring smoother studying curves. PPO's compatibility with deep neural networks helps its integration into complicated environments, enhancing its versatility. Thru its iterative approach, PPO iteratively refines guidelines, converging closer to ideal answers. Inside the landscape of reinforcement mastering, PPO shines as a cornerstone algorithm, riding improvements in independent structures and selection-making techniques [4].

Organization of the paper: The work's introduction and an outline of its mechanism are the main topics of Sect. 1. Existing research has been evaluated in Sect. 2, the problem statement and objectives are discussed in Sect. 3. Section 4 discusses the proposed approach and the operation of the deep reinforcement learning algorithm. The experimental results are described in Sect. 5. The final section, Sect. 6, summarizes the main conclusions and talks about possible enhancements.

2 Related Work

The optimized deep reinforcement model learned about feature award features, action-selection regulations, RMSprop for learning charge regulation, algorithms for mobile robotic nearby path creating plans, and specific PPO initialization procedures. It has been demonstrated to perform better than earlier algorithms across a number of metrics. RMSprop is used by DRL for dynamic adaptive mastering rate adjustments. A 50×50 randomized item grid environment with three sets of grid environments each with an obstacle grid that made up 20%, 30%, and 40% of the total grid cells was the dataset used for the analysis. In simulation trials, the performance of the set of rules is assessed in three distinct labyrinth scenarios using a range of metrics [1]. The Morris Water Maze is simulated in dimensions to train deep reinforcement mastery advertisers. Additionally, self-sufficient navigation technique classification has been achieved and the outcomes are contrasted with experimental documentation. The internal representations of artificial agent neural networks are examined, and biologically feasible tasks are suggested [2, 17]. In a real-world environment, it supports the TurtleBot3 burger robot's capacity. The study used genuine benchmarks, such as grid maps and GIS data from the Beijing 2008 road network. Future studies could investigate improved methods for experience transfer and reuse, enabling robots to adapt more effectively to shifting tasks and environments [5, 13]. Robots can navigate unexplored environments autonomously by utilizing unprocessed sensor information and ongoing reinforcement learning techniques. This approach enhances their ability to adapt and operate efficiently in changing situations through continuous learning processes. The suggested method effectively tackles scenarios with numerous obstacles and demonstrates good performance and unseen environments. The methods have excellent performance and ease of adaptation. The study made use of a dataset of training and testing situations designed for mobile robot navigation trials. The Pioneer 3-DX robot platform and V-REP simulator were used to create these environments, which featured both straightforward and intricate scenarios [9, 10, 12].

3 Problem Statement and Objectives

The task entails creating a customized the hardware platform that meets the unique needs of the robot and implementing an advanced software system that uses reinforcement learning to make decisions in real time. Software development includes developing an intricate set of rules that use reinforcement learning to understand concepts so the robot can react in real time to a variety of real-world situations. Through a complete strategy that addresses both hardware and software difficulties for a reliable and intelligent solution, this challenge seeks to advance the field of autonomous robots.

Main Objectives are:

1. to develop and apply the technique of reinforcement algorithms for the autonomous decision-making of the JetBot control system.
2. To navigate real-world situations and avoid obstacles by using an autonomous robotic device.
3. To develop a well-organized circuit design that enables the algorithm, control system, and sensor array to integrate seamlessly.
4. To Determine shortest path and make good decisions by using an autonomous robot to efficiently achieve the needed position in the environment.

4 Proposed Method

4.1 Reinforcement Learning

An agent learns to make decisions within the subfield of reinforcement learning inside machine learning by interacting with its environment and pursuing preset goals. Generally speaking, an entity being taught, or a reinforcement learning agent, may sense and interpret its environment, act, and learn by making mistakes [14, 16]. Theoretical reinforcement learning models are often constructed using Markov features. According to the Markov attribute, a system's lower state is only related to its current state and not to its past state [15]. A quaternion (S, A, P, R) is used to represent the Markov decision (MDP) model. S represents all the different situations we could encounter when making decisions [6]. The list A contains the options available to us. P indicates the likelihood of changing circumstances. R provides us with the outcome of our actions, which brings us to the subsequent scenario. Return evaluation and algorithm usability are given priority in current reinforcement learning, which is based on the MDP paradigm. The goal of this effort is to solve the problem of learning ability, namely how to go somewhere faster and safely. This is all about the reinforcement learning, and the technique of reinforcement learning is inspired by the identical query as TRPO: how are we able to take the most important possible improvement step on a policy the use of the information we currently have, without stepping to date that we by chance motive overall performance collapse? wherein TRPO attempts to resolve this problem with a complicated 2D-order approach, PPO is a circle of relatives of first-order techniques that use some different hints to hold new policies near antique. PPO techniques are considerably less complicated to put into effect and empirically seem to carry out at the least as well as TRPO used for this research and execution of it.

4.2 Deep Reinforcement Learning

A class of machine learning algorithms known as reinforcement learning (RL) lies in the middle between supervised and unsupervised learning [7]. Since it doesn't only rely on a set of labeled training data, it can't be considered supervised learning, but it also can't be considered unsupervised learning because our reinforcement learning agent is being trained to maximize a reward. The agent's primary objective is to figure out what the "correct" course of action is in different situations. The machine would never outperform a human at chess if we were to train it to mimic human behavior; after all, it would only be copying what the human would do. Therefore, by definition, we are unable to teach the machine using supervised learning. Is it possible to have an agent play a game by itself, though? Yes, that is the application of reinforcement learning. A kind of machine learning technique called reinforcement learning uses trial and error to figure out how to tackle a multi-level problem. To make a series of decisions, the machine is educated using real world circumstances [8], For the things it does, it is paid in incentives or penalties. Its objective is to maximize the overall gain. We take the maximum possible action in the state the agent is in to determine the value of a particular state (s). Selecting the course of action that will optimize value is the agent's goal. Consequently, the reward of the optimal action (a) in state (s) must be added, together with a multiplier (Y), which represents the discount factor that reduces the reward over time. Each time the agent acts, it returns to the subsequent state.

4.3 Proximal Policy Automation

The use of Proximal Policy Optimization (PPO) is the goal to reduce the number of policy updates at each training epoch in order to increase the training stability of the policy [9].

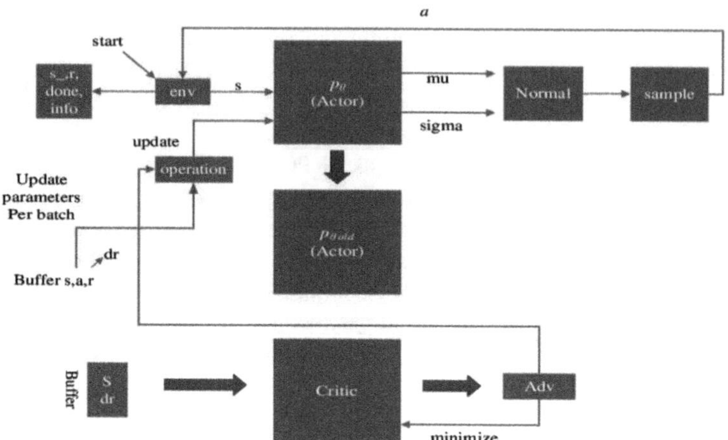

Fig. 1. Working of proximal policy algorithm

The complete working of PPO algorithm as shown in Fig. 1 PPO is inspired by the identical query as TRPO: how are we able to take the most important possible improvement step on a policy the use of the information we currently have without stepping to date that we by chance motive overall performance collapse? wherein TRPO attempts to resolve this problem with a complicated 2d-order approach, PPO is a circle of relatives of first-order techniques that use some different hints to hold new policies near antique [10]. PPO techniques are considerably less complicated to put into effect, and empirically seem to carry out at the least as well as TRPO.

The Jetson Nano serves as the JetBot's central processing unit in most hardware configurations. The computational capacity required for real-time picture processing and decision-making is provided by this single-board computer [11]. The main sensory input for the robot is provided via a camera module that is attached to the Jetson Nano. Depending on the exact needs and project budget, this camera may be either a USB camera or a csi camera module.

The Fig. 2 explains the fully assembled jetbot which is ready to perform the operation. The Motor drivers are used to convert the decisions made by the JetBot into motion. The Jetson Nano and the motors that operate the robot's wheels or tracks are interfaced by these motor drivers. They interpret the inputs from the Jetson Nano into commands that move the robot's motors in one direction or another [14]. The JetBot's chassis, which supports every component and ensures stability and mobility, makes up its physical structure. It should be lightweight and agile at the same time as being strong enough to endure the demands of different settings. The Jetson Nano and motors need a steady

Fig. 2. Fully assembled jetbot

power source to function properly. This could incorporate external power sources, power banks, or batteries, depending on the system's power requirements.

A WiFi module is built into the JetBot to enable wireless communication and remote control. This eliminates the requirement for physical tethering and enables operators to manage the robot remotely while also receiving real-time video feeds and sensor data. The JetBot's actions are determined by control logic, regardless of whether it is traversing its surroundings on its own or reacting to commands from a distance. To accomplish the intended movement, each motor's speed and direction must be coordinated. For improved perception and navigation, sensor integration may go beyond the camera module to incorporate other sensors like LiDAR, IMUs, or ultrasonic sensors.

The Robot Operating System (ROS) is frequently employed as the JetBot software stack's complexity management framework. By enabling developers to build separate nodes that interact via topics, services, and actions, ROS simplifies modular. OpenCV and other computer vision libraries are essential for processing the photos that the camera module takes. By enabling functions like obstacle avoidance, lane tracking, and object detection, these libraries give the JetBot the capacity to sense and comprehend its surroundings. High-level commands, like "move forward" or "turn left," are translated into low level signals that operate the motors through the use of motor control logic. The main channel of communication between the various Jetson Nano software modules is through ROS topics. For instance, the camera module publishes picture data to a ROS topic that modules handling object identification or image processing can subscribe to it. Other communication protocols, like TCP/IP or MQTT, can be used in addition to ROS to interface with external systems or devices. These protocols make it easier for components of the larger robotics ecosystem to integrate and work together.

5 Results and Discussions

The performance and result of the project and the implementation of proximal policy optimization for the automation of the jetbot and to perform the other autonomous functions. Have a look at the images. In Figs. 3 and 4 represents the movements of the bot, and Figs. 5, 6 and 7 are representing the collision avoidance feature. Of course, A thorough grasp of the system architecture and design of a JetBot using the Jetson Nano is the outcome of the explanation given. This dissection covers the hardware as well as the software, describing the parts, how they work together, and what each does to make a working robot. In Figs. 8 and 9 we can see how the bot is reaching its destination.

The JetBot is primarily dependent on the Jetson Nano, an incredibly potent single-board computer, for data processing and decision-making. The Jetson Nano has a camera module attached to it that acts as the robot's eyes, taking pictures of its surroundings. These choices are translated into motion by motor drivers, which manage the robot's wheels or tracks. The system is kept operational by a dependable power source, and the chassis provide structural support. The JetBot is primarily dependent on the Jetson Nano, an incredibly potent single board computer, for data processing and decision-making. The Jetson Nano has a camera module attached to it that acts as the robot's eyes, taking pictures of its surroundings.

These choices are translated into motion by motor drivers, which manage the robot's wheels or tracks. The system is kept operational by a dependable power source, and the chassis provides structural support. The operating system, usually Ubuntu or JetPack OS, offers the framework for program development. Modular software architecture is made easier by ROS, allowing for intercomponent communication.

Camera data is processed by computer vision libraries like OpenCV for purposes like object detection and navigation. Movement is coordinated by motor control logic, and perception and navigation are improved by sensor integration. For the JetBot to function, communication is essential between software modules as well as externally with humans or other devices. WiFi-based wireless connection allows data transfer and remote control. Software modules can communicate with each other through ROS topics, and external systems can be interfaced with other protocols like TCP/IP or MQTT.

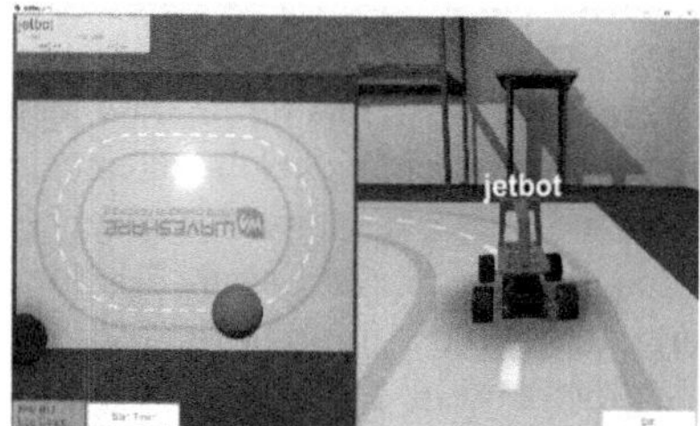

Fig. 3. A Jetbot train environment

Fig. 4. Jetbot ready to move forward

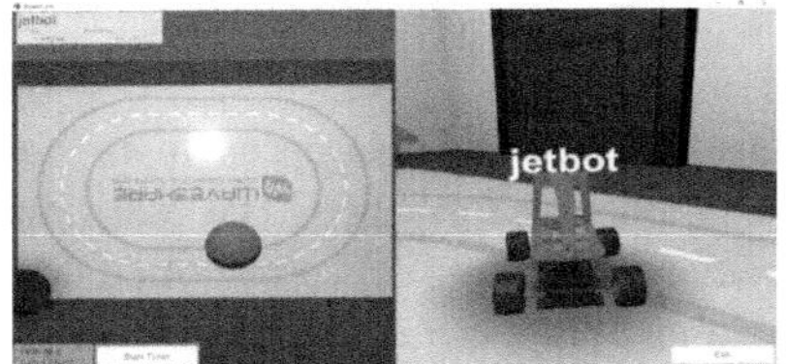

Fig. 5. Jetbot moving forward without collision

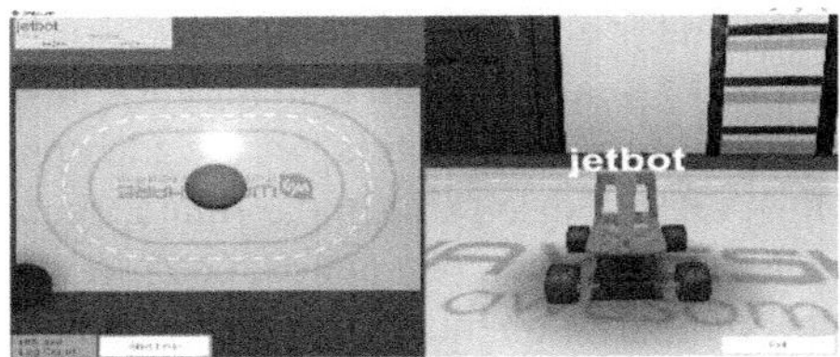

Fig. 6. Jetbotbot taking decision and changing path

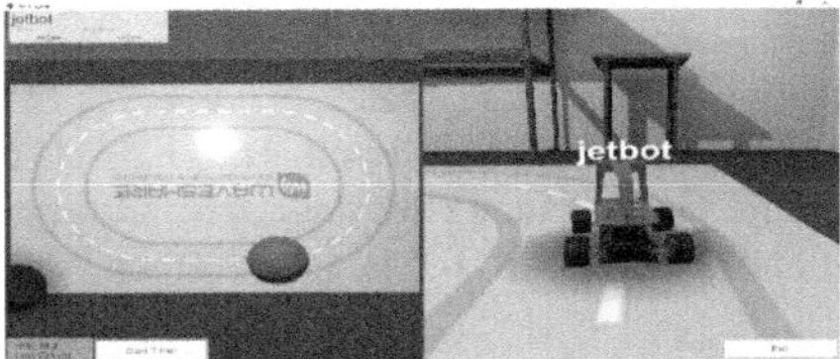

Fig. 7. Jetbot travelling in the specified path

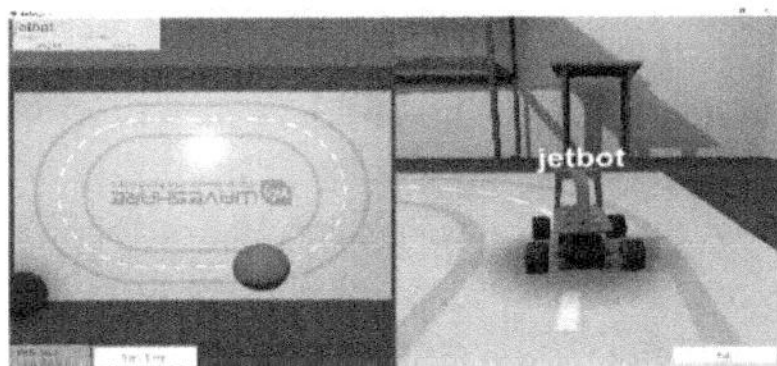

Fig. 8. Jetbot reaching to the destination by optimal route

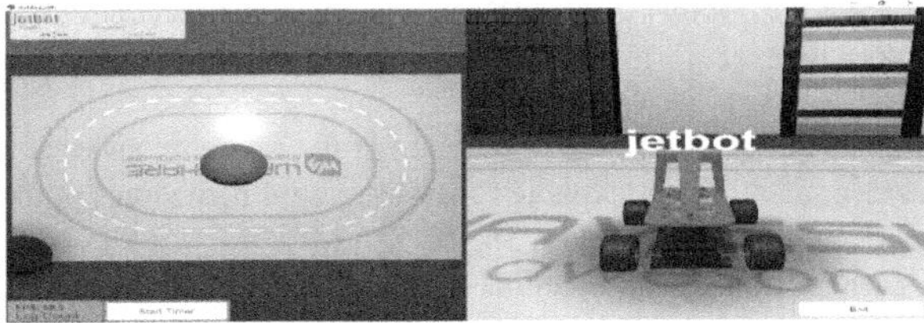

Fig. 9. Jebot changing its path to the real pat

6 Conclusion and Future Scope

The main work is finding the path-finding approach for autonomous robots, which enhances the suggested method's performance. The robot may continuously enhance its comprehension of its environment and make informed decisions through this iterative process, ultimately leading to the selection of the most effective path. Also, this bot includes the feature of collision avoidance, which is very helpful in the autonomous feature. The JetBot's wireless communication features allow it to share data and be remotely controlled, increasing its usefulness in a variety of applications. In the end, the JetBot symbolizes a flexible platform for robotics research, instruction, and real-world applications, representing the union of state-of-the-art technology with creative engineering. In order to improve upon the existing configuration, we are attempting to incorporate additional functions into the NVIDIA Jetson Bot in the future, such as a camera-equipped bot. The JetBot accomplishes autonomy, perception, and movement by combining sensors, motor drivers, and communication modules with the computational capacity of the Jetson Nano.

References

1. Kuutti, S., Bowden, R., Jin, Y., Barber, P., Fallah, S.: A survey of deep learning applications to autonomous vehicle control. IEEE Trans. Intell. Transp. Syst. **22**(2), 712–733 (2021)
2. Yao, Z., Liang, X., Jiang, G.-P., Yao, J.: Model-based reinforcement learning control of electrohydraulic position servo systems. IEEE/ASME Trans. Mechatron. **28**(3), 1446–1455 (2023)
3. Chae, H., Kang, C.M., Kim, B., Kim, J., Chung, C.C., Choi, J.W.: Autonomous braking system via deep reinforcement learning. In: Proceedings of the IEEE 20th International Conference on Intelligent Transportation Systems (ITSC), pp. 1–6 (2017)
4. Zhu, Z., Zhao, H.: A survey of deep RL and IL for autonomous driving policy learning. IEEE Trans. Intell. Transp. Syst. **23**(9), 14043–14065 (2022)
5. Alzubaidi, L., et al.: Review of deep learning: concepts, CNN architectures, challenges, applications, future directions. J. Big Data **8**(1), 1–74 (2021)
6. Sarker, I.H.: Deep learning: a comprehensive overview on techniques, taxonomy, applications and research directions. Soc. Netw. Comput. Sci. **2**(6), 420 (2021)
7. Fan, J., Wang, Z., Xie, Y., Yang, Z.: A Theoretical analysis of deep Q-learning. In: Proceedings of the 2nd Conference on Learning for Dynamics and Control, pp. 486–489 (2020)
8. Schulman, J., Wolski, F., Dhariwal, P., Radford, A., Klimov, O.: Proximal policy optimization algorithms. arXiv:1707.06347 (2017)
9. Antonio, G.-P., Maria-Dolores, C.: AIM5LA: a latency-aware deep reinforcement learning-based autonomous intersection management system for 5G communication networks. Sensors **22**(6), 2217 (2022)
10. Zhang, Y., Sun, P., Yin, Y., Lin, L., Wang, X.: Human-like autonomous vehicle speed control by deep reinforcement learning with double Q-learning. In: Proceedings of the IEEE Intelligent Vehicles Symposium (IV), pp. 1251–1256 (2018)
11. Li, G., et al.: Deep reinforcement learning enabled decision-making for autonomous driving at intersections. Autom. Innov. **3**(4), 374–385 (2020)
12. Khalifa, A.A., Ramesh, G., Deekshith, R., Harsha, K., Shreyas, J.: Forecasting financial frontiers: real-time insights in stock price prediction through LSTM, linear regression, and sentiment analysis. In: 2024 International Congress on Human-Computer Interaction, Optimization and Robotic Applications (HORA), pp. 1–6. IEEE (2024)

13. Karthikeyan, P., Chen, W.-L., Hsiung, P.-A.: Autonomous intersection management by using reinforcement learning. Algorithms **15**(9), 326 (2022)
14. Li, G., Yang, Y., Li, S., Qu, X., Lyu, N., Li, S.E.: Decision making of autonomous vehicles in lane change scenarios: deep reinforcement learning approaches with risk awareness. Transp. Res. Part C: Emerg. Technol. **134**, 103452 (2022)
15. Ramesh, G., et al.: Hybrid manifold smoothing and label propagation technique for Kannada handwritten character recognition. Front. Neurosci. **18**, 1362567 (2024)
16. Sharma, S., Tewolde, G., Kwon, J.: Lateral and longitudinal motion control of autonomous vehicles using deep learning. In: Proceedings of the IEEE International Conference on Electro Information Technology (EIT), pp. 1–5 (2019)
17. Ramesh, G., Jeswin, Y., Rao, D.R., Suhaag, B.R., Uppoor, D., Kiran Raj, K.M.: Real-time object detection and tracking using SSD MobilenetV2 on JetBot GPU. In: 2024 IEEE International Conference on Distributed Computing, VLSI, Electrical Circuits and Robotics (DISCOVER), pp. 255–260. IEEE (2024)

An Extreme Learning Machine-Based Approach for Geriatric Depression Prediction

K. P. Impana[1(✉)], Pushpalata[1], R. Raksha[1], K. B. Vikhyath[2], and Meenakshi V. Patel[3]

[1] Department of Computer Science and Engineering, JSS Academy of Technical Education, Bengaluru, India
impanaraj@gmail.com, mppvin@sjce.ac.in, raksha@jssstuniv.in
[2] Department of Computer Science and Engineering, Dr. H N National College of Engineering, Bengaluru, India
[3] Department of General Medicine, Apollo BGS Hospital, Mysore, India

Abstract. This research investigates application of advanced machine learning (ML) algorithms within predicting geriatric depression, focusing on a cohort from Kaggle. The study involved analysis of various demographic, clinical, and socio-economic factors among elderly patients, enabling development of a robust predictive model for depression. Machine learning models were trained and evaluated on precision, accuracy, recall, f1-score after rigorous preprocessing, normalization, and data splitting. The recommended approach's Extreme Learning Machine (ELM) has maximum recall = 1.0000, accuracy = 0.9692, precision = 0.7912, and f1-score = 0.8834 compared to all other existing approaches. Findings indicate that ELM achieved greatest ability for prediction, highlighting its potential as established diagnostic aid in order to detect depression in geriatric patients. This model can assist healthcare practitioners in resource-constrained settings, ultimately enhancing mental health treatment as well as quality of life (QoL) for elderly population. Results of this study underscores the significance of AI-driven predictive analytics in addressing geriatric mental health concerns and contribute to broader literature on applications of ML in healthcare.

Keywords: Depression · geriatric patients · Machine learning (ML) · Extreme learning machine (ELM)

1 Introduction

Ageing is characterized as biological process of deteriorating with age [1]. Ageing denotes accumulation of alterations that increase probability of mortality in humans. Medawar, 1952. Ageing is characterized as a steady loss in physiological function, including a reduction in productivity over time [2]. Ageing is a change involving dynamic biological, physiological, psychological, behavioral, environmental, and social factors. Indeed, rising age is primary risk factor for several chronic illnesses in humans. According to WHO, active ageing involves process of optimizing opportunities for engagement,

J. Shreyas et al. (Eds.): CODE-AI 2025, CCIS 2690, pp. 90–101, 2026.
https://doi.org/10.1007/978-3-032-19321-6_9

health, as well as safety to improve people's QoL [3]. In 2017, the worldwide population of individuals aged 60 years and older reached 962 million, having almost doubled since 1980. By 2050, its projected that 79% of people aged 60 and over would reside in developing nations [4]. By 2050, almost 80% of elderly population will reside in emerging nations. The World Population Prospects 2017 aged over 60 years are illustrated in below Fig. 1.

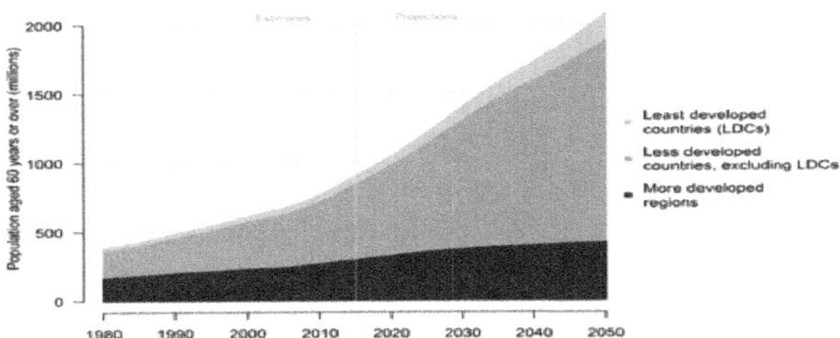

Fig. 1. World Population Prospects 2017 [4].

2 Geriatric Population in India

India has 81 million elderly individuals. People over 60 who are considered geriatrics are increasing, particularly in India. Population of geriatric individuals will continue to rise, resulting in a greater number of older adults compared to current figures. The nation requires more healthcare and supplementary medical service facilities and resources to accommodate 76.6 million individuals over age of 60, representing over 7.7% of India's total population [5]. There is a gradual rise in population individuals in India who are 60 years of age or older. Figure 2 illustrates the demographics of population ageing in India, highlighting trends and disparities.

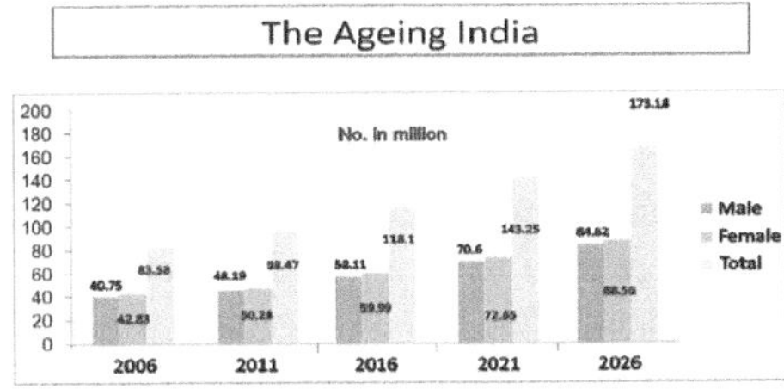

Fig. 2. Population India, Census 2011 [6].

Old age is correlated with considerable decline in physical, social, and cognitive areas [7]. Aged population has a range of health issues, including physical decline, physiological ailments, mental illnesses, emotional challenges, and social difficulties, prompting to seek targeted medical treatment. Mental health issues in the geriatric population can be precisely diagnosed and efficiently treated if assistance is sought promptly; yet, they are mistakenly regarded as a component of normal aging [6]. The rapid rate of demographic aging is a significant public health challenge for geriatric mental health in India. Geriatricians consistently encounter the challenge of evaluating patients' health condition while diagnosing diseases and syndromes, as they must consider numerous connected symptoms (health risk factors) [7].

Latest advancements were largely enabled by increasing prevalence of machine learning in medical data science. As subset of AI, ML allows computers to be taught on data to discern patterns and provide predictions [8]. This method is ideal for creating algorithms that assess patient's likelihood to possess a condition via examination of extensive medical data. Upon completion of training, these algorithms may be evaluated on novel data to measure their efficacy outside training context. Numerous machine learning approaches exist, with two most prevalent being supervised and unsupervised ones. In supervised learning, data are annotated with intended output.

Primary contribution of this research is, the development and assessment of an effective ML model tailored for geriatric depression prediction. By incorporating a comprehensive set of variables, including socio-demographic, clinical, and personal health factors, the study presents a holistic approach to predicting mental health outcomes in elderly patients.

- This work is significant as it addresses a pressing healthcare issue—depression in aging population using advanced machine learning techniques. The findings have practical implications for healthcare practitioners, as they offer a non-invasive, data-driven approach to identifying high-risk individuals, which can facilitate timely intervention and improved patient outcomes.
- Furthermore, study contributes to broader AI and healthcare literature by demonstrating viability of ML to improve geriatric mental health diagnosis precision.

This paper's remaining structure is as follows, Sect. 3 presented literature survey; Sect. 4 indicates methodology and Sect. 5 and 6 represents the research findings and conclusion of the proposed work.

3 Literature Review

AI techniques for geriatric depression prediction are examined in this overview of the literature. Conventional models such as Support Vector Machines (SVM), Random Forest (RF), Naïve Bayes (NB), and Logistic Regression (LR) are extensively utilized. Tran et al. [9] utilized logistic regression to examine risk factors for depression, attaining an accuracy of 79.8% and identifying loneliness and frailty as significant predictors. Zheng et al. [10] employed Random Forest with SMOTE on a sample of 3,959 seniors, achieving an AUC of 0.957, highlighting the importance of class balance. Sabouri et al. [13] conducted a comparison of six machine learning models, finding that support vector machine (SVM) and logistic regression (LR) achieved the highest accuracy at 83.32%.

Advancements in deep learning have contributed to the prediction of depression. Mirza et al. [19] developed an artificial neural network model that achieved an accuracy of 94.79%, surpassing traditional machine learning methods. Rajawat et al. [14] employed convolutional neural networks combined with Fusion Fuzzy Logic for facial expression analysis, achieving an accuracy of 94.3%. Choi et al. [15] developed a wearable system based on a multi-layer perceptron, attaining a recall rate of 82.7% in remote monitoring applications. Recent research incorporates various AI models or multimodal data. Lee et al. [16] integrated Random Forest (RF), Artificial Neural Networks (ANN), and Support Vector Machines (SVM), with SVM attaining an accuracy of 96.9%. Song et al. [12] employed SHAP in conjunction with machine learning, finding that the random forest model achieved the highest performance with an AUC of 0.81. Cho et al. [18] introduced a hybrid LASSO-ML model, attaining an AUC-ROC of 0.903, which underscores the effectiveness of feature selection. Table 1 gives the comparison of different studies carryout for geriatric depression.

Table 1. Depression prediction studies comparison.

Findings	Dataset	Models used	Best model	Performance (Accuracy/AUC)
Tran et al. [9]	976 elderly patients	Logistic regression	LR	79.8% accuracy
Zheng et al. [10]	3,959 elderly (China)	RF,SVM, KNN	RF with SMOTE	AUC = 0.957
Sabouri et al. [13]	Kaggle dataset	RF,SVM, KNN,LR, DT, NB	SVM & LR	83.32% accuracy
Mirza et al. [19]	40 elderly patients	ANN	ANN	94.79% accuracy
Rajawat et al. [14]	Facial expression dataset	CNN + Fuzzy Logic	CNN-FFL	94.3% accuracy
Choi et al. [15]	Wearable band data	MLP	MLP	Recall = 82.7%
Lee et al. [16]	8,628 hypertensive patients	RF,ANN, SVM	SVM	96.9% accuracy
Song et al. [12]	797 elderly	ML + SHAP	RF	AUC = 0.81
Cho et al. [18]	9,488 community patients	LASSO + ML	Hybrid LASSO-ML	AUC = 0.903

Traditional machine learning models (SVM, RF, LR) provide interpretability and efficiency, although need manual feature selection. Deep learning models (ANNs, CNNs) excel with extensive datasets, identifying nonlinear markers of depression. Hybrid methodologies (e.g., SHAP-ML, CNN-FFL) reconcile precision and interpretability, hence augmenting clinical applicability. Research limitations encompass insufficient multimodal data, dependence on surveys, and constrained deep learning interpretability, hence demanding Explainable AI (XAI) methodologies such as SHAP and LIME. Limited sample sizes underscore the necessity for more extensive databases. Future investigations should examine CNNs, LSTMs, and the integration of multimodal AI, alongside implementing real-world trials for AI-facilitated depression screening in clinical environments.

4 Research Methodology

This study on geriatric depression prediction using machine learning involved several systematic steps. Initially, data was collected from Kaggle, covering demographic, clinical, and socio-economic features. After obtaining informed consent, these variables were recorded, standardized, and normalized to ensure uniformity. To make model creation and evaluation easier, dataset was subsequently separated into training as well as testing subunits. To find patterns linked to depression in elderly individuals, variety of ML algorithms received training on training data. Models have been assessed employing precision, accuracy, recall, as well as f1-score metrics, comparing their effectiveness in identifying depressive cases. Model that had best performance measures had been picked as proposed solution for accurate and reliable prediction of geriatric depression. Figure 3 Depicts the proposed framework of this study.

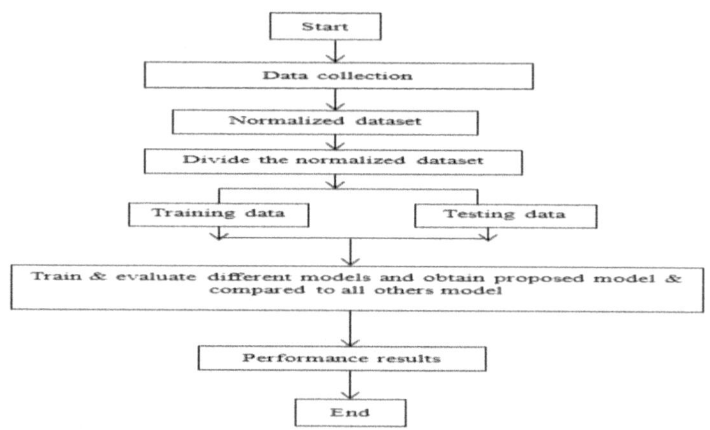

Fig. 3. Proposed framework.

This work used Kaggle data, including demographic and clinical characteristics, to construct and test machine learning models for geriatric depression prediction. The features were chosen from literature and psychiatric consultations. The dataset was separated into training and testing sets after normalisation. To find the optimal model, accuracy, precision, recall, and F1-score were tested. Medical applications must balance false positives and negatives, according to the study. AI-driven depression diagnosis may enhance geriatric mental health by detecting and treating depression early.

4.1 Data Collection

Kaggle, collected data on sex, literacy, age, place of residence, recent loss, marital status, work status, personal income (PI), socioeconomic status, history of anxiety, family history of depression, history of depression, addiction in any form, diabetes, hearing, vision, mobility, hypertension (HTN), as well as insomnia. These elements were chosen after a comprehensive literature study. The dataset contained 304 records and 39 features, covering demographic, psychological, and social factors. Psychological features (happiness, boredom, helplessness) were the strongest predictors.

4.2 Proposed Methodology

Extreme Learning Machine (ELM) ANNs may address several classifying problems. Extreme Learning Machine (ELM) is a cutting-edge training approach for single hidden-layer feed-forward networks (SLFNs). ELM is a common feed-forward neural network for classification, clustering, compression, regression, small estimation, and pattern recognition. These models learn much faster than back propagation-trained networks. Back propagation is a common learning approach in feed-forward neural networks that determines gradients from output to input. Back propagation still has numerous issues. Due to rationalising biases and weights after each iteration, training can be time-consuming. In order to achieve maximum accuracy, the model ignores weight size, lowering output quality. Local minima affect back propagation learning's effectiveness [20]. ELM is a feed-forward neural network that solves weight and bias updating problems. Beyond minimising training error, this strategy aims to achieve optimal weight criteria, improving model efficiency. ELM architecture is depicted in Fig. 4.

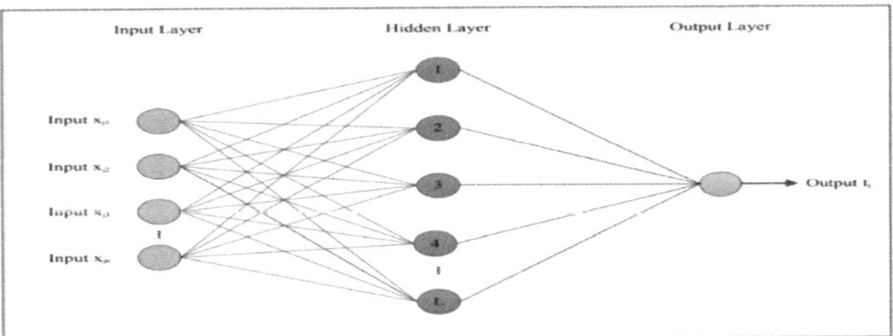

Fig. 4. Diagram of ELM [21].

Algorithm of ELM

The following is description of ELM algorithm:

1. N: total number of training-samples
2. $\tilde{N}$: number of hidden layer units;
3. n, m: Dimension of input and output layers;

4. $(x_j, t_j), j = 1,2, \ldots N$: Training-sample, where $x_j = (x_{j1}, x_{j2}, \ldots, x_{jn})^T \in R^n$, and $t_j = (t_{j1}, t_{j2}, \ldots, t_{jn})^T \in R^m$.
   To obtain the total output matrix, combine each output vector in a row.

$$T = \begin{bmatrix} t_1^T \\ t_2^T \\ \vdots \\ t_N^T \end{bmatrix} = \begin{bmatrix} t_{11} & \cdots & t_{1m} \\ \vdots & \ddots & \vdots \\ t_{N1} & \cdots & t_{Nm} \end{bmatrix}$$

5. $O_j j = 1,2, \ldots, N$: actual output vector corresponding to label t_j.
6. $W = (w_{ij})_{N\times n}$ a weight-matrix among input layer and hidden layer, wherein vector $w_i = (w_{i1}, w_{i2}, \ldots, w_{in})^T$ related to the ith row of W represents a wright-vector connecting ith unit and input unit of the hidden layer.
7. $b = (b_1, b_2, \ldots, bN)^T$: offset vector, bi represent the threshold of ith hidden layer unit.
8. $\beta = (\beta_{ij})_{N\times m}$: a weight-matrix among hidden layer and output layer, wherein vector $\beta_i == (\beta_{i1}, \beta_{i2}, \ldots, \beta_{in})^T$ corresponding to the ith row β symbolizes a weight vector that joins hidden layer's ith unit and output-layer unit. Following is one way to write matrix β in rows.

$$\beta = \begin{bmatrix} \beta_1^T \\ \beta_2^T \\ \vdots \\ \beta_N^T \end{bmatrix} = \begin{bmatrix} \beta_{11} & \cdots & \beta_{1m} \\ \vdots & \ddots & \vdots \\ \beta_{N1} & \cdots & \beta_{Nm} \end{bmatrix}$$

9. $g(x)$: The excitation functions. Sigmoid, tribas, sin, hardlim, as well as radbas functions are 5 categories of excitation functions that are frequently employed. Consider step 6 and 8, we can get

$$H\beta = T$$

where, T represents transposition of T and H represents hidden layer's output. The weight matrix values of β are calculated using least squares approach in order to achieve unique solution with least amount of error.

$$\beta = H + T$$

Pseudo code of the proposed method as given below:

1. Randomly assign the input weight w_i and biases b_i,
2. Calculate the hidden-layer output matrix H,
   where $H = \{h_{ij}\}$ (i = 1,….,N and j = 1,….,K) and $h_{ij} = g(w_j.x_i + b_j)$;

3. Calculate the output weights matrix as $\beta = H + T$,
   where H is the generalised inverse of matrix H^{+}

ELM was chosen for its exceptional accuracy, swift training abilities, strong generalization, and proficiency in handling intricate relationships. The evaluated models demonstrated optimal performance. The achieved accuracy is 96.92%, exceeding that of KNN, SVM, NB, MLP, and AdaBoost. Recall: 100% (outstanding in recognising patients with depression). Balanced f1-score: 0.8834, indicating a commendable equilibrium between accuracy and recall.

5 Results and Discussion

K-NN, MLP, ELM, SVM, as well as Naïve Bayes models had their initial run. In an attempt to improve ELM's accuracy, authors changed quantity of nodes in models' concealed layer. There were nodes within buried layer as little as 50 and as many as 250. Changing number of nodes within hidden layer of ELM improved performance resulting in better outcomes. Table 2 gives brief summary of results, while Fig. 5 depicts a graphical representation.

Table 2 makes it abundantly evident because when 200 hidden layer nodes are selected, ELM reaches its maximum accuracy. Consequently, there are instances in which concealed layer's node count is set at 250, 150, 100, as well as 50. It is clear that, in comparison to other nodes in concealed layer, recall & F-score values are higher for 200 as well as 150 hidden layer nodes.

Table 2. Performance analysis for ELM with different hidden layers.

Nodes in hidden layer	50	100	150	200	250
Accuracy	0.9341	0.9451	0.9560	0.9692	0.9648
Precision	0.8947	0.8929	0.8851	0.7912	0.7294
Recall	0.9855	0.9868	1.0000	1.0000	0.9118
F1 score	0.9379	0.9375	0.9390	0.8834	0.8105

With 200 hidden layer nodes, the ELM model's accuracy was at its best (0.969), followed by 250 nodes (0.9648), 150 nodes (0.956), 100 nodes (0.945), and 50 nodes (0.934). By dividing processing over several neurones, increasing the number of hidden nodes improves computing performance. But after a while, complexity increases and performance suffers. The model achieved an F1-score of 0.8834, recall of 0.099, accuracy of 0.9692, and precision of 0.7912 at 200 nodes. Accuracy decreased little with 250 nodes (0.9648). This demonstrates that although increasing the number of hidden nodes boosts efficiency, going above the ideal number raises computational burden and degrades performance and model efficacy.

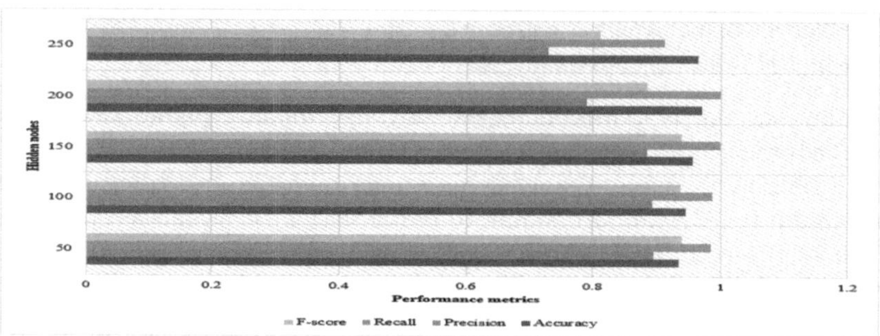

Fig. 5. Comparison graph of proposed method with performance metrics and other hidden nodes.

5.1 Comparison of Proposed Method with Other Classification Models

In a solitary setting, utilizing an ELM classifier of 200 nodes, this section compares effectiveness of different classifiers that of other classifiers. It should be mentioned that while 20% of dataset's tuples are designated for testing, remaining 80% are designated for training. Evaluation measures including F-score, recall, accuracy, as well as precision were utilized to create comparisons. While Fig. 6 provides a visual representation of results, Table 3 provides a succinct summary of findings.

Table 3. Performance of different ML models.

ML models	Ada boost	KNN	NB	MLP	SVM	ELM
Accuracy	0.9298	0.9064	0.8480	0.8304	0.9298	0.9692
Precision	0.9375	0.9211	0.8000	0.9765	0.9464	0.7912
Recall	0.9545	0.9375	0.9796	0.7545	0.9464	1.0000
F-score	0.9459	0.9292	0.8807	0.8513	0.9464	0.8834

Accuracy values in Table 3 are compared, and it was evident that ELM had highest accuracy (0.9692) and perceptron has lowest (0.8304). When compared to all other classifiers, ELM classifier additionally possesses highest recall value (1.00). The results of the aforementioned analysis clearly reveal that ELM-based model has highest classification accuracy, followed by K-NN, SVM, NB, AdaBoost, as well as perceptron models.

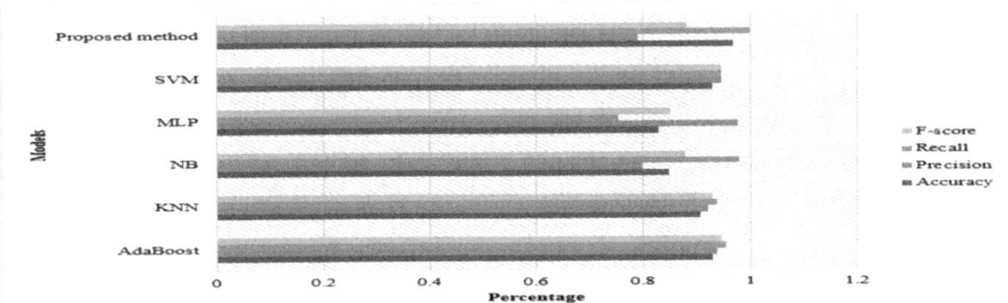

Fig. 6. Comparison graph of proposed method with other existing method.

Proposed method's comparison graph with other current methods is displayed in above Fig. 6 the results of the AdaBoost approach were 0.929 for accuracy, 0.9375 for precision, 0.9545 for recall, and 0.9459 for f1-score. K-NN approach yielded f1-score of 0.9292, accuracy of 0.906, precision of 0.9211, and recall of 0.9375. With recall 0.7545, precision 0.9765, and f1-score 0.8513, MLP approach achieved lowest accuracy of any method, at 0.8304. Maximum recall, accuracy, precision, as well as f1-score of suggested approach (ELM) were 0.9692, 1.00, and 0.7912, respectively. Consequently, it is clear that suggested approach outperforms all other approaches.

6 Conclusion and Future Work

Based on aged datasets, we used machine learning models to predict geriatric depression. Our study demonstrated that ELM outperforms other models in f1-score, accuracy, recall, and precision, making it a promising therapeutic tool. This model accurately predicts depressive symptoms, which helps reduce the course of depression in the elderly. Clinical decision support systems (CDSS) in Electronic health records (EHR) can help clinicians diagnose geriatric depression using the ELM model. Nursing homes and assisted living facilities may use it for mental health examinations. Telemedicine and mHealth applications can use the model to enable seniors self-assess their mental health or obtain AI-driven advice. Public health authorities track depression quantitatively to help legislators distribute mental health funds. Although useful, the study's 304 records may restrict generalizability. Demo-graphic imbalance or self-reported psychological disorders might cause biases. Clinicians have trouble understanding black-box ELM predictions. Larger, more diverse datasets, bias reduction, and XAI approaches like SHAP or LIME increase model openness and clinician trust. Study CNNs for facial expression analysis and LSTMs for depression monitoring to improve feature extraction. Multimodal AI may improve accuracy by combining clinical notes, wearable sensor data, and social media sentiment analysis. Hospital and mental health clinical trials must evaluate model performance, usability, and patient and healthcare professional acceptability. These findings improve early diagnosis, personalised treatment, and geriatric mental health.

References

1. Williams, G.C.: Pleiotropy, natural selection, and evolution of senescence: evolution 11, 398–411 (1957). Sci. Aging Knowl. Environ. **2001**(1), cp13–cp13 (2001)
2. Lopez-Otin, C., et al.: HallMarks of aging. J. Eur. PMC **153**(6), 1194–1217 (2013)
3. WHO: National institute on aging: global health and aging. NIH Publ. **11–7737**, 2011 (2011)
4. United Nations. World Population Prospects.: Revision, key findings and advance table Newyork: United Nations, Department of Economic and social Affairs PD (2017). http://esa.un.org/unpd/wpp/Publication/files/WPP2017
5. WHO.: WHO department of health statistics and information system of health systems and Innovation cluster, WHO press, Geneva (2015). https://www.who.int
6. BKPAI, SRS census.: Demographics of population aging in India: trends and differentials, 1, United Population Fund (UNFPA), New Delhi (2011)
7. Urry, H.L., Gross, J.J.: Emotion regulation in older age. Curr. Dir. Psychol. Sci. **19**(6), 352–357 (2010)
8. Cho, G., Yim, J., Choi, Y., Ko, J., Lee, S.-H.: Review of machine learning algorithms for diagnosing mental illness. Psychiatry Investig. **16**(4), 262 (2019)
9. Tran, T., Tan, Y.Z., Lin, S., Zhao, F., Ng, Y.S., Ma, D., Ko, J., Balan, R.: Exploring key factors influencing depressive symptoms among middle-aged and elderly adult population: a machine learning-based method. Arch. Gerontol. Geriatr. 105647 (2024)
10. Zheng, Y., Zhang, C., Liu, Y.: Risk prediction models of depression in older adults with chronic diseases. J. Affect. Disord. **359**, 182–188 (2024)
11. Song, Y., Zhang, D., Wang, Q., Liu, Y., Chen, K., Sun, J., Shi, L., et al.: Prediction models for postoperative delirium in elderly patients with machine-learning algorithms and SHapley Additive exPlanations. Transl. Psychiatry **14**(1), 57 (2024)
12. Song, Y., Qian, L., Sui, J., Greiner, R., Li, X.M., Greenshaw, A.J., Liu, Y.S., Cao, B.: Prediction of depression onset risk among middle-aged and elderly adults using machine learning and Canadian longitudinal study on aging cohort. J. Affect. Disorder. **339**, 52–57 (2023)
13. Sabouri, Z., Gherabi, N., Nasri, M., Amnai, M., Massari, H.E., Moustati, I.: Prediction of depression via supervised learning models: performance comparison and analysis. Int. J. Online Biomed. Eng. **19**(9) (2023)
14. Rajawat, A.S., Bedi, P., Goyal, S.B., Bhaladhare, P., Aggarwal, A., Singhal, R.S.: Fusion fuzzy logic and deep learning for depression detection using facial expressions. Procedia Comput. Sci. **218**, 2795–2805 (2023)
15. Choi, J., Lee, S., Kim, S., Kim, D., Kim, H.: Depressed mood prediction of elderly people with a wearable band. Sensors **22**(11), 4174 (2022)
16. Lee, C., Kim, H.: Machine learning-based predictive modeling of depression in hypertensive populations. Plos One **17**(7), e0272330 (2022)
17. Trivedi, N.K., Tiwari, R.G., Witarsyah, D., Gautam, V., Misra, A., Nugraha, R.A.: Machine learning based evaluations of stress, depression, and anxiety. In: 2022 International Conference Advancement in Data Science, E-learning and Information Systems (ICADEIS), pp. 1–5. IEEE (2022)
18. Cho, S.-E., Geem, Z.W., Na, K.-S.: Predicting depression in community dwellers using a machine learning algorithm. Diagnostics **11**(8), 1429 (2021)
19. Mirza, M.: Detecting depression in elderly people by using artificial neural network. Elderly Health J. (2020)
20. Dang, W., et al.: A semi-supervised extreme learning machine algorithm based on new weighted kernel for machine smell. Appl. Sci. **12**(18), 9213 (2022)
21. Huang, G.-B., Wang, D.H., Lan, Y.: Extreme learning machines: a survey. Int. J. Mach. Learn. Cybern. **2**, 107–122 (2011)

Methods of Machine Learning for Classifying Human Emotions Based on Electroencephalogram Signals Under Different Audio-Visual Stimuli

G. S. Shashi Kumar and Kiran Poojary(✉)

Department of Electronics and Communication, Manipal Institute of Technology, Manipal Academy of Higher Education, Manipal, Karnataka, India
shashi.gs@manipal.edu, kiran1.mitmpl2024@learner.manipal.edu

Abstract. This study main focus is on applying machine learning classifiers, such as support vector machines (SVM) and k-nearest neighbor (k-NN) algorithms, to categorize human emotions using electroencephalogram (EEG) signals. A classifier learns input features from a dataset by using a specific technique and set of tuning parameters to create a classification model. A two-dimensional emotion model (valence-arousal) was used to evaluate the emotion state that was produced in individuals from the Database for Emotion Analysis using Physiological Signals (DEAP). After that, a new input in an unknown dataset is predicted to belong to the same class using the model. It has been shown that when applied to EEG classification, k-NN and SVM can effectively differentiate between features in an EEG dataset. However, a number of EEG applications showed different results. Emotions are analyzed using SVM and KNN classifiers using EEG signals. The gamma-trained model outperforms KNN in SVM classification, according to our results; however, this advantage vanishes when selecting features using PCA.

Keywords: Electroencephalogram · Emotion recognition · Physiological signal · Valance · Arousal

1 Introduction

Human-to-human communication is greatly influenced by emotions. Emotional ties between humans and computers have become a major concern for HMIs and BCIs due to the extensive usage of technology in today's world [1]. Robots must therefore be able to comprehend and react to emotions in order to improve upon this without human input. Several studies on automatically detecting emotions have already been conducted. They come in two varieties. The first method uses facial features or speech patterns that are easily recognized [2, 3]. Without using physical contact, these audio-visual methods enable us to ascertain an individual's emotions, preventing distress.

The methods described may be more vulnerable to parameter changes in some situations, even though errors are possible. The autonomic nervous system causes instantaneous changes in the physiological signals used by the other appendage roaches, such

J. Shreyas et al. (Eds.): CODE-AI 2025, CCIS 2690, pp. 102–113, 2026.
https://doi.org/10.1007/978-3-032-19321-6_10

as skin conductance, pulse respiration, EEG and ECG signals, etc. [4, 5]. Compared to audio-visual methods, physiological signal responses offer greater details and information to illustrate emotional states. It has been shown that physiological markers, including EEG data, can provide valuable information on emotional triggers [6]. Interest in the connection between EEG signals and emotions has increased dramatically since Davidson et al. [7] found that the EEG of the frontal brain is linked to both happy and negative emotional states.

The 2D version of emotion proposed by Davidson et al. [8] shows that the 2D plane on which emotions are represented is covered by two axes: the arousal and valence axes. The arousal axis illustrates the spectrum of activation levels, ranging from relaxed to anxious.

Bos used the international affective digitized sound system (IADS) and international affective picture system (IAPS) to elicit emotions and Fisher's discriminant analysis (FDA) as a classification model in a related study [9, 11]. Data from a different study shows that a KNN and SVM classifier can achieve an accuracy of about 60% for two different emotional states. Takahashi et al. showed how to use EEG signals to recognize sentiments using movie images as stimuli [10], with a recognition rate of 41.7% for different emotional states.

The average classification accuracy was 68% for valance and 65% for arousal from alpha and beta bands. Although there has been a lot of research on the identification of emotions from EEG, more study is required to find other traits and categorization methods that would enhance recognition skills.

In the first chapter, the project "Development of SVM/KNN based protocol for classification of EEG recorded with different Audio-Visual Stimuli" is presented. The second chapter discusses the concepts that underlie emotion as well as the biological bases of emotion generation. The third chapter will describe the steps taken to transform all of the data into an EEG dataset and then identify these emotions using classification models. Finally, the fourth and fifth chapters analyze the results and conclusions.

2 Literature Survey

This chapter discusses the four most widely accepted theories of emotion and the unique biological processes that underpin each. Emotions are strong feelings that come from one's surroundings, mental condition, or relationships with other people.

Emotions are composed of physiological, cognitive, and behavioral responses to stimuli. As a result, there are multiple theories of emotion. There are a lot of theories about emotions. For the purposes of this debate, we will be discussing four different theories of emotion. The first theory of emotion, called the James-Lange hypothesis, is seen in Fig. 1. It postulates that our ability to sense bodily reactions is what gives rise to our emotional experiences. Therefore, holding a pet cat evokes enjoyment [12].

In accordance with the James-Lange hypothesis, holding the pet cat causes a physiological reaction that may increase heart rate. In the brain, some neurotransmitters become more prevalent. There might be someone holding a cat and smiling. Notable is the James-Lange theory of emotion, which holds that understanding this response leads to happiness.

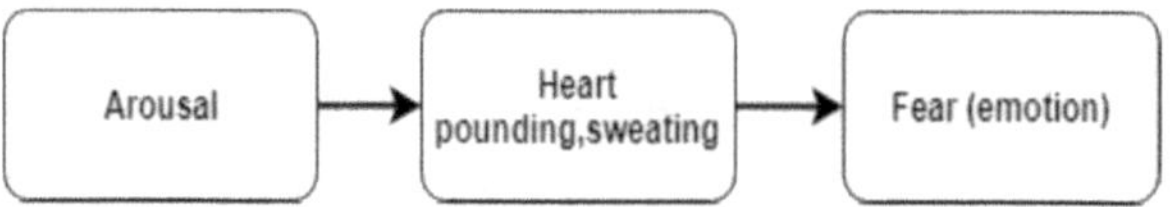

Fig. 1. The James-zange theory of emotion

The Cannon-Bard theory of emotion was developed by two intellectuals who disagreed with the James-Lange theory. They thought that the idea that emotion was triggered by physiological processes was not entirely true [13], such as the idea that fear causes the heart to beat faster. But long runs are also followed by races. Those with racing hearts might become afraid if emotions were evoked by a physiological response. Figure 2 illustrates the Cannon-Bard concept that they put out.

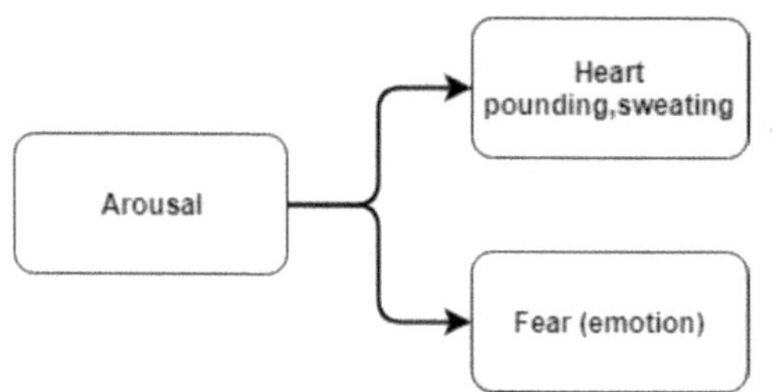

Fig. 2. The Cannon-Bard theory of emotion

Another significant theory that maintains that both physiological and cognitive responses can contribute to the formation of an emotion is the Schachter-Singer hypothesis, which is depicted in Fig. 3.

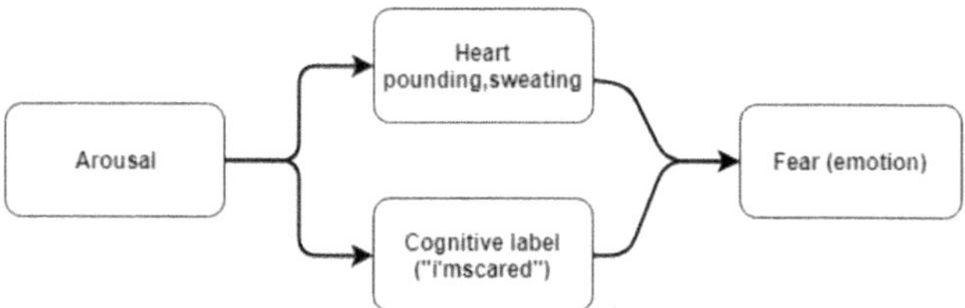

Fig. 3. The Schachter-Singer theory of emotion

Another notion is the Lazarus theory. It was proposed that the cognitive assessment of the experience determines the mood [15]. There are similarities between it and the Schachter-Singer hypothesis. Using the petting of a cat as an example, someone who has touched a cat in the past and may have been bitten by it or experienced a bad interaction with it may characterize a situation as frightening. However, both individuals react differently and feel different feelings, as shown in Fig. 4.

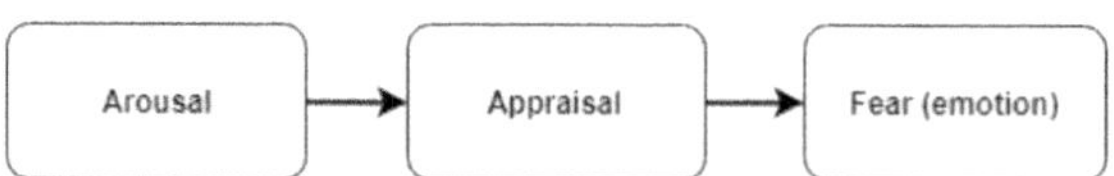

Fig. 4. The Lazarus theory of emotion

The amygdala, hypothalamus, thalamus, and hippocampus. The Limbic system is made up of these four structures. The spinal cord reaches all the way down our back to the area around our tailbone, as seen in Fig. 5. For the senses, the thalamus acts as a kind of sensory relay station. Therefore, all of our senses—hearing, touching, tasting, and seeing—pass through many brain nerves ends before reaching the thalamus, which is situated in each of our two hemispheres. The thalamus directs these sensations into particular areas of the cortex and other parts of the brain. The things we see, hear, and touch have a big impact on how we feel.

Nevertheless, scent is the sole human sense that does not pass through the thalamus [16]. Instead, it passes through a particular region of the brain via a unique signal channel that starts in the nose. Sometimes, certain odors can evoke powerful memories and take us back in time, which helps to explain why this part of our brain is so closely linked to other parts that regulate emotions. Thus, it is evident that our emotions depend heavily on our senses.

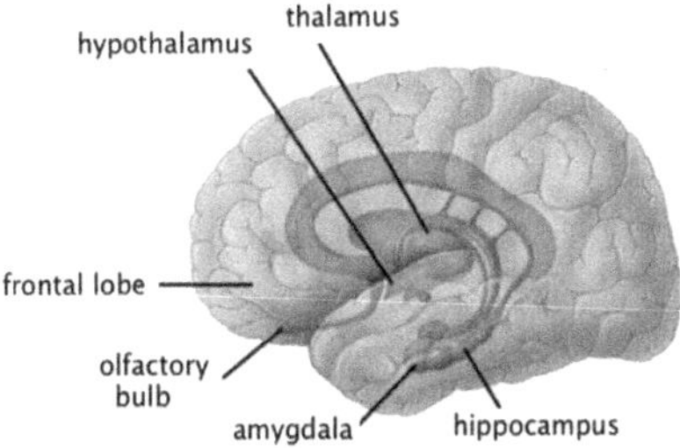

Fig. 5. Cross-section of the brain showing the limbic system. *Courtesy* https://webspace.ship.edu/cgboer/limbicsystem.html

The amygdala is the name given to the aggressive part of the brain. The condition is actually known as Kluver-Bucy syndrome, and it is caused by bilateral amygdala injury [17]. A person with the condition may act quite impulsively. They don't care about the risks involved with their behavior. They consequently act recklessly and take chances.

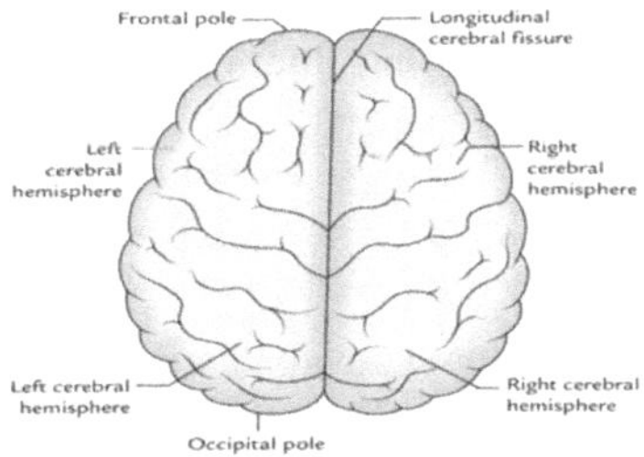

Fig. 6. Superior view of the cerebrum. *Courtesy* https://neupsykey.com/cerebrum

Emotions are primarily controlled by the cerebral cortex of the brain. The cerebral cortex can be divided and organized in a number of ways. So, we'll look at a few from an emotional standpoint. The hemispheres of the brain can be used to conceive it. Figure 6 shows a picture of the left and right hemispheres. In fact, scientists found that the left half of the brain is more electrically active while feeling positive emotions, and the right side is more electrically active when feeling negative emotions [18]. Some kids preferred to be gregarious. Some kids made the decision to isolate themselves by spending more time by themselves. EEG data showed that the children's left hemisphere was more active when they were playing in groups and were more outgoing situation that came about as a result of bilateral amygdala damage [17, 19].

In the Circumplex model of emotions, Russell proposed that the emotions may be viewed as a dispersed 2D space, with valence and arousal acting as its 2-dimension lines [20]. Valence represents the horizontal line, arousal the vertical line, and an average excitation level and middle valence the middle line (Fig. 7). The Circumplex model is used to convey emotion at different levels of valence and excitation, or at a middle level of one or both of these elements. Investigating the clues behind emotive language, facial expressions, and other emotional states was the most common use of these models [21]. Depending on the level of pleasure and excitement, this model might depict a variety of emotions that are elicited or directed by external stimuli, as illustrated in Fig. 7.

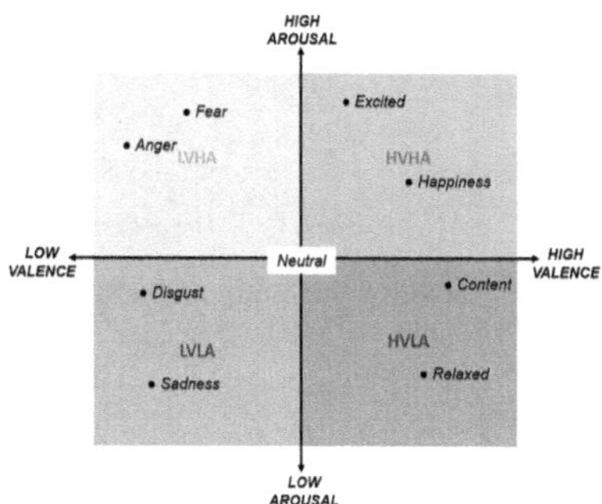

Fig. 7. The circumflex model of emotions indicates anger and calmness

3 Methodology

The creation of the dataset, pre-processing, filtering, feature extraction, and classification are all thoroughly explained in this chapter.

3.1 Pre-processing

There are two types of DEAP datasets available: partially pre-processed and unpre-processed [23]. Partially pre-processed data that was chosen for the investigation. A down sample of the data was made to 128 Hz. With frequencies between 4.0 and 45.0 Hz filtered and outliers and artifacts eliminated.

3.2 EEG Channel Selection

There are 32 patients in the dataset, including both male and female individuals. In addition to 40 sets of channels (32 EEG channels and 8 physiological signals), each patient had 40 separate trials (music videos) with their own EEG data [24–26].

3.3 Choosing a Frequency Band

According to some research, the Gamma Wavelength is where the most complex emotions are felt. Since the project's categorization goal involves classifying peacefulness and fury. Selecting alpha or beta wavelengths makes sense. Higher total energy activation in the beta wavelength is even indicated in a work by Kao FC [27]. As a result, bands in the Alpha, Beta, and Gamma range were chosen for the anger and tranquillity classification.

3.4 Feature Selection

Mean

Mean is the sum total number of samples divided by the number of samples.

$$\overline{\mathrm{x}} = \sum \frac{x_n}{n} \tag{1}$$

Sample maximum and minimum

The maximum and minimum, are the largest observation and smallest observation possible in a given sample

$$\mathrm{f(x)} = \mathrm{ax}^2 + \mathrm{bx} + \mathrm{c} \tag{2}$$

where a, b, and c are $\neq$ zero

Set the derivative equal to zero and Solve for x.

Insert the solved value of x into the original function.

Variance

Variance measures the spread of values from their average value.it is formulated by Eq. 3

$$S^2 = \frac{\sum (X - \overline{X})^2}{N - 1} \tag{3}$$

Standard deviation

A quantity expressing the value of the members of a group's difference from the mean value for the group.

$$\sigma = \sqrt{\frac{1}{N}\sum_{i=1}^{N}(x_i - \mu)^2} \tag{4}$$

Interquartile range

The interquartile range is the difference between the upper values and lower values in each set of data. It is also known as IQR and is a useful measure of variability spread in the data.

$$\text{IQR} = \text{Upper limit-Lower limit} \tag{5}$$

Median absolute deviation

The median absolute deviation is a way to describe variation in a dataset. Median absolute deviation (MAD) of a data set is the average distance between each data value and the median. It's another way to describe variation.

$$\text{MAD} = \text{median}\ (|\ \text{x} - \text{median(x)}|) \tag{6}$$

Range

The area of variation between upper and lower limits on a particular scale.

$$\text{Range} = \max - \min \tag{7}$$

Skewness

Skewness (Sk) is a measure of the lack of symmetry. A data set is said to be symmetric if it looks the same from the right and left of the center point.

$$\text{Sk} = \frac{(mean - median)}{Std.deviation} \tag{8}$$

Kurtosis

Kurtosis is a measure of whether the data are heavy-tailed or light-tailed relative to a normal distribution. Data sets with low kurtosis tend to have light tails or lack of outliers the ones with high kurtosis tend to have heavy tails or outliers. A uniform distribution would be the extreme case.

$$kurtosis = \frac{1}{(n-1)(n-2)(n-3)}(n(n+1)\sum(x_i - \overline{x})^4 - 3(\sum(x_i - \overline{x})^2)^2) \tag{9}$$

where,

n = no. of layer thickness measurements for the layer
x_i = individual layer thickness measurements for the layer
$\overline{x}$ = mean layer thickness.

Shannon's Entropy

It measures the average information required to identify random samples from that dataset.

$$H(X) = -\sum_{i=0}^{N-1} p_i \log_2 p_i \tag{10}$$

pi is the probability of i. To calculate log2 from another log base (e.g., log10 or loge):

$$Log_2(n) = \frac{\log_b(n)}{\log_b(2)} \tag{11}$$

Hjorth parameters

A few general characteristics of an EEG signal are called the Hjorth parameters.
They are mobility and complexity.

Mobility

This estimates the mean frequency. It is the root of the first derivative variance to the signal itself. It is calculated as shown

$$Mobility = \sqrt{\frac{\mathrm{var}(\frac{dy(t)}{dt})}{\mathrm{var}(y(t))}} \tag{12}$$

Here, var means variance.

Complexity

This represents a frequency change. The similarity of a pure sine wave is compared with the signal and if it's similar in nature, the value meets 1[hjorth original]. As shown in Eq. 13.

$$Complexity = \frac{Mobility(\frac{dy(t)}{dt})}{Mobility(y(t))} \tag{13}$$

Together, these two parameters along with variance categorize the EEG patterns in terms of randomness, and amplitude.

Higuchi fractal dimension

The structural complexity of a signal is denoted by a ratio called Fractal dimension (FD). Fractals are self-identical structures which repeat in nature. Similar to mapping a bigger area with the use of smaller lines. Higuchi [27–29] found a new way to find the fractal dimension values. To find it with the Higuchi algorithm, the main time series data (N) is decomposed into shorter time series, and it is described in Eq. 14.

$$X_k^m; X(m), X(m+k), X(m+2k), ..., X\left(m + \left[\frac{N-m}{k}\right].k\right)(m = 1, 2, ..., k), \tag{14}$$

X = primary time sequence
M = index based on 'k'
$X_k{}^m$ = smaller series.
We need to calculate the Lm(k) with the HFD to find the length m of the subsequence. The curve length for the interval k, (L(k)), is the average value over k sets of L(k).
L(k) and FD are related by Eq. 15

$$< \mathrm{L(k)} > \propto \mathrm{k}^{-\mathrm{D}} \tag{15}$$

After that, the fractal dimension is obtained by logarithmic plotting between different k and its associated L(k).

Higher Order Crossings (HOC)

Zero-order crossings are the number of times it will cross the x-axis. When a filter is applied to any time series, the oscillation changes; hence its zero crossing counts also. So, the zero crossing counts obtained from corresponding filter sequences is the HOC sequence.

HOC is obtained as follows:

1. Apply a filter and count the zero crossing
2. Apply another filter to the filtered signal and count the zero crossing again
3. Repeat Step-2 ‘z’ number of times.

4 Results & Discussion

In this study, electroencephalograms (EEG) and peripheral physiological signals were obtained while 32 patients, both male and female, viewed videos lasting 40 min. Multiple time-series characteristics in certain frequencies across an array of pertinent electrodes were extracted using MATLAB from down-sampled (to 128 Hz), pre-processed, and segmented versions of this data. Both SVM and KNN models were used to categorize these EEG signals into two classes—angry and calm—based on a fresh dataset that matched the channels and attributes. All three frequencies were compared for both models. The models under comparison are examined for their correctness, relevance, and potential drawbacks.

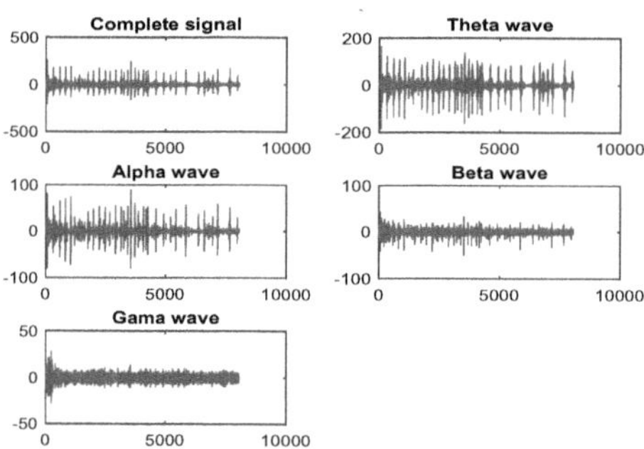

Fig. 8. A typical example of EEG stimulated by Calmness inducing Stimuli

A typical EEG signal produced by a calming stimulus is shown in Fig. 8 along with a breakdown of its individual frequencies, including alpha, beta, and gamma. A typical EEG signal produced by an anger-inducing stimulus at the same frequencies is also displayed in Fig. 9.

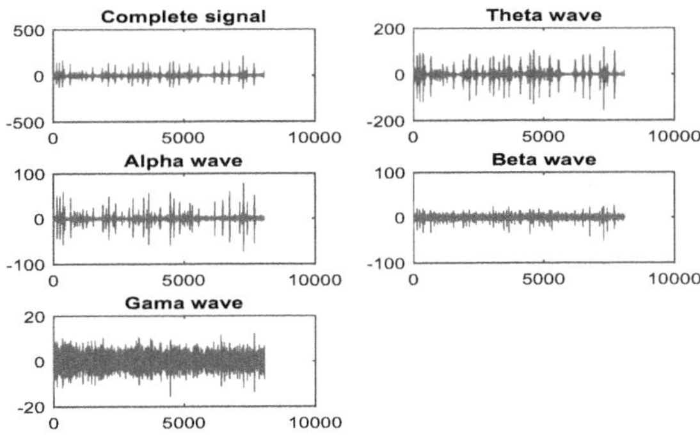

Fig. 9. A typical example of EEG stimulated by Anger-inducing Stimuli

The most pertinent frequencies are compared, as well as the accuracy of the two models. Similarly, as indicated in Tables 1 and 2, KKN and SVM models were trained for frequencies alpha, beta, and gamma with PCA both turned on and off.

Table 1. Accuracy of both models versus their respective frequencies (no cross-validation)

EEG wave	SVM		KNN	
	Without PCA (%)	With PCA (%)	Without PCA (%)	With PCA (%)
Alpha	98.40	62.50	98.40	98.4
Beta	98.40	77.4	98.4	100
Gamma	98.40	100	98.4	100
Raw	87.20	66.2	79.4	62.3

Table 2. Accuracy of both models versus their respective frequencies (10-Fold cross-validation)

EEG wave	SVM		KNN	
	Without PCA (%)	With PCA (%)	Without PCA (%)	With PCA (%)
Alpha	51.60	60.10	57.80	51.60
Beta	43.80	51.60	50	57.80
Gamma	48.40	60.90	54.70	60.90
Raw	23.87	26.22	37.40	35.00

5 Conclusions

EEG signal analysis of emotions is reported utilizing SVM and KNN classifier designs. The signals are captured using various visual and auditory stimuli. In conclusion, this paper's findings show that it is possible to map long duration stimuli to their corresponding psychological processes, even though there is a lot of research in the field of

EEG emotion recognition systems that is based on mapping onto a specific trigger to the stimuli. However, despite our current progress, more research is still required. These techniques can be used for everything from military research to targeted advertisements and marketing. According to our findings, the gamma-trained model's performance for SVM classification is higher than that of KNNs, attaining an accuracy of 60.9%; however, this difference is eliminated when choosing features from PCA. These findings also show that gamma wavelengths have a better accuracy than alpha wavelengths, which have a little lower accuracy (60.1%) and the lowest accuracy (31.6%) for raw data.

References

1. Picard, R.W.: Perceptual user interfaces: affective perception. Commun. ACM **43**(3), 50–51 (2000)
2. Petrushin, V.: Emotion in speech: recognition and application to call centers. In: Artificial Neural Networks in Engineering, pp. 7–10 (1999)
3. Black, M., Yacoob, Y.: Recognizing facial expressions in image sequences using local parameterized models of image motion. Int. J. Comput. Vis. **25**(1), 23–48 (1997)
4. Kim, K., Bang, S., Kim, S.: Emotion recognition system using short-term monitoring of physiological signals. Med. Biol. Eng. Comput. **42**(3), 419–427 (2004)
5. Brosschot, J., Thayer, J.: Heart rate response is longer after negative emotions than after positive emotions. Int. J. Psychophysiol. **50**(3), 181–187 (2003)
6. Chanel, G., Kronegg, J., Grandjean, D., Pun, T.: Emotion assessment: arousal evaluation using EEG's and peripheral physiological signals. In: Multimedia Content Representation, Classification and Security Lecture Notes in Computer Science, pp. 530–537 (2006)
7. Davidson, R., Fox, N.: Asymmetrical brain activity discriminates between positive and negative affective stimuli in human infants. Science **218**(4578), 1235–1237 (1982)
8. Davidson, R., Schwartz, G., Saron, C., Bennett, J., Goleman, D.: Frontal versus parietal eeg asymmetry during positive and negative affect. Psychophysiology **16**(2), 202–203 (1979)
9. Oude Bos, D.: EEG-based emotion recognition. In: The Influence of Visual and Auditory Stimuli (2007)
10. Takahashi, K.: Remarks on emotion recognition from multi-modal bio-potential signals. In: 2004 IEEE International Conference on Industrial Technology. IEEE ICIT 04 (2004)
11. Nie, D., Wang, X.-W., Shi, L.-C., Lu, B.-L.: EEG-based emotion recognition during watching movies. In: 2011 5th International IEEE/EMBS Conference on Neural Engineering (2011)
12. Lange, C.G.E., James, W.E.: The emotions, vol. 1 (1922)
13. Cannon, W.B.: The James-Lange theory of emotions: a critical examination and an alternative theory. Am. J. Psychol. **39**(1/4), 106–124 (1927)
14. Schachter, S., Singer, J.: Cognitive, social, and physiological determinants of emotional state. Psychol. Rev. **69**(5), 379 (1962)
15. Lazarus, R.S.: Progress on a cognitive-motivational-relational theory of emotion. Am. Psychol. **46**(8), 819 (1991)
16. Shepherd, G.M.: Perception without a thalamus: how does olfaction do it? Neuron **46**(2), 166–168 (2005)
17. Preliminary analysis of functions of the temporal lobes in monkeys. J. Neuropsychiatr. Clin. Neurosci. **9**(4), 606–620 (1997)
18. Alfano, K.M., Cimino, C.R.: Alteration of expected hemispheric asymmetries: Valence and arousal effects in neuropsychological models of emotion. Brain Cogn. **66**(3), 213–220 (2008)
19. Lane, R.D., Nadel, L., (eds.): Cognitive Neuroscience of Emotion. Oxford University Press

20. Posner, J., Russell, J.A., Peterson, B.S.: The circumplex model of affect: an integrative approach to affective neuroscience, cognitive development, and psychopathology. Dev. Psychopathol. **17**(03) (2005)
21. Russell, J.A.: A circumplex model of affect. J. Pers. Soc. Psychol. **39**(6), 1161–1178 (1980)
22. Ismail, W.A.W., Hanif, M., Mohamed, S.B., Hamzah, N., Rizman, Z.I.: Human emotion detection via brain waves study by using electroencephalogram (EEG). Int. J. Adv. Sci. Eng. Inf. Technol. **6**(6), 1005 (2016)
23. Koelstra, S., et al.: DEAP: a database for emotion analysis using physiological signals. IEEE Trans. Affect. Comput. **3**(1), 18–31 (2012)
24. Alotaiby, T., El-Samie, F.E.A., Alshebeili, S.A., Ahmad, I.: A review of channel selection algorithms for EEG signal processing. EURASIP J. Adv. Signal Process. **2015**(1) (2015)
25. Hamann, S.: Mapping discrete and dimensional emotions onto the brain: controversies and consensus. Trends Cogn. Sci. **16**(9), 458–466 (2012)
26. Kassam, K.S., Markey, A.R., Cherkassky, V.L., Loewenstein, G., Just, M.A.: Identifying emotions on the basis of neural activation. Plos One **8**(6) (2013)
27. Shetty, H., Sampathila, N., Shashi Kumar, G.S., Behere, R.: Interactive self assessment toll for analysis of emotional status. Indian J. Sci. Technol. **10**(16) (2017). https://doi.org/10.17485/ijst/2017/v10i16/95314
28. Kumar, G.S., Sampathila, N., Shetty, H.: Neural network approach for classification of human emotions from EEG signal. In: Engineering Vibration, Communication and Information Processing, pp. 297–310. Springer LNEE (2019)
29. Bairy, G.M., Bhat, S., Eugene, L.W.J., Niranjan, U.C., Puthankattil, S.D., Joseph, P.K.: Automated classification of depression electroencephalographic signals using discrete cosine transform and nonlinear dynamics. J. Med. Imaging Health Inform. **5**(3), 635–640 (2015)

Optimizing Handwritten Alphabet Recognition: A Comparative Study of ML Algorithms for Assistive Applications

Hajara Sabnam Kareem Navaz(✉) and Archana Pandita

Amity University Dubai, Dubai, UAE
hajarasabnamkareemN@amitydubai.ae

Abstract. Handwritten Alphabet Recognition plays a crucial role in assistive technologies, enabling individuals with disabilities to interact with digital systems, thereby facilitating communication, education, and accessibility across various domains. This paper presents a comparative analysis of the performance of four algorithms—SVM, KNN, Decision Tree, and Random Forest—in recognizing and accurately classifying handwritten alphabets. The SVM model was trained with four kernels (RBF, Linear, Polynomial, Sigmoid), KNN with varying n_neighbors, Decision Tree with different max_depth values, and Random Forest with varying n_estimators. Results evaluated via prediction accuracy, precision score, recall score and F1 score, demonstrated that SVM with RBF Kernel outperformed other models with a testing accuracy of 97.25% followed by Random Forest (96.25% with n_estimators = 200) and KNN (95.20% with n_neighbors = 5). These findings highlight the effectiveness of machine learning models, particularly SVM, in building accurate handwritten alphabet recognition systems with applications in assistive technologies and accessibility tools.

Keywords: Handwritten alphabet recognition · Support vector machine (SVM) · K-nearest neighbor (KNN) · Decision tree · Random forest · Multiclass classification

1 Introduction

Handwritten alphabet recognition is a field within computer vision and machine learning that focuses on identifying and processing handwritten characters through software. Scanned images of these alphabets are analyzed using image processing techniques, extracting unique features like shape, curvature, and structural patterns, which are then classified by machine learning models. Recognizing handwritten alphabets and characters is a task of transforming a language represented in its spatial form of graphical marks into its symbolic representation. The goal here is to identify input characters correctly which are then analyzed for automated process systems [1]. While this technology is already applied in various domains, its true potential lies in enhancing accessibility and inclusivity. Assistive Technology, as defined by the World Health Organization, is a broad term that encompasses systems, services, products, methodologies, and strategies

J. Shreyas et al. (Eds.): CODE-AI 2025, CCIS 2690, pp. 114–126, 2026.
https://doi.org/10.1007/978-3-032-19321-6_11

designed to minimize or eliminate restrictions caused by disabilities or incapacities. The Assistive Technology Industry Association further describes these products and services as items, equipment, hardware, or software aimed at assisting people with disabilities [2]. These tools improve independence, enhance communication, and make technology more accessible. The recognition of handwritten alphabets is an integral part of this landscape, offering solutions that aids in the connection between individuals with disabilities and modern digital tools. By enabling users to interact with technology more comfortably and efficiently, assistive tools promote inclusion and allow people to lead more independent lives, contributing to greater social participation and well-being. Handwritten alphabet recognition in general is a fundamental technology that is applied across many domains, but particularly in assistive tools, it proves invaluable by helping individuals with visual, motor, or learning disabilities. The elements of written expression require a set of complicated skills that go beyond the act of holding a pencil and putting words on paper and encompass the complex interaction among physical, cognitive, and sensory systems. Most students on the autism spectrum are likely to have difficulties with written expression that impact their academic performance across subject matter areas [3]. When we look at assistive technologies for visually impaired people, handwritten character recognition extends their ability to read any document, regardless of its status of being printed or not as it aids in converting handwritten text written on mobile and tablet devices into computer encoded text [4]. Handwritten Alphabet Recognition helps bridge the gap between traditional written communication and digital platforms, allowing people with visual impairments, motor disabilities, or limited mobility to interact with devices more easily. By enabling the recognition of handwritten text, these systems provide more inclusive and accessible ways to convert handwritten inputs into digital formats, offering functionalities like reading assistance, text-to-speech conversion, and even enabling users to write and communicate freely. In future, character recognition might serve as a key role to create a paperless environment that helps the visually impaired people to gain enormous number of educational materials [5]. Efficiently converting handwritten text into Braille for visually impaired individuals is a highly impactful application enabled by handwritten character recognition. When handwriting recognition methods become operational, this also saves us from using the keyboard and allows us to type and draw in a much more natural way [6]. Furthermore, handwriting difficulties are a significant concern for students, often leading to referrals for occupational therapy. Research highlights that occupational therapists frequently recommend technology-based solutions, such as keyboard or dictation strategies, to support students facing such challenges, with decisions influenced by factors like equipment cost and available support in schools [7]. The assistive applications for handwritten alphabet recognition can range from something as significant and impactful as helping people with disabilities to everyday conveniences such as allowing users to quickly jot down numbers and names for contacts compared to inputting the same information via the onscreen keyboard [8]. The importance of handwritten alphabet recognition in assistive technology is therefore very evident in its ability to support personalized communication. People who have lost the ability to speak or type can still convey their thoughts and ideas through writing, ensuring that they remain connected with others. Despite significant advancements in handwritten character recognition for numeric datasets, alphabet-based recognition remains a complex multiclass classification problem due to the larger character set and inter-class similarities. Although there have been substantial developments in pattern recognition and classification, the identification of handwritten characters continues to pose challenges [9]. For these systems to be practical, the recognition model must be highly accurate. Any errors in interpreting handwritten input can lead to frustration, confusion, and a loss of confidence in using the technology. In assistive technologies, such as speech-to-text systems or communication devices for those with speech impairments, accurate recognition ensures that users can convey their thoughts effectively. Inaccurate recognition may lead to misinterpretation of the user's intent, which could have serious consequences in critical situations. Furthermore, a high level of accuracy in recognition models enhances the overall user experience by providing seamless and efficient communication. The system should be able to handle a wide variety of handwriting styles, accounting for different speeds, shapes, and forms of writing, without compromising on performance. For assistive applications to be truly helpful, they must be dependable, reducing the need for constant corrections and promoting a smoother interaction between the user and the device. These issues highlight the need for robust algorithms capable of generalizing across varied samples while maintaining high accuracy.

2 Literature Review

Several works have been done in building and evaluating machine learning models for recognizing handwritten alphabets and they clearly outline the effectiveness of various algorithms for the same. Mustafa S. Kadhm and Asst. Prof. Abd Ali Nayif Hassan in their paper on "Handwriting Word Recognition Based on SVM Classifier", proposed a model to recognize handwritten words and achieved the best accuracy of 96.317% based on several feature extraction methods and SVM classifier [10]. Shamim et al., in their paper on "Handwritten Digit Recognition Using Machine Learning Algorithms", concluded that the Multilayer Perceptron classifier gave the most accurate results with minimum error rate followed by Support Vector Machine, Random Forest Algorithm, Bayes Net, Naive Bayes, j48, and Random Tree respectively [11]. Norhidayu binti Abdul Hamid et al., performed a detailed comparison of SVM, KNN and MLP models to classify handwritten text and concluded that KNN and SVM predict all the classes of dataset correctly with 99.26% accuracy [12]. U Ravi Babu et al., in their paper on "Handwritten Digit Recognition Using Structural, Statistical Features and K-nearest Neighbor Classifier", tested a total of 5000 numeral images of handwritten digits using KNN and achieved an overall accuracy of 98.42% [13]. Yevhen Chychkarov et al., in their paper on "Handwritten Digits Recognition Using SVM, KNN, RF and Deep Learning Neural Networks" evaluated various models, ranging from KNN to CNN, and reported an accuracy of 97.5–98.5% on the MNIST test sample after proper tuning of the models. Additionally, it was observed that the SVC classifier achieved the best accuracy for MNIST image recognition using the RBF kernel, with a regularization parameter set to at least 50 [14]. Riyadi et al., in their paper on "Classification of Alphabetic Handwriting of Pre-toddlers using Support Vector Machine Method" developed a SVM model with linear kernel to classify the handwritten alphabets and the model achieved a maximum accuracy of 86.33% [15]. Nadine et al., in their paper on "Isolated Handwriting

Recognition via Multi-Stage Support Vector Machines" implemented a multi-stage SVM approach to classify handwritten letters, achieving an average accuracy of 91.8% through 4-fold cross-validation [16]. Koushik et al., in their paper on "Handwriting Analysis for Classification of Human Personality" investigated various machine learning models for handwritten letter classification. Among these, the SVM model achieved an accuracy of over 95%, with KNN performing comparably at more than 94% [17]. Tsehay et al., in their paper 'Handwritten Digits Recognition with Decision Tree Classification: A Machine Learning Approach,' proposed a decision tree model for recognizing handwritten digits, achieving an accuracy of 79.8% [18]. Munish et al., in their paper on "Performance evaluation of classifiers for the recognition of offline handwritten Gurmukhi characters and numerals: a study" developed several machine learning models to classify the characters. Among these models, Random Forest classifier outperformed others, achieving a recognition accuracy of 87.9% on a dataset of 13,000 samples [19]. Table 1 presents a detailed comparative analysis of a few additional literature work.

Table 1. Comparative analysis of related literature review

Methodology	Limitations	Literature gap	Conclusion	References
Support vector machines (SVM) for handwritten digit recognition [20]	Focused solely on digit recognition; did not address alphabet characters	Lack of application to handwritten alphabet recognition and assistive technologies	High accuracy in digit recognition, demonstrating SVM's effectiveness	Dixit et al. [20]
Convolutional neural networks for Arabic handwritten character recognition [21]	Concentrated on Arabic script; limited exploration of other scripts and assistive applications	Did not investigate the impact on assistive technologies or compare with other ML algorithms	CNNs showed high accuracy for Arabic characters, indicating potential for other scripts	Shams et al. [21]
Hybrid CNN-SVM model for handwritten character recognition [22]	Emphasized model performance without considering real-time application	Lacked focus on deployment in assistive tools and comparison with simpler models	The hybrid model improved recognition rates over individual models	Kamal et al. [22]
K-nearest neighbors and SVM [12]	Limited to digit classification; did not address alphabet recognition	Did not explore the role of these algorithms in assistive technology contexts	SVM outperformed KNN with 99.26% accuracy	Hamid et al. [12]

(continued)

Table 1. (*continued*)

Methodology	Limitations	Literature gap	Conclusion	References

3 Materials and Methods

This research paper analyses and compares the effectiveness of four machine learning algorithms, SVM, K-NN, Decision Tree and Random Forest in correctly identifying handwritten alphabets through features that specify the co-ordinates of the alphabet. In machine learning, classification is the problem of classifying a new instance into pre-identified set of categories. These classification algorithms primarily are divided into two types: binary and multiclass. Binary classification is classifying instances into one of two classes, while multiclass classification is classifying instances into one of three or more distinct classes [23, 24]. The research incorporates systematic steps of data preprocessing, hyperparameter tuning, and performance evaluation using established metrics such as accuracy, precision, recall, and F1 score. Each algorithm was trained and tested under optimized configurations, with particular attention paid to the comparative analysis of prediction accuracy. This methodology provides a structured approach to addressing the challenges of handwritten alphabet recognition while exploring opportunities for enhancing performance in real-world applications.

Handwritten character recognition is an expansive research area that already contains detailed ways of implementation which include major learning datasets, popular algorithms, features scaling and feature extraction methods [20]. The dataset chosen for this study is a secondary dataset where each row represents an image of a handwritten alphabet in the form of certain co-ordinates. Using some basic image processing, these images have been converted into m x n pixels by the primary data collector, where m and n depend on the size and resolution of the original image of the alphabet. Each pixel contains numeric values, where higher values denote the presence of an 'ink' and the pixel with value 0 denotes its absence, implying nothing has been written in that pixel block. A pixel is called 'on' if it contains a positive numeric value, else it is called 'off'. Upon capturing geometric and pixel-based characteristics of the images of the handwritten alphabets, 16 features were derived for each image, and are included in this dataset. Table 2 provides a detailed summary of the derived features, generated using the function, data.describe().

Table 2. Dataset description

Feature name	Data type	Description	Mean	Standard deviation
Letter	Object	Target variable	–	–
xbox	int64	x-coordinate of the top left corner of the bounding box	4.023550	1.913212

(*continued*)

Table 2. (*continued*)

Feature name	Data type	Description	Mean	Standard deviation
ybox	int64	y-coordinate of the top left corner of the bounding box	7.035500	3.304555
Width	int64	Width of the bounding box	5.121850	2.014573
Height	int64	Height of the bounding box	5.37245	2.26139
onpix	int64	Total number of pixels with ink in the bounding box	3.505850	2.190458
xbar	int64	Mean of x-coordinates of all 'on' pixels	6.897600	2.026035
ybar	int64	Mean of y-coordinates of all 'on' pixels	7.500450	2.325354
x2bar	int64	Variance of x-coordinates of all 'on' pixels	4.628600	2.699968
y2bar	int64	Variance of y-coordinates of all 'on' pixels	5.178650	2.380823
xybar	int64	Mean correlation between x and y coordinates of 'on' pixels	8.282050	2.488475
x2ybar	int64	Mean of x*x*y	6.45400	2.63107
xy2bar	int64	Mean of x*y*y	7.929000	2.080619
xedge	int64	Mean edge count from left to right of bounding box	3.046100	2.332541
xedgey	int64	Correlation of xedge with y	8.338850	1.546722
yedge	int64	Mean edge count from bottom to top	3.691750	2.567073
yedgex	int64	Correlation of yedge with x	7.80120	1.61747

The features listed in Table 2 include pixel intensity distributions, stroke direction metrics, and structural characteristics, providing a robust representation of the handwritten alphabets. The character images are based on 20 different fonts and each letter within these 20 fonts were randomly distorted to produce a file of 20,000 unique stimuli and each stimulus has been converted into corresponding numerical features which were scaled to fit a range of 0–15. The dataset is well-balanced, containing equal samples for each alphabet class, ensuring unbiased model training and evaluation. The relevance of this dataset in real-world scenarios makes it a good dataset for evaluating machine learning algorithms for assistive technologies and automated systems.

Pre-processing is an initial step in machine learning which focuses on improving the input data by reducing unwanted impurities and redundancy [20]. In this multi-class classification task, the required Python libraries—NumPy, Pandas, Matplotlib, and Seaborn—were first imported. After understanding the dataset using data.info() and data.describe(), numerical features were filtered for further analysis and processing. To visualize the distribution of these numerical features, box plots were created using Matplotlib and Seaborn. These box plots provide a clear way to identify the range, median, and potential outliers in numerical features, making it easier to compare distributions across multiple groups. Figure 1 shows the box plot visualization of the features of the dataset for outlier detection.

The box plot visualizations show a significant number of outliers present in the dataset for each feature but since every individual measurement and stimuli for each feature is very crucial in enabling the model to understand the geometric shapes of the alphabets, these outliers were deemed significant to improve the model's accuracy and hence, they were retained and not explicitly handled as typical outliers. The dataset columns were then divided into features (x) and the target variable (y).

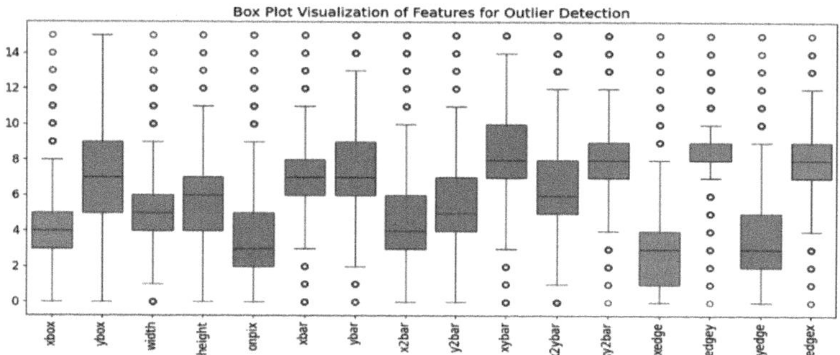

Fig. 1. Box plot visualization for outlier detection

Here, the column 'letter' was identified as the target variable. Following this, the dataset was split into training and testing dataset in an 80:20 ratio with 80% of the data allocated for training and remaining 20% for testing. Once the target variable was separated from the features, the categorical values were converted into numerical values using LabelEncoder. This is done to reduce complexity and enable the model to process data more efficiently as most of the machine learning models operate well only on numerical data. Following this, the categorical values were mapped to numerical values, creating a clear connection between the unique class labels and their corresponding numerical labels.

Then the features were scaled using StandardScaler, ensuring that they have the properties of a normal distribution with a mean of 0 and a standard deviation of 1.

Standardization, also known as z-score normalization helps reduce the dimensionality is calculated for each feature using the following formula:

$$z = \frac{x - \mu}{\sigma} \tag{1}$$

This is done to ensure that the features are brought to a common scale, thereby preventing features with larger ranges from dominating the distance-based models. It improves model performance and convergence during training.

4 Results and Discussion

The performance of the SVM algorithm with different kernels (RBF, Linear, Polynomial, and Sigmoid) is illustrated in Fig. 6, showing the performance of each kernel in terms of testing accuracy. This comparison allows for the identification of the kernel that achieves the highest accuracy for handwritten alphabet recognition. Figure 7 presents the comparison of the accuracies of KNN Classifier with varying values of n_neighbors (3, 5, 7, 9, 11). This graph highlights how different values of n_neighbors affect the model's accuracy and helps determine the optimal value for classification. Figure 8 presents the performance Decision Tree classifier with varying maximum depths in terms of both training and testing accuracies. This graph provides insights into how tree depth influences the model's ability to correctly classify handwritten alphabets, helping to determine the best tree depth value. Figure 9 illustrates the accuracies of the Random Forest classifier with different numbers of trees in the forest. The results of evaluation of the performance of top-performing model for each algorithm is presented in Table 3, with detailed comparison of their accuracy metrics, precision score, recall score and F1 score.

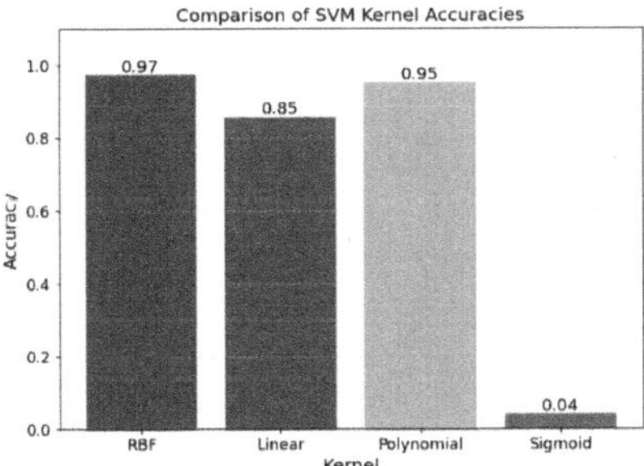

Fig. 6. Comparison of SVM Kernel accuracies

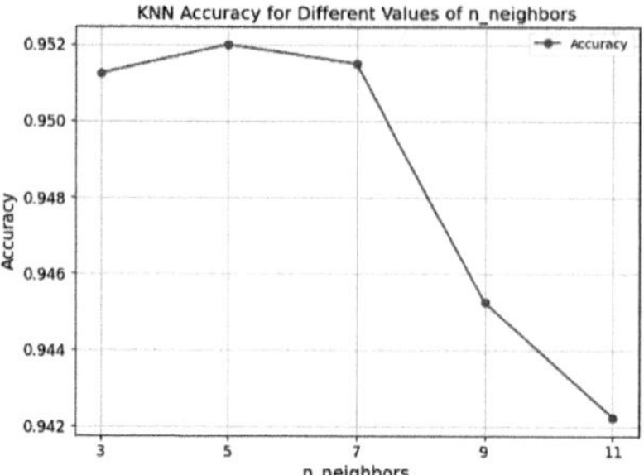

Fig. 7. KNN accuracy

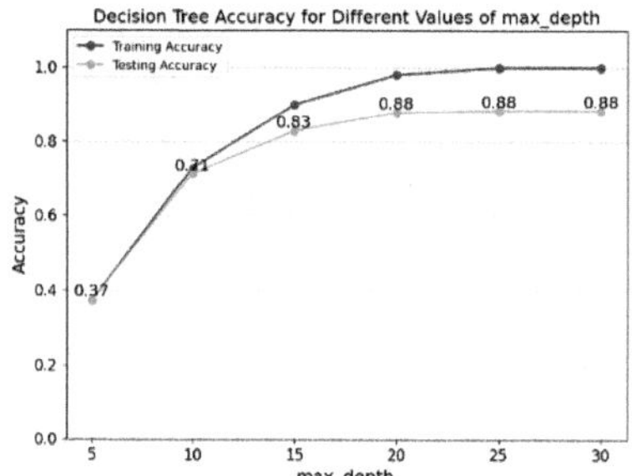

Fig. 8. Decision tree accuracy

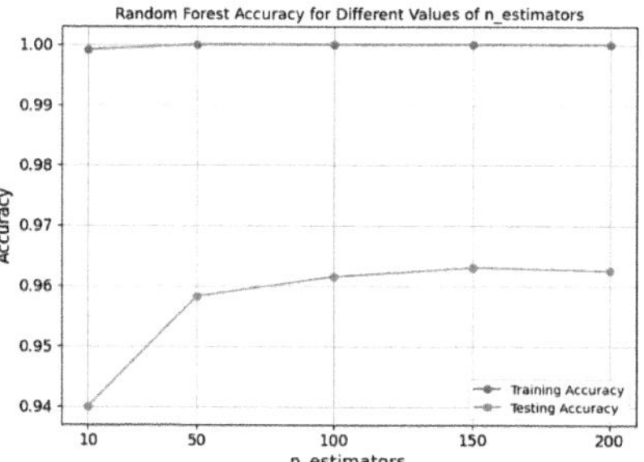

Fig. 9. Random forest accuracy

Table 3. Comparison of the performance of top-performing models of each algorithm

Algorithm	Accuracy (Training) (%)	Accuracy (Testing) (%)	Precision (%)	Recall (%)	F1 score (%)	Comments
SVM (RBF Kernel)	99.04	97.25	97.21	97.19	97.18	Best performance after hyperparameter tuning
KNN (n_neighbors = 5)	97.52	95.20	95.19	95.14	95.12	Good performance, but slightly lower than SVM with RBF kernel
Decision Tree (max_depth = 15)	89.87	82.90	84.11	82.78	83.05	Lower accuracy compared to Random forest
Random Forest (n_estimators = 200)	100	96.25	96.28	96.14	96.16	High accuracy, best performing after SVM

In the comparison of various machine learning algorithms for the given classification problem, the Support Vector Machine (SVM) with the RBF kernel demonstrated the highest test accuracy of 97.25% after hyperparameter tuning, establishing it as the top-performing model. This indicates the RBF kernel's suitability for capturing complex patterns in the data. Further, its high accuracy also proves its ability to model non-linear decision boundaries and perform very well in high-dimensional space compared to other kernels. Hence, this makes RBF kernel more adaptable and makes it easier to achieve great results without extensive tuning, and therefore makes it the most optimal kernel for this multiclass classification problem. While linear and polynomial kernels work well for these kinds of problems, they struggle slightly with complexity and non-linearity. The reason for poor performance of Sigmoid kernel here is possible due to its inability in handling complex decision boundaries in high-dimensional data. Sigmoid kernel is not positive semi-definite, which can cause instability in the optimization process and it also struggles with capturing intricacies of non-linear relationships. Following SVM with the RBF kernel model, the Random Forest algorithm closely followed with an accuracy of 96.25% using 200 estimators, demonstrating its robustness and generalizability. While K-Nearest Neighbors (KNN) with k = 5 achieved a test accuracy of 95.2%, it slightly underperformed compared to SVM and Random Forest. The Decision Tree algorithm, with a max depth of 15, attained a lower test accuracy of 82.90%, highlighting its relatively limited performance for this task compared to ensemble methods like Random Forest. The probable reason for poor performance of Decision Tree compared to other models could be due to possible overfitting. This comparative analysis emphasizes the effectiveness of SVMs in handling complex data when paired with appropriate kernel selection and hyperparameter tuning. While Random Forest and KNN remain viable alternatives, their computational efficiency and suitability for specific applications should be considered.

5 Conclusion and Future Work

This study demonstrates the effective application of machine learning algorithms for solving classification problems. As we can see from the result, SVM with RBF kernel displayed the highest accuracy and emerged as the most reliable model, showcasing its ability to handle complex dataset. Random Forest and KNN proved to be a robust alternative, offering strong and consistent results while Decision Tree highlighted the importance of depth constraints in achieving competitive performance. Overall, the findings reinforce the value of model selection and hyperparameter optimization in achieving optimal results for classification tasks. These models can be extended to develop intelligent systems that support voice-to-text or text-to-speech applications, creating inclusive environments for people with visual or motor impairments. Furthermore, the integration of advanced deep learning techniques, such as Convolutional Neural Networks (CNNs), could enhance the performance of handwriting recognition systems by leveraging their ability to handle complex features and patterns. Additionally, exploring ensemble methods that combine multiple models, such as SVM and Random Forest, could improve prediction robustness and accuracy.

Disclosure of Interests. The authors have no competing interests to declare that are relevant to the content of this article.

References

1. Bojja, P., Bojja, P., Velpuri, N.S.S.T., Pandala, G.K., Polavarapu, S.D.L.R.S.: Handwritten text recognition using machine learning techniques in application of NLP. Int. J. Innov. Technol. Explor. Eng. **9**(2), 1394–1397 (2019)
2. De Freitas, M.P., Piai, V.A., Farias, R.H., Fernandes, A.M.R., De Moraes Rossetto, A.G., Leithardt, V.R.Q.: Artificial intelligence of things applied to assistive technology: a systematic literature review. Sensors **22**(21), 8531 (2022)
3. Coffin, A.B., Myles, B.S., Rogers, J., Szakacs, W.: Supporting the writing skills of individuals with autism spectrum disorder through assistive technologies. In: Autism and Child Psychopathology Series, pp. 59–73. Springer (2015)
4. Computational Approaches to Assistive Technologies for People with Disabilities. Springer (n.d.)
5. Anitha, M., Elangovan, D.: Efficient conversion of handwritten text to braille text for visually challenged people. In: Lecture Notes on Data Engineering and Communications Technologies, pp. 941–949. Springer (2019)
6. Zeki, U., Karanfiller, T., Yurtkan, K.: Handwriting and voice recognition application for students who need special education. Online J. Sci. Technol. **10**(3), 96–110 (2020)
7. Freeman, A.R., MacKinnon, J.R., Miller, L.T.: Assistive technology and handwriting problems: what do occupational therapists recommend? Can. J. Occup. Ther. **71**(3), 150–160 (2004)
8. Vijaya, D., Panda, M.R., Babu, D.A., Yadav, S.L., Nidhya, D.: Machine learning algorithm to detect handwritten character recognition. Migr. Lett. (2023)

9. Sarkhel, R., Das, N., Saha, A.K., Nasipuri, M.: A multi-objective approach towards cost-effective isolated handwritten Bangla character and digit recognition. Pattern Recogn. **58**, 172–189 (2016)
10. Kadhm, M.S., Hassan, A.K.A.: Handwriting word recognition based on SVM classifier. Int. J. Adv. Comput. Sci. Appl. **6**(11) (2015)
11. Shamim, S.M., Miah, M.B.A., Sarker, A., Rana, M., Jobair, A.A.: Handwritten digit recognition using machine learning algorithms. Indonesian J. Sci. Technol. **3**(1), 29 (2018)
12. Hamid, N.A., Sjarif, N.N.A.: Handwritten recognition using SVM, KNN and neural network. arXiv:1702.00723 (2017)
13. Babu, U.R., Chintha, A.K., Venkateswarlu, Y.: Handwritten digit recognition using structural, statistical features and K-nearest neighbor classifier. Int. J. Inf. Eng. Electron. Bus. **6**(1), 62–68 (2014)
14. Chychkarov, Y., Serhiienko, A., Syrmamiikh, I., Kargin, A.: Handwritten digits recognition using SVM, KNN, RF and deep learning neural networks. In: CMIS-2021: the Fourth International Workshop on Computer Modeling and Intelligent Systems, CEUR-WS, vol. 2864 (2021)
15. Riyadi, S., Gunawan, P., Ihsan, A.: Classification of alphabetic handwriting of pre-toddlers using support vector machine method. In: 2023 IEEE International Conference on Communication, Networks and Satellite (COMNETSAT), pp. 153–157. IEEE (2023)
16. Hajj, N., Awad, M.: Isolated handwriting recognition via multi-stage support vector machines. In: 2012 6th IEEE International Conference on Intelligent Systems (IS), pp. 152–157. IEEE (2012)
17. J, B., S, K., Suraj, A., K, N., Venkitesan, P.: Handwriting analysis for classification of human personality. In: 2024 Second International Conference on Advances in Information Technology (ICAIT), vol. 1, pp. 1–6. IEEE (2024)
18. Assegie, T., Nair, P.: Handwritten digits recognition with decision tree classification: a machine learning approach. Int. J. Electr. Comput. Eng. (IJECE) **9**(5), 4446–4451 (2019)
19. Kumar, M., Jindal, M., Sharma, R., Jindal, S.: Performance evaluation of classifiers for the recognition of offline handwritten Gurmukhi characters and numerals: a study. Artif. Intell. Rev. **53**, 2075–2097 (2019)
20. Dixit, R., Kushwah, R., Pashine, S.: Handwritten digit recognition using machine and deep learning algorithms. Int. J. Comput. Appl. **176**(42), 27–33 (2020)
21. Shams, M., A, A., Z, W.: Arabic handwritten character recognition based on convolution neural networks and support vector machine. Int. J. Adv. Comput. Sci. Appl. **11**(8) (2020)
22. Kamal, M., Shaiara, F., Abdullah, C.M., Ahmed, S., Ahmed, T., Kabir, M.H.: Huruf: an application for Arabic handwritten character recognition using deep learning. In: 2022 25th International Conference on Computer and Information Technology (ICCIT), pp. 1131–1136. IEEE (2022)
23. Apao, N.J., Feliscuzo, L.S., Sta. Romana, C.L.C., Tagaro, J.A.S.: Multiclass classification using random forest algorithm to prognosticate the level of activity of patients with stroke. Int. J. Sci. Technol. Res. **9**(4) (2020)
24. Silva-Palacios, D., Ferri, C., Ramírez-Quintana, M.J.: Improving performance of multiclass classification by inducing class hierarchies. Procedia Comput. Sci. **108**, 1692–1701 (2017)
25. Doğan, U., Glasmachers, T., Igel, C.: A unified view on multi-class support vector classification. J. Mach. Learn. Res. **17**, 1–32 (2016)
26. Nair, K., Date, H.B., Shelke, K.P., Bangar, S.N., Sanas, U.G.: Handwritten digit recognition using SVM. J. Emerg. Technol. Innov. Res. **10**(4) (2023)
27. Hamdan, Y.B., Sathish, N.: Construction of statistical SVM based recognition model for handwritten character recognition. J. Inf. Technol. Dig. World **3**(2), 92–107 (2021)

28. Preece, S.J., Goulermas, J.Y., Kenney, L.P.J., Howard, D.: A comparison of feature extraction methods for the classification of dynamic activities from accelerometer data. IEEE Trans. Biomed. Eng. **56**(3), 871–879 (2008)
29. Shah, K., Patel, K.S.: Study of multiclass classification techniques. IARJSET **11**(2) (2024)
30. Breiman, L., Friedman, J.H., Olshen, R.A., Stone, C.J.: Classification and Regression Trees. 2nd edn. Routledge (2017)
31. Chaudhary, A., Kolhe, S., Kamal, R.: An improved random forest classifier for multi-class classification. Inf. Process. Agric. **3**(4), 215–222 (2016)

Enhancing Stress Detection Accuracy Using Support Vector Machine Classifiers with Multimodal Data

Britney Biju(✉) and Archana Pandita

Amity University, Dubai, UAE
britneybiju22@gmail.com

Abstract. In today's fast-paced world, stress has evolved into a widespread societal challenge, fueled by work deadlines, academic pressures, financial uncertainty, and the omnipresent influence of social media. Detecting stress levels is critical, as unmanaged stress can lead to severe physical and mental health issues, impacting both individuals and communities. Real-time stress detection offers the promise of early intervention, helping to mitigate long-term harm and improve overall well-being. This study investigates the performance of Support Vector Machines (SVM) with polynomial (POLY) and radial basis function (RBF) kernels for stress detection using a multimodal dataset. The dataset integrates personality traits, behavioral metrics, and physiological signals, providing a robust foundation for analysis. Results indicate that SVM models are highly effective in detecting stress, with the RBF kernel achieving an impressive accuracy of 98%, outperforming the POLY kernel's 94%. These findings underscore the capability of SVM, particularly non-linear kernels, in handling complex multimodal data to identify stress levels. Despite these promising results, challenges such as the subjectivity in categorizing stress levels and the variability in physiological signals across individuals remain. The report also highlights the potential for future advancements, including the integration of real-time monitoring systems and more nuanced stress categorization techniques, to further enhance stress detection and intervention strategies.

Keywords: Stress Detection · Support Vector Machine (SVM) · Multimodal Stress Detection · Machine Learning Classification · Behavioral and Physiological Features

1 Introduction

Stress has become an invisible force shaping modern life, impacting our health, productivity, and overall quality of life in ways often unnoticed until it reaches a tipping point. Stress, often considered a silent killer, is linked to numerous physical and mental health issues, including heart disease, anxiety disorders, depression, and a weakened immune system [1, 2]. The early detection of stress can help mitigate its adverse effects, leading to timely interventions and better well-being. This is where stress detection

J. Shreyas et al. (Eds.): CODE-AI 2025, CCIS 2690, pp. 127–138, 2026.
https://doi.org/10.1007/978-3-032-19321-6_12

plays a crucial role, enabling researchers and developers to build reliable systems for recognizing stress in individuals. Stress detection techniques, ranging from monitoring physiological indicators like heart rate and cortisol levels to analyzing behavioral and psychological patterns, are essential tools in identifying stress early [3, 4]. To better understand and predict stress, researchers require datasets that encompass various aspects of human behavior, personality traits, and physiological signals. This dataset includes diverse features, such as personality scores (Openness, Conscientiousness, Extraversion, Agreeableness, Neuroticism), sleep metrics (sleep and wake times), activity data (call duration, number of calls and SMS), physiological responses (skin conductance), and movement patterns (accelerometer readings, mobility radius and distance). Additionally, it contains perceived stress scores (PSS) and stress classifications, making it ideal for stress analysis and detection tasks. Despite the availability of various stress detection methods, there remains a significant gap in accurately classifying stress levels due to the diverse and multifaceted nature of stress indicators. Traditional approaches often fail to integrate diverse data types, such as physiological signals, personality traits, and behavioral metrics, into a unified detection framework. Moreover, the efficiency and accuracy of machine learning models, particularly in handling nonlinear and complex patterns in stress data, require further exploration. This report addresses these challenges by leveraging Support Vector Machine (SVM) classifiers with Polynomial (POLY) and Radial Basis Function (RBF) kernels to analyze a comprehensive stress dataset. These SVM kernels are particularly suited to managing nonlinear relationships within data, making them ideal for detecting stress from diverse features, including personality scores, sleep metrics, activity patterns, and physiological signals. By comparing the performance of these kernels, this report aims to identify the most effective approach for classifying stress levels, ultimately contributing to the development of more accurate and reliable stress detection systems. The dataset provides a multidimensional approach to stress analysis by incorporating psychological, physiological, and behavioral data. Key psychological metrics such as personality traits are paired with physiological measures like skin conductance and accelerometer readings, as well as behavioral indicators such as phone usage and mobility patterns. This multimodal structure enables the development of accurate and holistic stress detection models, as stress often manifests through a combination of these factors [5]. By including features such as the Perceived Stress Score (PSS) and Stress Class, the dataset serves as a robust training ground for machine learning models. These labels allow researchers to build predictive systems capable of classifying stress levels with high precision. Such advancements hold the potential for real-time stress monitoring applications, paving the way for proactive stress management solutions [5, 6]. Furthermore, the dataset captures individual differences in stress responses through its personality metrics and behavioral data. This insight is critical for creating personalized intervention strategies, as stress experiences and coping mechanisms vary widely among individuals. By analyzing these factors, systems can generate tailored recommendations, such as improving sleep quality or adopting relaxation techniques, to address unique stress management needs effectively [7]. The dataset's relevance extends to wearable and smartphone integration, as many features—such as skin conductance, accelerometer readings, and screen-on time—can be measured using these devices. This makes it particularly suitable for non-invasive, real-time stress detection applications, aligning with the current trend of utilizing everyday devices for health monitoring. Such applications have the potential to bring stress management tools to a broader audience, enhancing accessibility and convenience [8]. Finally, the dataset's combination of psychological and physiological data contributes significantly to public health research. By elucidating the relationship between behavior, personality, and stress, it provides valuable insights into the broader impact of stress on different populations. These findings can inform public health initiatives aimed at reducing stress and improving overall well-being through targeted strategies [9].

2 Literature Review

Stress detection has gained significant attention due to the growing concerns about its impact on both physical and mental health. Stress is a natural reaction to challenges but chronic stress is linked to various health issues, including cardiovascular diseases, depression, and anxiety [10]. Traditional methods for stress detection, such as self-reported questionnaires and physiological monitoring, are widely used but have limitations, including subjectivity, inconsistency, and delayed responses [11]. To overcome these challenges, research has increasingly focused on automated stress detection systems that leverage physiological, behavioral, and psychological data. Physiological signals such as heart rate variability (HRV), skin conductance, and electroencephalography (EEG) are commonly used for real-time stress detection. These signals provide valuable insights into autonomic nervous system responses to stress, as they change rapidly in response to stressors. For example, heart rate variability has been shown to decrease under stress, while skin conductance increases with arousal [12]. EEG signals, which measure brain activity, have been used to detect stress by identifying changes in brainwave patterns, such as a shift toward theta and alpha waves during stress [13]. However, using these physiological signals alone may not fully capture the complexity of stress responses, and combining them with behavioral data can provide a more holistic understanding. In addition to physiological indicators, behavioral signals such as smartphone usage, sleep patterns, and activity levels have proven to be valuable in predicting stress. For instance, research by Sano & Picard (2013) showed that changes in smartphone use, such as increased texting frequency and the number of social media interactions, were linked to higher stress levels [14]. Similarly, sleep disturbances, which are often a consequence of stress, can serve as a predictor for stress detection [15]. However, the use of single-modal data in stress detection systems has limitations. These methods may overlook the interaction between various factors and their cumulative effect on stress, underscoring the need for multimodal approaches that combine physiological, behavioral, and environmental data. Machine learning (ML) plays a pivotal role in modern stress detection by enabling the automatic identification of patterns from complex and large datasets. Several machine learning algorithms, such as Support Vector Machines (SVM), Random Forests, and k-Nearest Neighbors (KNN), have been applied to classify stress levels based on physiological and behavioral data [16, 17]. These models are designed to capture non-linear relationships between input features, such as heart rate, sleep data, or smartphone usage patterns, and stress outcomes.

SVM has been one of the most widely used algorithms in stress detection due to its ability to handle high-dimensional data and perform well even with small sample sizes.

Table 1. Comparative analysis of stress detection using machine learning

S.No	Data sources	ML model	Methodology	Accuracy	Limitation
[24]	Physiological signals from PhysioNet database	KNN, SVM	Signal processing, feature extraction, classification	99%	Limited to driving scenarios, processing time increases with more sensors
[25]	EEG signals, Fast Fourier transform (FFT)	Support Vector Machine (SVM), Naïve Bayes (NB)	Frontal lobe EEG spectrum analysis, feature extraction	98.21%	One gender research, one modality, and one lobe
[26]	Statistical dataset, Electrocardiogram, Electromyogram, Galvanic Skin Response, Heart Rate, Respiration	Decision Tree, Naïve-Bayes, K-Nearest Neighbor	Statistical data analysis, Machine learning algorithms comparison	100%	-
[16]	Dataset from 206 students of Jaypee institute of information technology Noida	SVM, Random Forest, Naïve Bayes, K-Nearest Neighbour	Cross-Validation applied for performance enhancement.	85.71% (SVM)	Lacks generalize to other demographics. Lacks all stress factors
[27]	SAID Dataset (Collected own dataset using ECG and GSR sensors)	Multilayer Perceptron (MLP), Decision Tree (DT), K-Nearest Neighbour (KNN), SVM, Deep Learning (DL)	Mean, Median, Standard deviation, Min, Max, Max Ratio, Min Ratio.	DT: 95 DL: 91 MLP: 86 KNN: 82 SVM: 79	Limited dataset to a specific age group (20-22 years)
[28]	Private dataset collected own dataset from BNPPGED	SVM	Used wearable physiological sensors like ECG, GSR	82%	Limited to a single machine learning model (SVM)

Sano & Picard (2013) used SVM to detect stress based on physiological and behavioral data from wearable sensors and smartphones, achieving an accuracy of 85% [14]. Despite its success, SVM's performance is sensitive to the choice of kernel functions and requires careful parameter tuning, making it computationally expensive for large-scale applications [17]. Deep learning models, particularly Long Short-Term Memory

(LSTM) networks, are effective for analyzing time-series data, such as physiological and behavioral signals. LSTMs excel at capturing long-term dependencies, essential for understanding stress progression. However, while they surpass traditional machine learning models in accuracy, LSTMs are computationally demanding and require large labeled datasets, which make them less suitable for real-time applications in certain cases [18, 19]. Psychological factors play a crucial role in an individual's response to stress and are important for accurate stress detection. Personality traits, such as those defined in the Big Five personality model—Neuroticism, Extraversion, Openness, Agreeableness, and Conscientiousness—are strongly associated with stress susceptibility. For example, individuals with high neuroticism tend to experience greater emotional instability and are more likely to be stressed in response to negative events [20]. Conversely, individuals high in conscientiousness exhibit better stress coping strategies and are less likely to experience chronic stress [21]. Emotional states, such as anxiety and depression, also affect stress responses. These emotional states can amplify an individual's stress reaction and alter their coping mechanisms. Including psychological assessments, such as questionnaires or personality tests, in stress detection models could improve the accuracy of predictions. However, integrating psychological data into machine learning models poses challenges due to the subjective nature of psychological assessments and the difficulty in quantifying these traits [7, 19, 23]. Moreover, there is a lack of large-scale, publicly available datasets that combine physiological, behavioral, and psychological data for stress detection, hindering progress in this area. Table 1 compares various studies on stress detection using machine learning algorithms, highlighting differences in accuracy, data sources, and limitations. For example, Machine Learning-based Signal Processing Using Physiological Signals achieved up to 99% accuracy with SVM, but it is limited to driving scenarios. Frontal Lobe Real-Time EEG Analysis reached 98.21% accuracy, though it focused on a single gender and brain region. Studies like Stress Detection Using Machine Learning Algorithms achieved 100% accuracy with a statistical dataset, but lacked generalizability. On the other hand, Mental Stress Detection in University Students achieved 85.71% accuracy, limited by its focus on a specific student population. Studies using wearable sensors and IoT, like Stress Detection Using Wearable Physiological Sensors, showed varied results, with the Decision Tree model achieving 95%, but faced limitations in signal variety and age group. The Machine Learning and IoT for Stress Detection study had lower accuracy (68% for SVM), highlighting challenges in dataset and model performance. These findings emphasize the need for more diverse datasets and refined models.

3 Materials and Methods

I used the Stress Detection Dataset [22]. This dataset includes features such as personality traits (Big Five), sleep patterns, phone usage (e.g., call duration, SMS count, screen time), and physiological signals (e.g., skin conductance and mobility). Since the original dataset did not include predefined stress levels, I manually categorized stress into three classes: low stress (1), medium stress (2), and high stress (3), based on relevant feature thresholds and distributions to enable classification and prediction tasks.

The dataset comprises various features aimed at analyzing stress levels among participants. Each participant is uniquely identified by a participant_id, ensuring individual data differentiation. The day column provides a timeline for data collection, enabling time-series analysis. Psychological stress is measured through the PSS_score (Perceived Stress Score), reflecting subjective stress perception, where higher scores indicate greater stress. Personality traits, based on the Big Five model—Openness, Conscientiousness, Extraversion, Agreeableness, and Neuroticism—offer insights into individual characteristics, including creativity, organization, sociability, empathy, and emotional stability. Sleep metrics like sleep_time (hours slept) and wake_time (wake-up time in 24-h format) are critical for assessing sleep patterns and their impact on stress. The PSQI_score (Pittsburgh Sleep Quality Index) evaluates sleep quality, with higher scores signifying poorer sleep. Behavioral metrics include call_duration (total minutes spent on calls), num_calls (number of calls), and num_sms (SMS messages sent), which reveal changes in communication patterns potentially linked to stress. Digital engagement is measured through screen_on_time (total hours the phone screen was active), while physiological responses are captured via skin_conductance, reflecting stress-induced sweating. Physical activity is analyzed through accelerometer data, indicating movement levels, and geographical activity is measured using mobility_radius (movement spread) and mobility_distance (total distance traveled). The target variable, Stress_class, categorizes stress levels into three groups: 1 (low stress), 2 (medium stress), and 3 (high stress). This dataset provides a comprehensive foundation for exploring the multifaceted influences on stress using machine learning models. The dataset was preprocessed by separating features (X) and the target variable (y). Data was split into training (70%) and testing (30%) subsets using the train_test_split function from scikit-learn. Support Vector Machine (SVM) is a supervised machine learning algorithm commonly used for classification, regression, and outlier detection tasks. It is particularly effective for solving problems involving non-linear relationships and high-dimensional datasets. The primary goal of SVM is to find the optimal hyperplane that best separates data points belonging to different classes in a dataset. The algorithm seeks to maximize the margin, which is the distance between the hyperplane and the nearest data points from each class, known as support vectors. The hyperplane serves as the decision boundary that separates the data into different classes. In a two-dimensional plane, this decision boundary is a straight line, while in higher dimensions, it becomes a hyperplane. Support vectors, the data points closest to the hyperplane, are critical because they determine its position and orientation. By maximizing the margin, SVM ensures better generalization and robustness in predictions. When the data is not linearly separable, SVM employs kernel functions to transform the data into a higher-dimensional space where a hyperplane can effectively separate the classes. Common kernel functions include the linear kernel for linearly separable data, the polynomial kernel for capturing polynomial relationships, and the radial basis function (RBF) kernel for non-linear relationships. The sigmoid kernel, often used in neural networks, is another option for mapping complex relationships. The architecture of Support Vector Machine is depicted in Fig. 1. Support Vector Machine (SVM) models with both polynomial and radial basis function (RBF) kernels were employed for classification to evaluate their effectiveness in capturing complex, non-linear relationships between features. The models were trained on the training data and subsequently evaluated on the testing data. To assess the classifiers' effectiveness, several performance metrics were used:

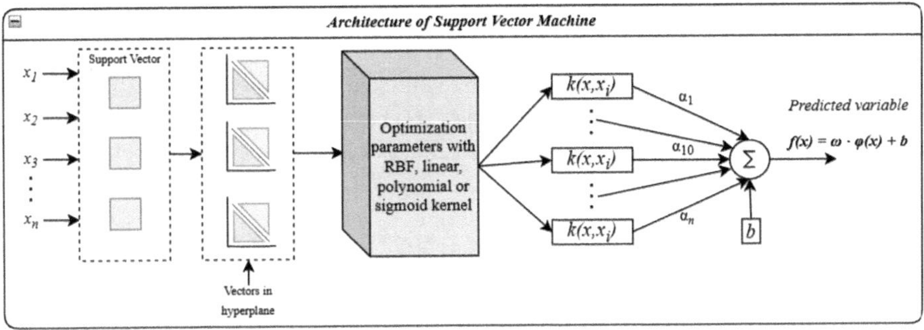

Fig. 1. Architecture of support vector machine (SVM)

1. Accuracy This metric measured the proportion of correctly classified instances, including low, medium, and high stress, out of all instances. Accuracy evaluated the overall effectiveness of the model in distinguishing between different stress levels by con-sidering true positives (TP) and true negatives (TN) while accounting for false posi-tives (FP) and false negatives (FN). (FN).

$$Accuracy = \frac{TP + TN}{TP + FP + TN + FN} \tag{1}$$

2. Precision: assessed the proportion of correctly predicted instances of a stress level (e.g., high stress) relative to all predictions made for that level. High precision indicated reduced false alarms (lower FP) for a given stress category, emphasizing the model's reliability in its classifications.

$$Precision = \frac{TP}{TP + FP} \tag{2}$$

3. Recall: measured the model's effectiveness in correctly identifying actual cases of a specific stress category (e.g., medium stress). This metric focused on minimizing FN by ensuring all relevant instances were accurately captured.

$$Recall = \frac{TP}{TP + FN} \tag{3}$$

4. F1-score: The F1-score provided a balanced evaluation by combining precision and recall, offering a single metric that accounted for both. It was especially valuable for im-balanced datasets, ensuring fair assessment across stress levels, including low, me-dium, and high categories.

$$F1 - Score = 2 \times \frac{Precision \times Recall}{Precision + Recall} \tag{4}$$

These metrics provided insights into the models' ability to correctly identify stress levels across different classes, with comparisons made between the polynomial and RBF kernels to determine the most suitable approach for this dataset.

4 Results

Figure 2 illustrates the correlations between features in the dataset and a key observation from Fig. 2 is the strong positive correlation (0.92) between PSS_score and Stress_class, suggesting that the Perceived Stress Scale score is closely tied to stress classification and may serve as a key predictor.

Table 2. Confusion matrix for SVM (Polynomial Kernel)

	Predicted values			
Actual values	Stress class	Low stress (1)	Medium stress (2)	High stress (3)
	Low stress (1)	159	11	0
	Medium stress (2)	15	299	9
	High stress (3)	2	16	289

Table 3. Classification report for SVM (Polynomial Kernel)

Stress class	Precision	Recall	F1-score	Support
Low stress (1)	0.90	0.94	0.92	170
Medium stress (2)	0.92	0.93	0.92	323
High stress (3)	0.98	0.96	0.97	407
Accuracy	0.94			900

Table 4. Confusion matrix for SVM (Radial Basis Function (RBF) Kernel)

	Predicted values			
Actual values	Stress class	Low stress (1)	Medium stress (2)	High stress (3)
	Low stress (1)	168	2	0
	Medium stress (2)	7	311	5
	High stress (3)	0	8	399

Based on the results shown in Table 2, the SVM model with a polynomial kernel correctly classified 847 out of 900 instances, achieving an overall accuracy of 94%. For Low Stress, 159 instances were accurately identified, while 11 were misclassified as Medium Stress and none as High Stress. As detailed in Table 3, for Low Stress, the precision is 90%, and recall is high at 94%, indicating that most Low Stress cases were

Table 5. Classification report for SVM (Radial Basis Function (RBF) Kernel)

Stress Class	Precision	Re-call	F1- score	Support
Low stress (1)	0.96	0.99	0.97	170
Medium stress (2)	0.97	0.96	0.97	323
High stress (3)	0.99	0.98	0.98	407
Accuracy	0.98			900

correctly identified. These results, as summarized in Tables 1 and 2, indicate that the model is highly effective, particularly in identifying High Stress cases, which is critical for timely interventions. The results presented in Table 4 demonstrate that the SVM model with an RBF kernel achieved excellent performance, correctly classifying 878 out of 900 instances for an overall accuracy of 98%. As shown in Table 5, for Low Stress, the precision is 96%, and recall is nearly perfect at 99%, indicating that almost all Low Stress cases were correctly identified. These results, as detailed in Tables 3 and 4, highlight the robustness of the RBF kernel in handling non-linear relationships, enabling the model to effectively distinguish stress levels across all categories. Table 6 provides a comparative analysis of the current study with two other research works on stress detection using machine learning and deep learning approaches.

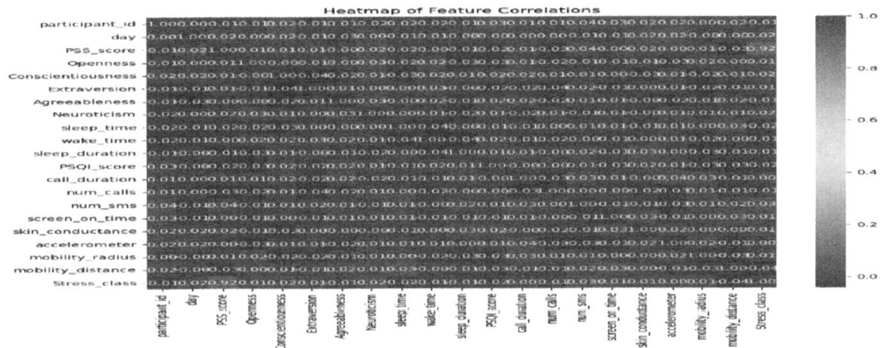

Fig. 2. Correlation heatmap: highlights feature relationships, with PSS_score strongly correlated (0.92) with Stress_class, indicating predictive relevance.

Table 6. Comparative analysis of current study and existing works

	Proposed study	[5]	[16]
Dataset used	Stress detection dataset from Kaggle	WESAD dataset	Perceived stress scale dataset (PSS)
ML models	SVM (Poly and RBF kernels)	KNN, SVM, AdaBoost, RF, ANN	SVM, Random forest, Naïve bayes, KNN

(continued)

Table 6. (*continued*)

	Proposed study	[5]	[16]
Methodology	Multimodal dataset with personality, behavioral, and physiological features;	Multimodal physiological data, deep learning, and ML for 3-class and binary classification	Stress detection among students, categorized stress before exams and internet usage
Accuracy	POLY: 94%	Up to 95.21% for ANN	85.71% (SVM)
Advantages	Multimodal dataset integrating personality traits, behavioral data, and physiological signals. Compares POLY and RBF kernels to optimize SVM performance	95.21% accuracy with a multimodal approach to physiological signals. Demonstrates the potential of ensemble learning techniques like AdaBoost	Integrates student-specific factors for context-specific detection. Includes comparative analysis of models like SVM, KNN, and Random forest
Disadvantages	Stress classes were manually categorized, which may introduce subjectivity. Focuses on a single dataset, limiting generalization to broader populations	Dataset focus is on physiological responses, limiting holistic stress detection approaches. No consideration of real-time	Dataset is limited to students, reducing generalizability to other populations. Achieves relatively low accuracy

5 Conclusion

The findings of this study confirm that SVM models are well-suited for the dataset used in stress detection. The RBF kernel performed exceptionally well, achieving an accuracy of 98%, making it highly effective in handling the non-linear relationships inherent in the multimodal dataset. Its ability to accurately distinguish between Low, Medium, and High Stress levels demonstrates its robustness in analyzing diverse features. The polynomial kernel, while still effective with an accuracy of 94%, was slightly less reliable due to a higher rate of misclassification. The dataset's inclusion of personality traits, behavioral patterns, and physiological signals provided a strong foundation for stress classification, and the SVM models capitalized on this multimodal nature effectively. However, the manual categorization of stress levels introduces a degree of subjectivity that could potentially affect the reliability of the results. Integrating these findings into real-time stress detection systems using wearable technology could make stress management tools more accessible and practical, paving the way for innovative health monitoring solutions. These steps would significantly contribute to the advancement of stress detection research and its practical applications.

References

1. Yaribeygi, H., Panahi, Y., Sahraei, H., Johnston, T.P., Sahebkar, A.: The impact of stress on body function: a review. EXCLI J. **16**, 1057–1072 (2017). https://doi.org/10.17179/excli2017-480
2. Dhabhar, F.S.: Effects of stress on immune function: the good, the bad, and the beautiful. Immunol. Res. **58**, 193–210 (2014)
3. Iqbal, T., Elahi, A., Redon, P., Vazquez, P., Wijns, W., Shahzad, A.: A review of biophysiological and biochemical indicators of stress for connected and preventive healthcare. Diagnostics. **11**(3), 556 (2021). https://doi.org/10.3390/diagnostics11030556
4. Frazier, S.E., Parker, S.H.: Measurement of physiological responses to acute stress in multiple occupations: a systematic review and implications for front line healthcare providers. Transl. Behav. Med. **9**(1), 158–166 (2019)
5. Bobade, P., Vani, M.: Stress detection with machine learning and deep learning using multimodal physiological data. In: 2020 Second International Conference on Inventive Research in Computing Applications (ICIRCA), pp. 51–57. IEEE (July 2020)
6. Chaptoukaev, H., et al.: Stressid: a multimodal dataset for stress identification. Adv. Neural Inf. Proces. Syst. **36** (2024)
7. Crum, A.J., Jamieson, J.P., Akinola, M.: Optimizing stress: an integrated intervention for regulating stress responses. Emotion. **20**(1), 120 (2020)
8. Gedam, S., Paul, S.: A review on mental stress detection using wearable sensors and machine learning techniques. IEEE Access. **9**, 84045–84066 (2021)
9. Jensen, C.G. et al.: Open and calm–a randomized controlled trial evaluating a public stress reduction program in Denmark. BMC Public Health, 15, 1–13 (2025)
10. McEwen, B.S.: Neurobiological and systemic effects of chronic stress. Chronic Stress. **1**, 2470547017692328 (2017)
11. McPartlin, D.A., O'Kennedy, R.J.: Point-of-care diagnostics, a major opportunity for change in traditional diagnostic approaches: potential and limitations. Expert. Rev. Mol. Diagn. **14**(8), 979–998 (2014)
12. Kim, H.G., Cheon, E.J., Bai, D.S., Lee, Y.H., Koo, B.H.: Stress and heart rate variability: a meta-analysis and review of the literature. Psychiatry Investig. **15**(3), 235 (2018)
13. Sanei, S., Chambers, J.A.: EEG signal processing. John Wiley & Sons (2013)
14. Sano, A., Picard, R.W.: Stress recognition using wearable sensors and mobile phones. In: 2013 Humaine Association Conference on Affective Computing and Intelligent Interaction, pp. 671–676. IEEE (Sept. 2013)
15. Harshini, V., Kuresan, H., Kancharlawar, O., Madhan, V.M., Balamithra, S.: Human stress detection in and through sleep using machine learning. In: 2024 International conference on recent advances in electrical, electronics, ubiquitous communication, and computational intelligence (RAEEUCCI), pp. 1–7. IEEE (Apr 2024)
16. Ahuja, R., Banga, A.: Mental stress detection in university students using machine learning algorithms. Procedia Comput. Sci. **152**, 349–353 (2019)
17. Guido, R., Ferrisi, S., Lofaro, D., Conforti, D.: An overview on the advancements of support vector machine models in healthcare applications: a review. Information **15**(4), 235 (2024)
18. Mane, S.A.M., Shinde, A.: StressNet: hybrid model of LSTM and CNN for stress detection from electroencephalogram signal (EEG). Results Control. Optim. **11**, 100231 (2023)
19. Phutela, N., Relan, D., Gabrani, G., Kumaraguru, P., Samuel, M.: Stress classification using brain signals based on LSTM network. Comput. Intell. Neurosci. **2022**(1), 7607592 (2022)
20. Hopkins, J., Lang, C., Glenn, D.E., Michalska, K.J.: Neuroticism predicts increased sensitivity in identifying negative facial affect in young adults. UC Riverside Undergrad. Res. J. **15**(1) (2021)

21. Chen, Y., Peng, Y., Ma, X., Dong, X.: Conscientiousness moderates the relationship between perceived stress and depressive symptoms among US Chinese older adults. J.S Gerontol. Ser. A Biomed. Sci. Med. Sci. **72**(suppl_1), S108–S112 (2017)
22. Stress Detection Dataset: Ashish Shende. Kaggle. (2024) https://www.kaggle.com/datasets/swadeshi/stress-detection-dataset
23. Hopkins, J., Lang, C., Díaz, D., Michalska, K.: Neuroticism predicts increased sensitivity in identifying negative facial affect in young adults. UC Riverside Undergrad. Res. J. Submit. **15** (2021). https://doi.org/10.5070/RJ515355194
24. Ghaderi, A., Frounchi, J., Farnam, A.: Machine learning-based signal processing using physiological signals for stress detection. In: 2015 22nd Iranian Conference on Biomedical Engineering (ICBME), pp. 93–98. IEEE (Nov. 2015)
25. AlShorman, O. et al.: Frontal lobe real-time EEG analysis using machine learning techniques for mental stress detection. J. Integr. Neurosci. **21**(1), 20 (2022)
26. Archana, V.R., Devaraju, B.M.: Stress detection using machine learning algorithms. Int. J. Res. Eng. Sci. Manag. **3**(8), 251–256 (2020)
27. Zainudin, Z., Hasan, S., Shamsuddin, S.M., Argawal, S.: Stress detection using machine learning and deep learning. J. Phys.: Conf. Series **1997**(1), 012019 IOP publishing (Aug. 2021)
28. Sandulescu, V., Andrews, S., Ellis, D., Bellotto, N., Mozos, O.M.: Stress detection using wearable physiological sensors. In: Artificial Computation in Biology and Medicine: International Work-Conference on the Interplay Between Natural and Artificial Computation, IWINAC 2015, Elche, Spain, Proceedings, part I 6, pp. 526–532. Springer International Publishing (2015).

Optimization of Fuzzy Decision Systems for Efficient Waste Collection, Recycling, and Pollution Reduction

Farhaanah, Asfiya Mahek Javeed(✉), Asha SunilKumar, and Rajesh Kumar Chandrawat

Manipal Academy of Higher Education, Dubai, United Arab Emirates
{farhaanah.mohamed,asfiya.mahek,Asha.sunilkumar, Rajesh.chandrawat}@dxb.manipal.edu

Abstract. Efficient waste management is a critical aspect of modern urban planning, aiming to reduce pollution and enhance sustainability through optimal waste collection schedules, recycling, and disposal practices. These ineptitudes lead to increased environmental degradation, higher operational costs, and missed resource recoveries. Traditional systems often struggle to accommodate the variability in waste production, traffic conditions, and recycling rates. This study proposes a sustainable model of a fuzzy decision support system (FDSS) to optimize waste collection schedules, enhance recycling efforts, and reduce pollution. This research focuses primarily on predicting the amount of time required for the collection of waste based on a gathered data set. The fuzzy logic-based model evaluates input factors such as waste generation rate, traffic congestion, weather conditions, and recycling capacities to produce optimal solutions for scheduling waste collection and recycling activities. By leveraging fuzzy logic's ability to handle uncertainty and imprecision, the proposed sustainable system can adapt to dynamic environmental conditions and improve the overall efficiency of waste management processes. Additionally, this approach encourages better resource allocation, minimizes landfill, and promotes eco-friendly waste disposal methods. The proposed model produces a crisp output which lies within the acceptable range of accuracy. Further, regression-based machine learning technique has been applied to significantly reduce RMSE and improving accuracy in garbage collection time.

Keywords: Waste management · Fuzzy Logic · Fuzzy Decision-Making System

1 Introduction

Urban waste management is increasingly challenging due to rising consumption, tourism, and population growth, which all drive up waste production. Traditional methods that rely on fixed schedules often fall short, resulting in issues such as overflowing bins, missed pickups, and ineffective recycling problems that not only hinder sustainability but also contribute to environmental degradation. These challenges are further

J. Shreyas et al. (Eds.): CODE-AI 2025, CCIS 2690, pp. 139–152, 2026.
https://doi.org/10.1007/978-3-032-19321-6_13

compounded by unpredictable factors like fluctuating population densities, migration, and seasonal changes, which in turn lead to higher operational costs and increased greenhouse gas emissions [1]. To address these challenges, innovative and flexible solutions are essential. Urban centers, especially dynamic cities like Dubai where pollution levels, tourist flows, and population densities change throughout the day, can benefit from integrating sustainable practices with data-driven technologies. Moreover, several studies have demonstrated the potential of fuzzy logic in waste management. For instance, Mishra et al. [2] introduced an innovative system to optimize the location selection for household waste recycling plants using Fermatean fuzzy sets. Similarly, Giel and Kierzkowski [3] presented a hierarchical fuzzy logic framework that aids municipalities and government agencies in evaluating garbage sorting and recycling processes for better resource allocation. Another study on the optimization of solid waste disposal using Fuzzy TOPSIS method effectively addressed by Mor and Ramachandran [4]. Additionally, Zafaranlouei et.al [5] employs a novel hybrid decision model based on fuzzy logic with fuzzy Z-numbers on sustainable waste Management Alternatives. A multi-criteria decision-making (MCDM) approach enabled experts to rank and choose the best recycling strategies Suvitha et al. [6]. Task scheduling, a critical component of these systems, shares many similarities with waste management, where the challenge lies in managing diverse and fluctuating input data—such as traffic patterns, weather conditions, and waste generation rates. A recent survey on fuzzy scheduling algorithms, particularly in distributed systems, highlighted by Abadi and Mansouri [7] for the flexibility and adaptability of fuzzy logic in handling uncertainties in dynamic environments like cloud, grid, and fog computing. Kokkinos et al. [8] mentioned that the waste collection scheduling benefits from optimizing factors such as waste generation, traffic congestion, recycling efficiencies and several other factors so it can tackle multiple areas at once. A notable application of fuzzy logic in waste management found in the bio-medical waste sector by Mali et al. [9] where decision-making processes often involve subjective judgment and complex variables. Thaseen et al. [10] explored the integration of IoT, genetic algorithms, and fuzzy logic to create more adaptive, real-time systems for waste collection, ultimately improving urban sustainability and reducing environmental impact. Another study by Nugraha et al. [11] aims to enhance waste transportation efficiency by integrating Internet of Things (IoT) technology with a Mamdani Fuzzy Logic algorithm. in complex systems involving uncertainty. Kabir et al. [12] analyzed the efficiency of the solid waste management system of Canadian provinces. The result of this study found integrates analytical hierarchy process to identify waste disposal sites. Abdulhasan et al. [13] integrated Geographic Information Systems (GIS) and the above-mentioned Analytical Hierarchy Process to identify waste disposal sites in Nasiriya, Iraq. Sathyamurthy [14] presents a fuzzy logic controller approach to particularly manage polymer wastes and formulates membership functions and rule bases to optimize polymer disposal and recycling processes. Marcelleon and Utama [15] suggested developing a decision support model combining Fuzzy Logic and simple additive weighing to assess the hazardousness of various waste types to factor environmental impact and human health risks. Oyebode and Abdulazeez [16] suggests a hybrid strategy by combining genetic algorithms with fuzzy logic to improve the supply chain network handling the solid waste in Lagos, Nigeria. Giel and Kierzkowski [3] implemented the fuzzy logic approach for waste management has potential to aid government and municipality corporations. Karadimas et al. [17] as suggested an adaptive fuzzy logic model to handle solid waste management. Despite the utility of fuzzy logic to address uncertainty in waste management and operational efficiency, the aforementioned study has significantly advanced waste management. To boost real-time sensitivity to dynamic elements like variability in waste generation rates, traffic conditions, and weather circumstances and fluctuating recycling efficiency, more advancements are required. There is still limited work has been done for recycling process optimization and pollution reduction. In this paper we proposed a regression based fuzzy logic model to optimize the waste collection time on variables including population size, pollution level, recycling rate, and garbage collection. We offer a robust framework for effectively allocating resources and scheduling trash collection by using fuzzy logic to manage uncertainty and machine learning for adaptive error minimization.

2 Data Table

The dataset used for analysis consists of synthetic values representing various waste collection scenarios. The input parameters, such as waste generation, population size, pollution level, and recycling rate, are assigned values based on logical assumptions rather than actual measurements. This allows us to systematically evaluate the performance of the fuzzy logic model and its enhancement through machine learning refinement (Table 1).

Table 1. Data table

Day	Time	Waste generation (kg/day)	Population density (persons/km^2)	Tourism activity (high/low)	Recycling infrastructure (Efficiency %)	Pollution level (High/Medium/Low)	Collection method	Recycling rate (%)	Collection time (hrs)
Monday	6:00 AM	8,000	4,500	Low	85%	Low	A	75%	2.5
Monday	12:00 PM	18,000	8,000	High	90%	Medium	A	80%	3
Monday	6:00 PM	22,000	10,000	High	95%	High	M	70%	4
Tuesday	9:00 AM	15,000	6,000	Medium	80%	Low	A	72%	2.5
Tuesday	3:00 PM	12,000	7,000	High	88%	Medium	A	78%	3
Tuesday	9:00 PM	10,000	5,500	Low	75%	Low	M	65%	2
Wednesday	7:00 AM	9,500	4,500	Low	83%	Low	A	76%	2.2
Wednesday	1:00 PM	17,500	8,500	High	90%	Medium	A	79%	3
Wednesday	7:00 PM	21,000	9,500	High	92%	High	M	71%	4.5
Thursday	8:00 AM	14,000	6,500	Medium	85%	Low	A	77%	3
Thursday	2:00 PM	18,500	9,000	High	89%	Medium	A	81%	3.5

(continued)

Table 1. *(continued)*

Day	Time	Waste generation (kg/day)	Population density (persons/km^2)	Tourism activity (high/low)	Recycling infrastructure (Efficiency %)	Pollution level (High/Medium/Low)	Collection method	Recycling rate (%)	Collection time (hrs)
Thursday	10:00 PM	12,000	5,200	Low	80%	Low	M	68%	2.5
Friday	7:00 AM	8,000	4,200	Low	70%	Low	A	70%	2
Friday	11:00 AM	15,500	7,500	Medium	85%	Medium	A	75%	3
Friday	5:00 PM	19,000	8,200	High	92%	High	M	79%	4
Saturday	8:00 AM	12,500	6,800	Medium	82%	Low	A	74%	2.5
Saturday	2:00 PM	18,000	9,500	High	88%	Medium	A	80%	3
Saturday	10:00 PM	14,000	7,000	Low	77%	Low	M	66%	2
Sunday	9:00 AM	10,000	5,800	Low	72%	Low	A	71%	2
Sunday	12:00 PM	16,000	8,000	High	90%	Medium	A	78%	3
Sunday	6:00 PM	21,000	9,200	High	93%	High	M	82%	4

3 Methodology

The goal of the study is to develop a fuzzy decision system that optimizes waste collection schedules, recycling efficiency, and pollution reduction in an urban metro city like Dubai, based on various factors such as waste generation rates, population density, tourism activity, recycling infrastructure, and pollution levels. The methodology outlines the steps for implementing a fuzzy system to handle these variables.

3.1 Fuzzification

Each input variable $x \in [a, d]$ (e.g., Waste Generation, Recycling Rate, Pollution Level) is converted into a fuzzy membership function using a Trapezoidal Membership Function:

$$\mu_A(x) = \begin{cases} 0, & 0 \le x < a \\ \frac{x-a}{b-a}, & a \le x < b \\ 1, & b \le x < c \\ \frac{d-x}{d-c}, & c \le x \le d \\ 0, & otherwise \end{cases} \tag{1}$$

where, a, b, c, d define the trapezoidal function parameters. $\mu_A(x)$ is the fuzzy membership value for input x. We apply this function to Waste Generation, Recycling Rate, and Pollution Level. The fuzzy membership functions are defined using triangular membership functions as:

$$\mu_B(x) = \begin{cases} 0, & 0 \le x < a \\ \frac{x-a}{b-a}, & a \le x < b \\ \frac{c-x}{c-b}, & b \le x \le c \\ 0, & x > c \end{cases} \tag{2}$$

The following fuzzy rules determine the collection time based on the fuzzy inputs.

$$\left\{ \begin{array}{c} If\ (W_G = High \wedge R_R = Low), \rightarrow\ C_T = High \\ If\ (W_G = Medium \wedge R_R = Medium), \rightarrow\ C_T = Medium \\ If\ (W_G = Low \wedge R_R = High), \rightarrow\ C_T = Low \end{array} \right\} \tag{3}$$

where, W_G is Waste Generation, R_R is Recycling Rate, and C_T is Collection Time. Each fuzzy input contributes to the collection time prediction through a weighted sum model:

$$S = \sum_{i=1}^{n} w_i \mu_A(Input\ Parameter) \tag{4}$$

That further modified as:

$$S = w_1 \cdot \mu Waste + w_2 \cdot \mu Recycling + w_3 \cdot \mu Pollution \tag{5}$$

where, w_1, w_2, w_3 are the weights assigned to each input variable. μ *Waste*, μ *Recycling*, μ *Pollution* are the fuzzy membership scores from Step 1. *S* represents the fuzzy output score. To obtain a real-valued prediction, we apply a defuzzification function:

$$Y = \sum_{i=1}^{n} \alpha_i X_i + \beta \tag{6}$$

where, *Y* is the predicted collection time. X_i represents the normalized values of the input variables (e.g., waste, population density, recycling rate, pollution). α_i are the coefficients (learned from data). β is a bias term ensuring realistic output constraints. A weighted formula is used for the waste collection time prediction:

$$T = \sum_{i=1}^{n} w_i \frac{T_{wc}(W, P, R, L)}{\mathrm{Max}(T_{wc}(W, P, R, L))} + C \tag{7}$$

That further modified as:

$$\begin{aligned} T = w_1 \cdot & \frac{T_{wc}(waste)}{Max(T_{wc}(waste))} \\ + w_2 \cdot & \frac{T_{wc}(\text{Population density})}{Max(T_{wc}(\text{Population density}))} \\ + w_3 . & \frac{T_{wc}(Recycling)}{100} \\ + w_4 \cdot & \frac{T_{wc}(Pollution)}{9} + C \end{aligned} \tag{8}$$

where, W = *Waste generation (kg/day)*, P = *Population density* (*persons/km*^{2}), R = *Recycling rate* (%), L = *Pollution level* $(1 - 9)$. w_1, w_2, w_3 are the weights assigned to each input variable. *C* is a bias term to avoid negative values. To ensure realistic results, we cap *T* between 1 and 10 and final time is computed as:

$$T_{final} = \max(1, \min(10, T)) \tag{9}$$

To Evaluate Prediction Accuracy, the Root Mean Square Error (RMSE) Is Calculated:

$$RMSE = \sqrt{\frac{1}{N} \sum_{i=1}^{N} \left(Y_{actual,i} - Y_{predicted,i}\right)^2} \tag{10}$$

where, N is the total number of data points. $Y_{actual,\, i}$ is the actual collection time for data point *i* $Y_{predicted,\, i}$ is the predicted collection time. A lower RMSE indicates better model accuracy. Linear regression optimization (ML enhancement) is done to refine the fuzzy model, we train a Linear Regression model:

$$Y = \beta_0 + \sum_{i=1}^{n} \beta_i X_i \tag{11}$$

where, *Y* is the optimized collection time. X_i represents the input variables. β_0 is the intercept. β_i are the regression coefficients (learned from training data). This step minimizes RMSE and ensures higher prediction accuracy.

4 Algorithm Implementation

To get efficiency in waste management, the waste collection time is targeted with respect to credentials like waste generation, population density, pollution level, and recycling rate. Fig. 1 shows the working of the model.

The following steps are followed as: (i) Input Processing: A dataset capturing waste generation patterns in Dubai is gathered across different time slots (morning, afternoon, evening). Key parameters include waste generation, population density, tourism activity, recycling efficiency, pollution levels, collection method, and collection time. (ii) Fuzzification: Crisp input values are converted into fuzzy sets using membership functions (e.g., Low, Medium, High). Using the eq. (1, 2), each variable like waste generation, population density, recycling efficiency, is represented by triangular or trapezoidal functions to handle uncertainty. (iii) Rule Base Formation: For collection time optimization, a set of fuzzy rules for three different categories namely low, medium, and high is employed using Eq. (3) to define relationships between inputs and outputs. The fuzzy inference engine evaluates rules by antecedent evaluation that matches inputs with fuzzy sets and consequent evaluation that determines fuzzy outputs based on rule strength. (iv) Defuzzification: Converts fuzzy output into actionable crisp values using Eq. (6) to derive the collection time in hours, recycling rate in %, and collection method. (v) Model Evaluation: The waste collection model evaluated in terms of predicted time using all the factors like waste generation, population density, recycling rate, and pollution level. The eq. (7, 8) this time is predicted, and the final time is computed by Max operation using Eq. (9). Then Root means square error of model is calculated through Eq. (10) to see the model efficiency. This error is further optimized in terms of higher prediction accuracy using an enhanced machine learning regression model by Eq. (11). (vi) System Testing & Validation: The model is tested against real-world data to refine rules, optimize membership functions, and ensure adaptability under varying conditions like high pollution or low tourist activity.

5 Results and Discussion

5.1 Computation of Membership Grade

To improve waste management performance, waste collection time is targeted based on factors such as generated waste, population size, pollution degree, and recycling rate. First the triangular and trapezoidal fuzzy membership functions are illustrated for each factor for low medium and high cases. Fig. 2 shows triangular degree function that the amount less than 10000 kg/day supports the low waste generation degree and amount more than 15000 kg/day reveals the membership grade for high waste generation. For the Medium waste generation this amount is distributed into two intervals in kg/day. The degree rise as waste generation amount is [5000, 15000] and it started decreasing in [15000, 25000]. While Fig. 3 depicts the trapezoidal membership degree that shows the modest waste with full degree is defined by the trapezoidal function for producing waste for amounts $\leq$5000 kg/day, progressively dropping to 0 between 5000 and 10,000 kg/day. Starting at 5000 kg/day, the medium category rises linearly to 10,000 kg/day, stays completely medium with 100% degree from 10,000 to 15,000 kg/day, and then starts to fall between 15,000 and 20,000 kg/day. Membership begins to rise to 15,000 kg/day for heavy waste generation, achieves full degree around 20,000 kg/day, and remains steady after this point. This structure efficiently captures the variety in waste generation levels and guarantees a seamless transition between categories.

Fig. 4 illustrates the triangular membership grade that a population density of less than 7,000 person/km^2 supports a low level of waste generation, while a density weight of more than 10,000 person/km^2 indicates a high population membership grade. This density is split into two intervals for medium population. The degree increased when the number of people in square kilometer area are between 3000 and 10000, then it began to decline between 10000 and 17000. Fig. 5 explains the level of pollution in trapezoidal membership grade it shows that It is fully included in the "Low" category when the pollution level is less than 3. The membership progressively drops from 3 to 5 as pollution rises, hitting zero at 5. This means that numbers higher than 5 are no longer considered "Low Pollution." As the number of members in "Medium Pollution" increases from 3 to 5, polluted levels in this range are partially classified as both "Low" and "Medium." "Medium" stays completely active from 5 to 7, meaning that pollution in this range is unquestionably classified as medium. The membership steadily decreases after level 7, enabling a smooth shift to "High Pollution." A low recycling capacity is shown by a regeneration rate of around 50%, and a high recycling performance is indicated by a rate above 50%, as Fig. 6 illustrates triangular function. The membership rating is allocated over two stages for medium recycling efficiency: it rises between [30, 70] and begins to fall between [70, 100]. While Fig. 7 shows the trapezoidal degree for recycle rate it is observed that the Low category is fully obtained by a rate of less than 30%. The membership steadily declines as it rises from 30% to 50%, hitting zero at that point, signaling the change to Medium. From 30% to 50%, membership in "Medium" remains in full degree between 50% and 80%, and then progressively drops from 80% to 100%, moving into High. Recycling rates of 100% or more are classified as "High," and membership progressively rises from 80% to 100%. With a collection period of less than 8 h denoting a low collection duration and a time beyond 13 h denoting a high collection membership grade, Fig. Fig. 8 depicts the triangle membership grade. The membership is divided into two periods for medium collection time: it rises between [5, 12] hours and begins to fall between [12, 17] hours. Similarly Fig. 9 illustrates the trapezoidal membership degree that exhibits the modest collection time with full degree is defined by the trapezoidal function for collecting with in $\leq$3 h/day progressively dropping to 0 between 3 and 5 h. Starting at 3 h, the medium category rises linearly to 5 h, and stays completely medium with 100% degree from 5 to 7 h, and then starts to fall between 7 h and 10 h/day.

Membership begins to rise to 5 h/day for high collection level, achieves full degree around 7 h/day, and remains steady after this point. This structure efficiently captures the variety in collection time and guarantees a seamless transition between categories.

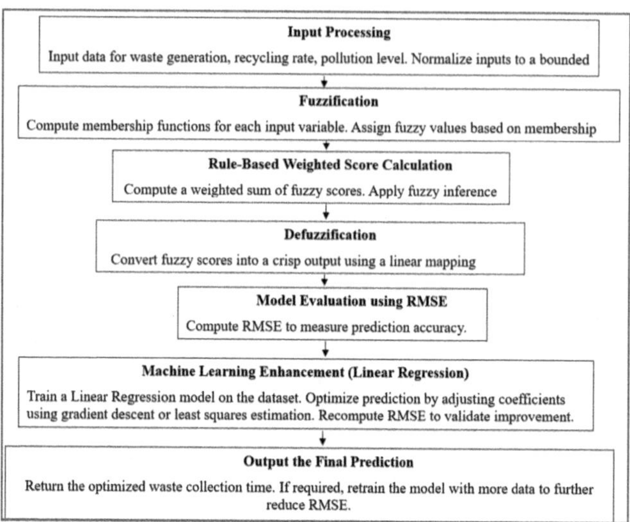

Fig. 1. Algorithm flow layout for designing fuzzy decision systems for effective waste collection routines

5.2 Prediction and Optimization of Collection Time

Following the acquisition of the fuzzy membership function for each case factor—that is, Low, Medium, and High—of all credential variables—that is, generated waste, population size, level of pollution, and recycling rate, the fuzzy scores are evaluated for all source according to its degree value. The collection time is determined using those scores and then optimized using a defuzzification model. The model is trained using a dataset of records that includes recycling rate, pollution level, population density, garbage generation, and real collection time. The fuzzy logic system is used by the function to calculate predictions, which are then contrasted with the dataset's actual values. Root Mean Squared Error (RMSE), which calculates the discrepancy between expected and actual values, is used to assess the initial forecasts. The forecasts are further refined using a linear regression model, which lowers the RMSE and increases forecast accuracy. It is observed from Fig. 9 that the fuzzy logic model predicts waste collection time depending on waste generation, population density, recycling rate, and pollution level by comparing real collection times with fuzzy predicted times for 4 data points. The normalized difference between the actual and anticipated (RMSE) values is 2.82 units, although trends are captured by the model forecasts that need to be improved, perhaps by integrating machine learning or modifying membership functions. Fig. 11 depicts models that run on smaller test sets and its observed that the enhancement over the original fuzzy model indicates that machine learning integration successfully improves the predictions, rendering the model extremely dependable for estimating collecting time. This allows us to evaluate the predictions of collecting time with a little large test case. Fig. 12 shows that the original fuzzy model produced an RMSE of 3.1561, which was a significant departure from real collection times, out of 12 test instances. The optimized model, however, dramatically decreased the error to 0.1649 after applying machine learning refinement, indicating considerably better alignment with actual data. This demonstrates how well fuzzy logic handles uncertainty when combined with machine learning-based changes for fine-tuning, resulting in forecasts of collection times that are more accurate and dependable (Fig. 10).

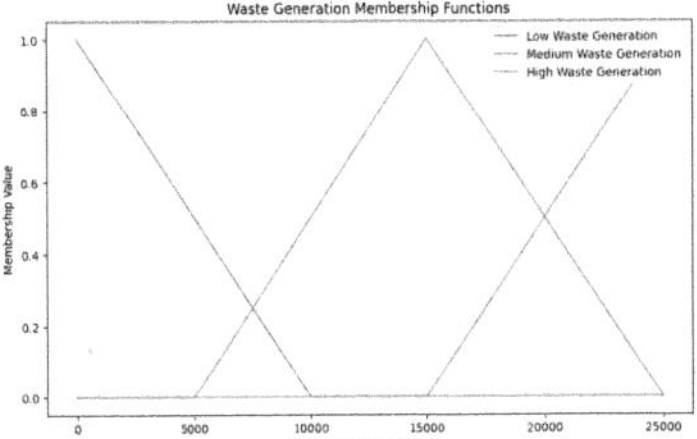

Fig. 2. Triangular fuzzy MG for waste generation

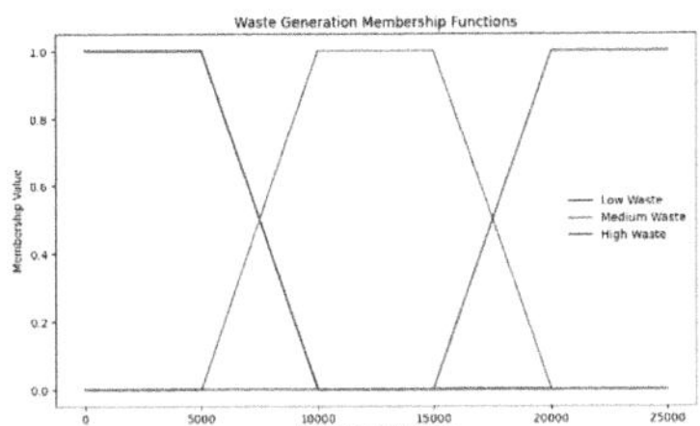

Fig. 3. Trapezoidal fuzzy membership grade for waste generation

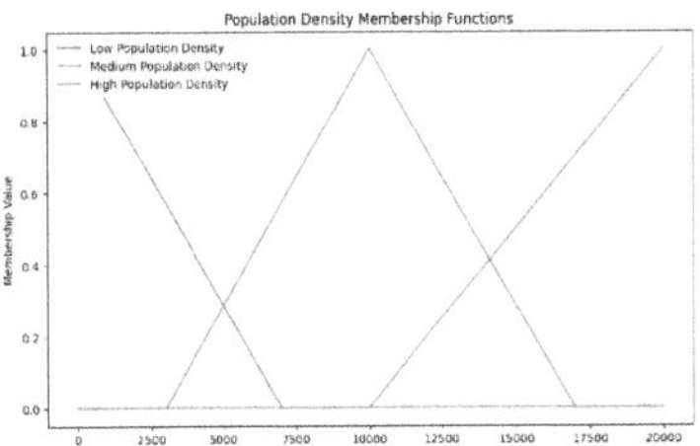

Fig. 4. Triangular fuzzy MG for population density

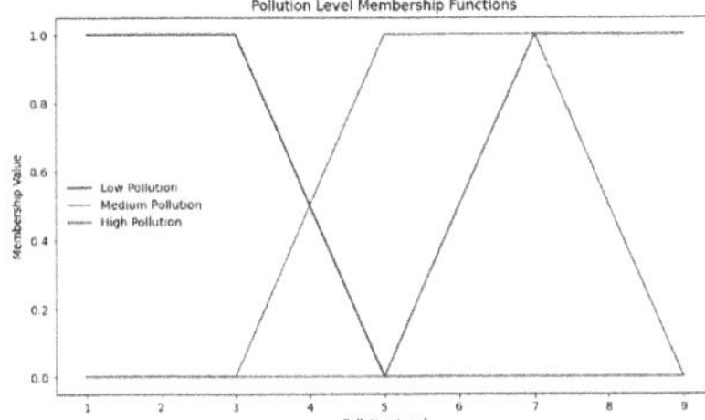

Fig. 5. Trapezoidal fuzzy MG for pollution level

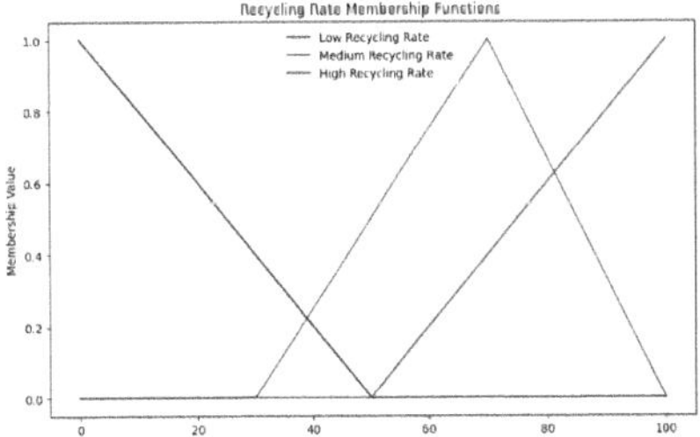

Fig. 6. Triangular fuzzy membership grade for recycling rate

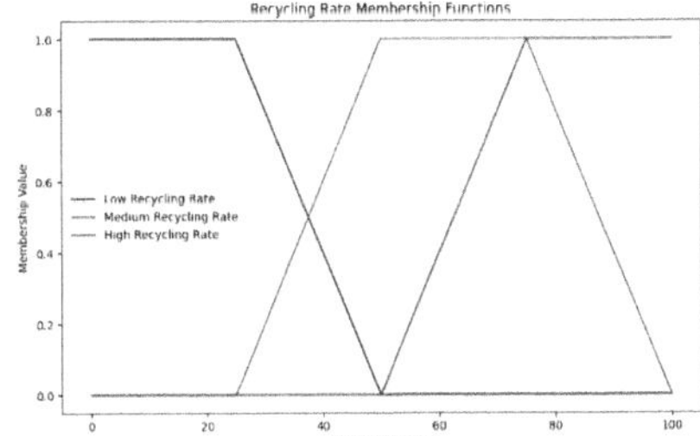

Fig. 7. Trapezoidal fuzzy MG for recycling rate

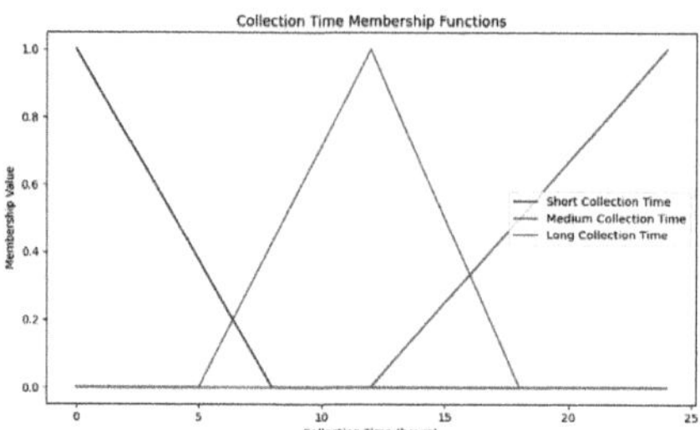

Fig. 8. Triangular fuzzy MG for collection time

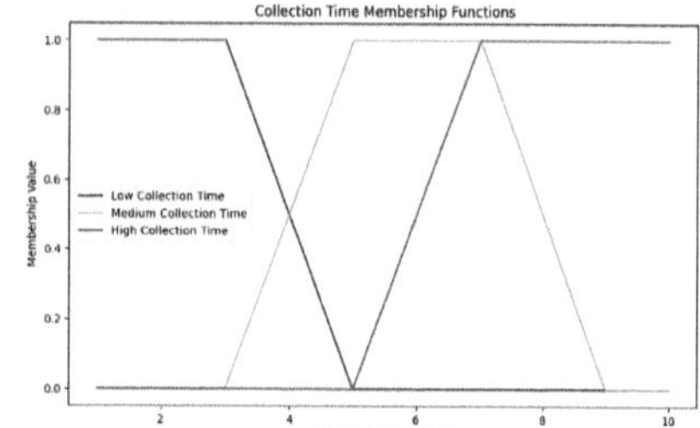

Fig. 9. Trapezoidal fuzzy MG for collection time

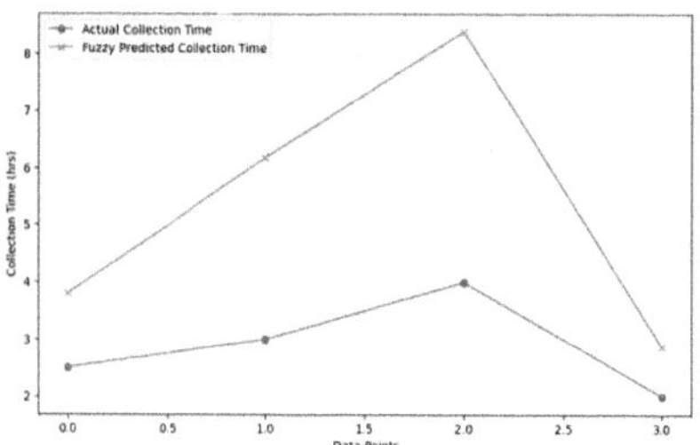

Fig. 10. Actual collection times versus fuzzy predicted times with Root Mean Squared Error (RMSE) = 2.82

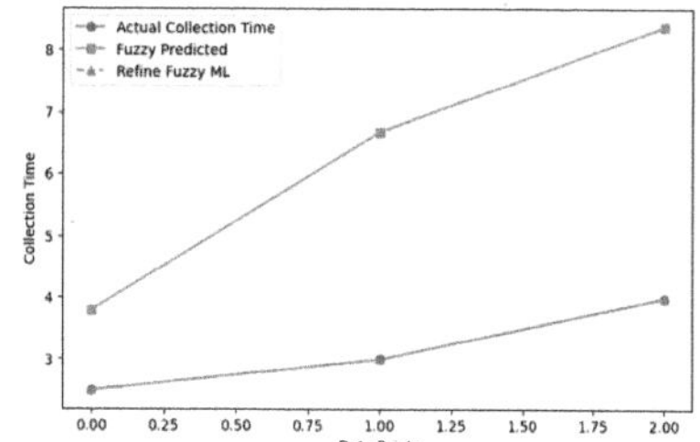

Fig. 11. Actual collection times versuss Refine fuzzy ML Model predicted times with Root Mean Squared Error (After ML Adjustment): 0.0000

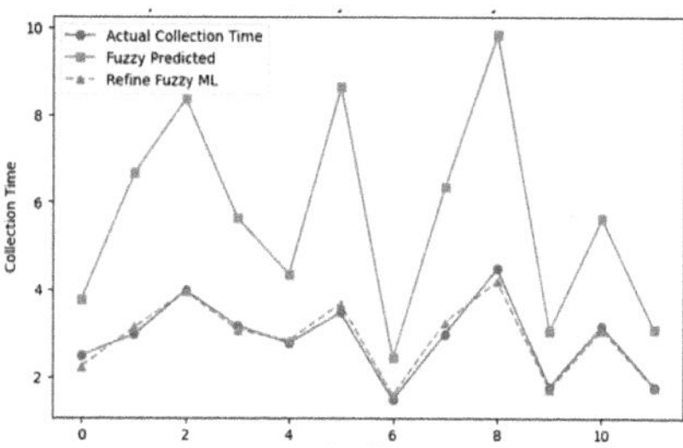

Fig. 12. Actual collection times vs. Refine fuzzy ML Model predicted times with Initial RMSE: 3.1561 Optimized RMSE (After ML Adjustment): 0.1649

6 Conclusion

In this paper the garbage collection time is predicted based on a number of contributing elements, such as the amount of pollution, population size, recycling rate, and created garbage, in order to improve waste management efficiency. The triangle and trapezoidal fuzzy degree functions that divide these elements into low, medium, and high categories are employed in the estimating process. Every situation's fuzzy score is calculated according to its degree, while a defuzzification model is used to calculate the collecting time. The accuracy of the fuzzy logic system's initial predictions is assessed by comparing them to the actual collection times using Root Mean Squared Error (RMSE).

The initial RMSE estimates show differences between expected and observed values, even if the fuzzy logic model reflects underlying patterns in trash collection time. Significant gains in predicted accuracy result in the use of machine learning-based adjustments to overcome this constraint. To fine-tune forecasts, the optimization method entails training a regression model using historical data, which includes trash creation, population density, pollution level, and recycling rate. The findings show a significant decrease in RMSE, demonstrating that the incorporation of machine learning improves the precision and dependability of estimates based on fuzzy logic. This method demonstrates the potential of hybrid models, which provide a strong framework for scheduling garbage collection and allocating resources efficiently by utilizing machine learning for adaptive error minimization and fuzzy logic for managing uncertainty. The limitation and Future work is that this model has computational complexity, lacks real-time adaptability, and depends on previous data. For better accuracy, scalability, and dynamic waste management, future research can combine behavioral aspects, deep learning, and IoT.

Disclosure of Interests. No conflict of interest

References

1. Amasuomo, E., Baird, J.: The concept of waste and waste management. J. Manag. Sustainability. **6**(88) (2016). https://doi.org/10.5539/jms.v6n4p88
2. Mishra, A.R., Rani, P., Saeidi, P., Deveci, M., Alrasheedi, A.F.: Fermatean fuzzy score function and distance measure based group decision making framework for household waste recycling plant location selection. Sci. Rep. **14** (2024). https://doi.org/10.1038/s41598-024-78158-z
3. Giel, R., Kierzkowski, A.: A fuzzy multi-criteria model for municipal waste treatment systems evaluation including energy recovery. Energies (Basel). **15** (2022). https://doi.org/10.3390/en15010031
4. Mor, D., Ramachandran, M.: Optimization of Solid Wastes Disposal Strategy by Fuzzy Topsis Method.
5. Zafaranlouei, N., Ghoushchi, S.J., Haseli, G.: Assessment of sustainable waste management alternatives using the extensions of the base criterion method and combined compromise solution based on the fuzzy Z-numbers. Environ. Sci. Pollut. Res. **30**, 62121–62136 (2023). https://doi.org/10.1007/s11356-023-26380-z
6. Suvitha, K., Narayanamoorthy, S., Pamucar, D., Kang, D.: An ideal plastic waste management system based on an enhanced MCDM technique. Artif. Intell. Rev. **57** (2024). https://doi.org/10.1007/s10462-024-10737-y
7. Z., J.K.A., Mansouri, N.: A comprehensive survey on scheduling algorithms using fuzzy systems in distributed environments. Artif. Intell. Rev. **57** (2024). https://doi.org/10.1007/s10462-023-10632-y
8. Kokkinos, K., Lakioti, E., Moustakas, K., Tsanaktsidis, C., Karayannis, V.: Sustainable medical waste management using an intuitionistic fuzzy-based decision support system. Sustainability (Switzerland). **16** (2024). https://doi.org/10.3390/su16010298
9. Mali, S.R., Ingole, N.W.: Student. P.G, Eco-Efficiency Optimization for Municipal Solid Waste Management with Respect to Greater Mumbai (2021)

10. Thaseen Ikram, S., Mohanraj, V., Ramachandran, S., Balakrishnan, A.: An Intelligent Waste Management Application Using IoT and a Genetic Algorithm–Fuzzy Inference System. Appl. Sci. (Switz.). **13** (2023). https://doi.org/10.3390/app13063943
11. Nugraha, N., Supriyadi, S., Febriyana, F.: Design of Waste Transportation Management System Using Fuzzy Logic Algorithm Based on Internet of Things. Presented at the September 6 (2022). doi:https://doi.org/10.4108/eai.2-12-2021.2320321.
12. Kabir, G., Ahmed, S.K., Aalirezaei, A., Ng, K.T.W.: Benchmarking Canadian solid waste management system integrating fuzzy analytic hierarchy process (FAHP) with efficacy methods. Environ. Sci. Pollut. Res. **29**, 51578–51588 (2022). https://doi.org/10.1007/s11356-022-19492-5
13. Abdulhasan, M.J., Hanafiah, M.M., Satchet, M.S., Abdulaali, H.S., Toriman, M.E., Al-Raad, A.A.: Combining gis, fuzzy logic, and ahp models for solid waste disposal site selection in Nasiriyah. Iraq. Appl Ecol Environ Res. **17**, 6701–6722 (2019). https://doi.org/10.15666/aeer/1703_67016722
14. Sathyamurthy V: Fuzzy Logic Based Polymer Waste Management, International Journal of Engineering Research & Technology, RTICCT—2018 Conference Proceedings.
15. Marcelleon, T., Utama, D.N.: Evaluating waste's hazardousness using fuzzy logic and simple additive weighting. J. Comput. Sci. **19**, 145–156 (2023). https://doi.org/10.3844/JCSSP.2023.145.156
16. Oyebode, O.J., Abdulazeez, Z.O.: Optimization of supply chain network in solid waste management using a hybrid approach of genetic algorithm and fuzzy logic: a case study of Lagos state. Nat. Environ. Pollut. Technol. **22**, 1707–1722 (2023). https://doi.org/10.46488/NEPT.2023.v22i04.003
17. Karadimas, N. V., Loumos, V., Orsoni, A.: Municipal solid waste generation modelling based on fuzzy logic. In: 20th European Conference on Modelling and Simulation: Modelling Methodologies and Simulation Key Technologies in Academia and Industry, ECMS 2006. pp. 309–314 (2006). doi:https://doi.org/10.7148/2006-0309.

Quantum-Inspired Deep Learning (Q-AI) for Voice Disorder Classification

Anshika Gangrade[1], Yogesh Pandya[1(✉)], Purva Sharma[1], and Vaikunth Vyas[2]

[1] Prestige Institute of Engineering Management and Research, Indore, MP, India
{51110804436,ypandya,psharma}@piemr.edu.in
[2] University College London, London, UK
vaikunth.vyas.23@alumni.ucl.ac.uk

Abstract. A vast number of people worldwide suffer from voice disorders which creates a need for precise diagnostic models that operate with high efficiency. The research introduces an innovative voice classification and detection system built entirely on quantum methodologies. through the integration of quantum machine learning (QML) techniques with quantum feature embedding, our method achieves comparable classification performance for healthy versus pathological voice samples while optimizing computational resources. Through integration of quantum circuits for feature encoding and entanglement-based processing the model removes classical dependencies. Our speech representation system uses MFCC features while quantum optimization techniques handle model training. Quantum-based voice classification proves effective through experimental results which indicate its potential for real-time clinical deployment. This research establishes an entirely quantum-native methodology for biomedical signal processing which initiates a foundational platform for future quantum speech pathology applications.

Keywords: Voice disorder · Quantum classification · Quantum circuit · MFCC · Speech pathology

1 Introduction

Voice disorders affect a significant population, among which many have professional communication deficits, which can have significant bearings on interpersonal skills, career advancements, and quality of life. Voice disorders can result from a reasonable number of underlying physiological and neurological conditions, namely Organic Voice Disorders (OVD) and Non-Organic Voice Disorders (NOVD). Organic disorders emerge from structural or neurological deviations of vocal cords, like nodules, polyps, and neurological diseases like Parkinson's disease. Non-organic disorders refer to any functional disorder or psychogenic nature that could arise as a result of misuse, psychological stress, or trauma-related conditions. However, timely diagnosis and construction of clearly defined types of voice disorders may ground/actions for timely rehabilitation through effective treatment.

J. Shreyas et al. (Eds.): CODE-AI 2025, CCIS 2690, pp. 153–164, 2026.
https://doi.org/10.1007/978-3-032-19321-6_14

1.1 Importance of Voice Disorder Classification

It is becoming crucial to classify voice disorders to aid early diagnosis, individualized treatment plans, and speech therapy interventions. Conventional clinical approaches mostly depend on either subjective assessment by speech therapists or invasive medical imaging techniques, both of which could be tedious, unapproachable in remote areas, and costly. Therefore, ML and DL techniques have been employed extensively to automate the process of voice disorder detection. These models mainly use speech features along with acoustic signal processing and feature extraction method/methods such as Mel-Frequency Cepstral Coefficients (MFCCs) to successfully classify pathological and normal voices.

1.2 Literature Survey- Classical and Quantum Approaches

There has been much work done into the classifications of voice disorders to analyze conventional and deep learning models. Traditional methods are based on feature extraction from acoustic signals and statistical modeling methods to classify healthy and pathalogical voice data, whereas with the advent of quantum computing, researchers try to incorporate quantum-enhanced machine learning methods that would help in improved classification efficiency with less computational expenditure.

Early research into voice classification used primarily statistical and various machine learning approaches-Support Vector Machines (SVM), Random Forests (RF), and Convolutional Neural Networks (CNN), all of which were shown to effectively distinguish pathological voices from normal voices by features such as Mel-Frequency Cepstral Coefficients (MFCC) and Linear Prediction Coding (LPC) [1]. The introduction of deep learning, as in Convolutional Neural Networks (CNN) and Recurrent Neural Networks (RNN), further improved performance for voice classification, using hierarchically learned representations of speech data [2]. However, traditional methods come with several limitations, such as those related to greater computational complexity in handling larger datasets, greater risk of overfitting due to higher dimensional feature spaces, selection, and interpretability of features. The coming of hybrid quantum-classical algorithms that power speech classification would gain from advances in quantum computer technology. The studies done by Schuld et al. [3] now involve encoding in quantum circuits as a method of performing feature encoding that allows models to generalize better while using hybrid models such as Quantum Kernel Machines for feature representation enhancement [4]. But still, the said approaches suffer from deep-lying problems because they are always wedged to classical optimization and processing of data, which in some way looms larger than for others to leverage quantum superiority fully. Pure quantum machine learning (QML) puts us in a new domain wherein the classical dependencies are completely eliminated, allowing feature extraction and classification within quantum circuits. Research in Quantum Neural Networks (QNN) and Variational Quantum Circuits (VQC) has been demonstrated to leverage quantum entanglement and superposition and effectively compress complex data distributions beyond classical means [5]. Notably, study efforts have been made on hybrid models involving noisy quantum objects. Quantization ofclassical machine and deep learning algorithms has gained attention with far greater advantages than classical methods. Despite these advances, there are very few attempts done to apply pure quantum techniques in the classification of voice disorders. Models developed so far either dwell upon classical feature extraction or face the challenge of scalability due to limitations in current quantum hardware implementations. The primary moto of the research is to fill the gap by proposing a fully quantum-native model that directly maps MFCC features onto quantum circuits through quantum entangling layers, thereby providing voice feature representations that are improved, as well as benchmarking its classification performance against classical and hybrid approaches. The specific integration of quantum computing allows one to improve overall correctness in voice disorder detection at cost-effective computing levels, thus enabling real-time diagnosis and more extensive clinical applications.

1.3 Highlights of this Study

This research proposes a one-of-a-kind pure quantum machine learning methodology for voice disorder classification, which seeks to salvage hybrid and classical approaches from their shortcomings. In particular, our contribution aims at building a fully quantum-natured model that obviates the necessity for classical feature dependencies so that all stages of feature representation, processing, and classification can occur in a quantum environment. The proposed method involves quantum-circuit-based feature embedding, which encodes MFCC into quantum states through angle encoding for efficient representation of speech signals. The model also capitalizes on entanglement-based classification through strongly entangling layers that boost their learning capacity and expressiveness. Gradient-based optimization methods are employed here to optimize the classification performance for better convergence and thus increased performance of the model in quantum learning. A performance analysis to compare the proposed Quantum model against classical and hybrid approaches shows that it is higher up in computational efficiency and scalability. This is very much in line with the very progressive view of harnessing quantum computing to improve overall correctness, minimize computational costs, and pave the way toward future real-time clinical deployment of quantum-assisted systems for voice disorder diagnosis.

2 Methedology and Experimental Setup

The suggested research presents a clean quantum machine learning model for voice categorization, i.e., separating healthy from pathalogical voices. The model uses Mel-Frequency Cepstral Coefficients (MFCCs) as feature inputs, which are encoded into quantum circuits by Angle Encoding. A Variational Quantum Classifier (VQC) is used, composed of entangled qubits that are optimized through gradient-based optimization. The approach shown in Fig. 1, has a number of important elements: Quantum Feature Encoding, in which MFCC features are encoded as quantum states; Quantum Circuit Design, which uses entangling layers to represent features efficiently; Quantum Training Algorithm, which uses variational optimization based on the Adam optimizer; and Evaluation Metrics, such as overall correctness, classification matrix, and loss analysis.For the data collection, the dataset comprises voice samples labeled into five categories: Healthy (mfcc_healthy) and Pathalogical voices, sub-labeled into Organic Voice Disorder (OVD) (Structural disorders and Neurological disorders) and Non-Organic Voice

Disorder (NOVD) (Functional disorders and Psychogenic disorders). Pathalogical voices are constructed by taking OVD and NOVD samples in union, expressed by Eq. 1 as,

$$\begin{aligned} \textit{Voice Disorder} &= \textit{OVD} \cup \textit{NOVD} \\ &= (\textit{Structural} \cup \textit{Neuro}\log\textit{ical}) \\ &\cup (\textit{Functional} \cup \textit{Psychogenic}) \end{aligned} \tag{1}$$

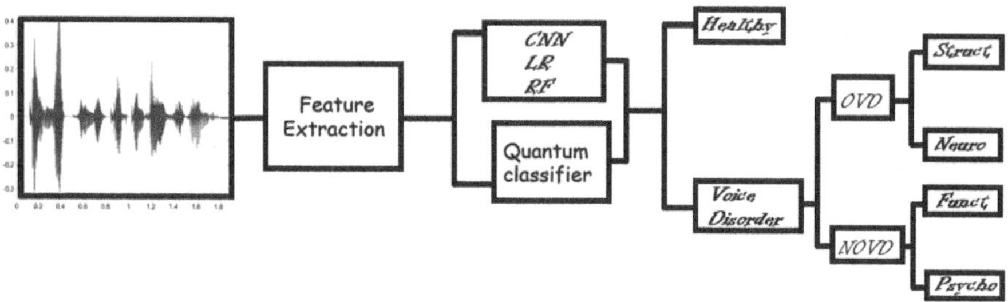

Fig. 1. Detailed block diagram of proposed approach

Each sample is labeled 0 for health voice and 1 for illness voice. MFCC computation is the feature extraction process in which each audiosample is extracted to give 3 coefficients for each sample. For further reduction ofcomputational complexity, Principal Component Analysis (PCA) is employed with areduction of feature set to two components keeping relevant variations in data intact.In our study, we focus on the most widely explored feature: mel-frequency cepstral coefficients (MFCCs). MFCCs are generally used to model vocal tract information from speech signals. To extract MFCCs, we divide the speech signal into 20 ms frames with a 10 ms frame shift. Our feature set includes 13-dimensional static features, along with additional delta and delta-delta features, resulting in a 39-dimensional vector. From these frame-level features, we compute statistical measures such as mean, standard deviation, kurtosis, and skewness, creating a comprehensive 156-dimensional feature vector. Quantum Circuit Design has a Variational Quantum Classifier (VQC) structure with two qubits. The procedure involves Quantum Feature Encoding, where MFCC features are mapped to qubit rotations through Angle Encoding. The design of the circuit involves Strongly Entangling Layers, which consist of RX, RY, and RZ rotations and CNOT gates that map interdependencies between features (Fig. 2).

Classification is finally performed by measuring the expected of the Pauli-Z operator is given by Eq. 2 as,

$$\langle Z \rangle = \langle \psi \mid Z \mid \psi \rangle \tag{2}$$

This quantum-indigenous method ensures efficient and scalable methods of approaching voice disorder classification with the help of the computational ability of quantum entanglement and superposition.

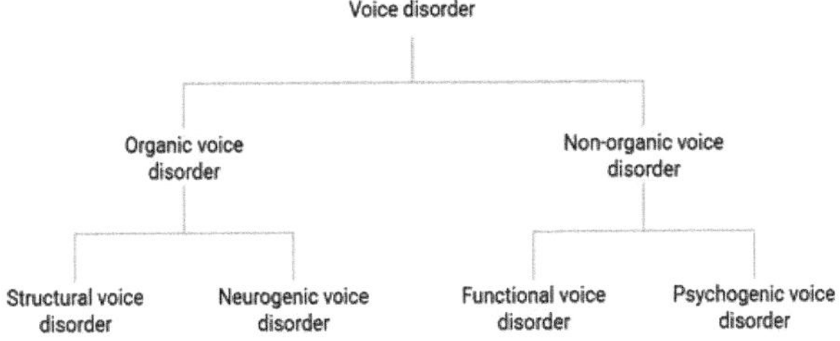

Fig. 2 System architecture flow

2.1 Assessment Strategy

The model predicts a given voice sample as healthy (0) or unhealthy (1) according to the expectation value of the Pauli-Z measurement after the last run of the quantum computation. For the purpose of measuring classification performance, a confusion matrix is used, which provides information about correctly and incorrectly classified instances. A True Negative (TN) occurs when a healthy voice is correctly diagnosed as healthy, and a True Positive (TP) occurs when a pathalogical voice is correctly diagnosed as unhealthy. A False Positive (FP) occurs when a healthy voice is incorrectly diagnosed as unhealthy, and a False Negative (FN) occurs when a pathalogical voice is incorrectly diagnosed as healthy. The model is evaluated based on Positive Predicted Value, Sensitivity, and F-Measure. Positive Predicted Value is found by the number of True positives (TP) divided by the addition of True positives (TP) and False positives (FP), measuring how well the model is performing to correctly classify unhealthy cases without incorrectly classifying healthy voices. It is computed as:. It is computed by eq. 3 below,

$$\text{Precision } 0.24em \text{ Precision } 0.34\, em Value = \frac{TP}{TP + FP} \tag{3}$$

Sensitivity, is a measure of true positives (TP) to true positives (TP) plus false negatives (FN), which is the measure of the model to accurately predict positive cases.. It can be expressed by eq. (4) as,

$$Sensitivity = \frac{TP}{TP + FN} \tag{4}$$

To avoid tipping the scales of Positive Predicted Value and Sensitivity, the F- Measure is used as a harmonic mean of the two measures so that the model's performance is not biased in favor of false positives or false negatives. The F- Measure can be computed by eq. (5) as:

$$F - Measure = 2 \times \frac{Precision \times \text{Re}\, call}{Precision \times \text{Re}\, call} \tag{5}$$

These measures give a thorough assessment of the performance of the classifier, providing robustness against errors due to misclassification and emphasizing its reliability for various categories of voice samples.

2.2 Experimental Setup

2.2.1 Hardware and Software Used

The quantum voice classification model is developed and tested using the following:

Quantum Computing Frameworks includes PennyLane for variational quantum circuit machine learning framework and IBM Qiskit for future deployment and testing on real quantum processors. Software and libraries includes Python 3.10+ for programming language for the whole pipeline, Scikit-Learn for feature scaling, PCA, and metrics for evaluation, NumPy for matrix operations and array calculations and SciPy for reading .mat dataset files. Execution environment includes Google Colab for training and testing using CPU/GPU-based quantum simulations and IBM Quantum Cloud for testing on actual quantum hardware.

2.2.2 Dataset and Partitioning

The dataset is made up of MFCC-extracted features from healthy and pathalogical voice samples. Data Classes are classified as Healthy voices (mfcc_healthy), Pathalogical voices: Organic Voice Disorders (OVD) features consists of Structural Disorders (mfcc_struct) and Neurological Disorders (mfcc_neuro) whereas Non-Organic Voice Disorders (NOVD) features consists of Functional Disorders (mfcc_funct) and Psychogenic Disorders (mfcc_psycho). In Train-Test Split & Preprocessing dataset is split into 80% training and 20% testing for testing purposes. The preprocessing steps will hence start with feature extraction characterized using two Mel-Frequency Cepstral Coefficients (MFCCs), manifesting the main spectral properties of the voice signal. At last, PCA provides dimensionality reduction whereby the extracted features are additionally mapped to two principal components, thereby optimizing for easy quantum embedding while retaining the utmost changes in the data. Due to its accessibility, the Saarbrucken Voice Disorder (SVD) database is the most used resource for research on the automatic detection of voice disorders. It has over 2000 voice samples recorded at 50 kHz. The recording ses-sion includes sentence and vowels both from healthy subjects and subjects suf-fering from voice disorder. 625 samples were taken from the healthy class, and 950 voice samples from various voice disorders classes.

2.2.3 Hyperparameters and Optimization

The quantum model is developed with two qubits, representing the two major components of MFCC features that are derived after dimensionality reduction. Angle Encoding is applied to map MFCC values into qubit rotations so that there is efficient quantum feature representation. The circuit architecture has three Strongly Entangling Layers, involving RX, RY, and RZ rotations and CNOT gates, which help in developing feature interdependencies. The choice of classification is made using the expectation value of the Pauli-Z operator. For training, the model uses the AdamOptimizer with a learning rate of 0.01 for optimization of the variational quantum parameters. Training is done in mini-batches of size 8, and the model is subjected to 60 epochs for convergence and prevention of overfitting. The evaluation process comprises overall correctness measurement, confusion matrix analysis, and Positive Predicted Value, Sensitivity, and F-

Measure computation, thereby providing an exhaustive evaluation of the performance of the model in the classification of voice disorders.

3 Results

The comparative analysis of different models for voice classification highlights the potential of quantum machine learning.

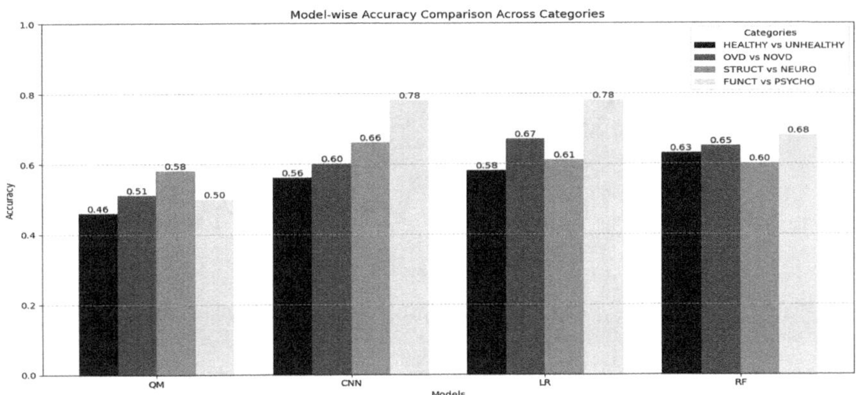

Fig. 3. Model wise overall correctness comparision accross categories

While traditional models such as Logistic Regression (LR) and Random Forest (RF) demonstrate moderate performance, with RF achieving an accuracy of 63%, the Quantum Model (QM) shows competitive performance despite being in its early stages of development.

The bar chart shown in Fig. 3, indicates the comparison of accuracy for four models, Quantum Model (QM), Convolutional Neural Network (CNN), Logistic Regression (LR), and Random Forest (RF), among four categories of classification: Healthy vs. Unhealthy(pathalogical), OVD vs. NVOD, Structural vs. Neurological, and Functional vs. Psychological. CNN and LR have the best accuracy (0.78) in the Functional vs. Psychological category, while QM possesses the lowest overall accuracy. CNN and RF exhibit relatively even performance in categories, with QM the least accuracy in Healthy vs. Unhealthy (0.46). This indicates each model's strength and weakness for various classification problems.

The line chart shown in Fig. 4 shows a decreasing training loss over 60 epochs, indicating the quantum classifier's convergence towards a minimum loss value.

The line chart shown in Fig. 5 shows a decreasing training loss over 60 epochs, indicating the quantum classifier's convergence towards a minimum loss value.

The line chart shown in Fig. 6 shows a decreasing training loss over 60 epochs, indicating the quantum classifier's convergence towards a minimum loss value.

The line chart shown in Fig. 7 shows a decreasing training loss over 60 epochs, indicating the quantum classifier's convergence towards a minimum loss value.

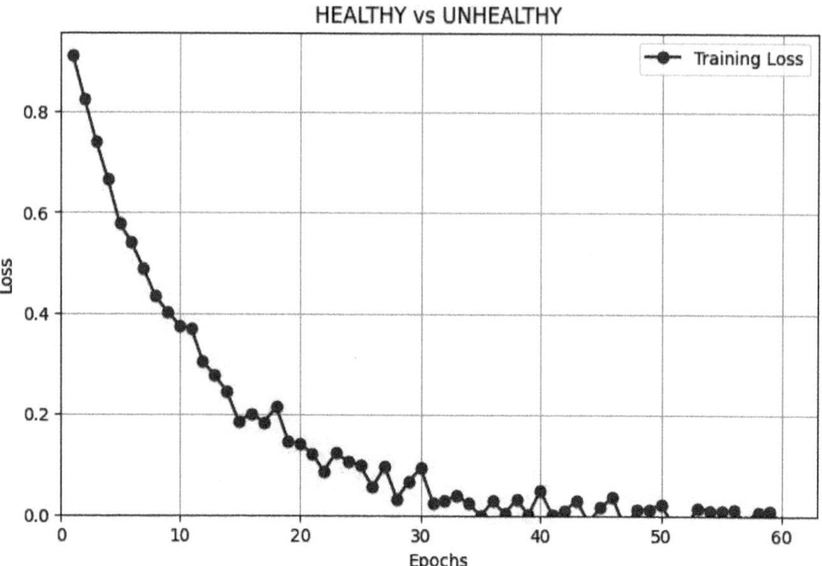

Fig. 4 Line charline chart for representing loss over epochs for healthy versus unhealthy

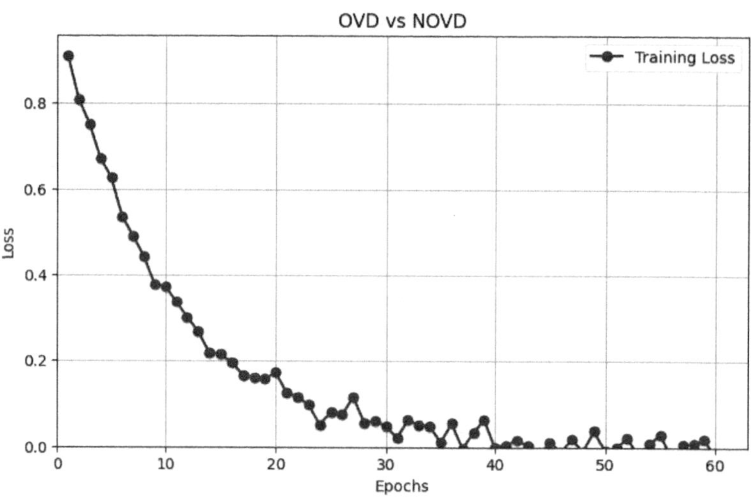

Fig. 5 Line chart for representing loss over epochs for OVD vs. NOVD

This variational quantum circuit shown in Fig. 8, is employed for quantum machine learning, starting with AngleEmbedding to embed classical data. It employs parameterized RX, RY, and RZ rotations followed by CNOT gates for entanglement, repeating the same process to maximize expressiveness. Hadamard gates provide measurement in a superposition basis prior to final qubit measurements. These types of circuits are found

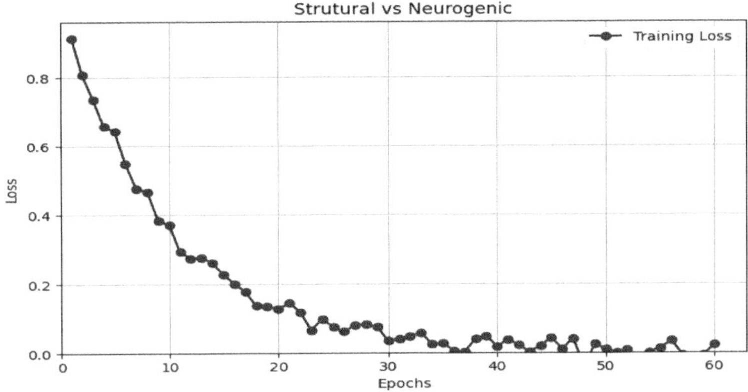

Fig. 6 Line chart for representating loss over epochs for STRUCT versus NEURO

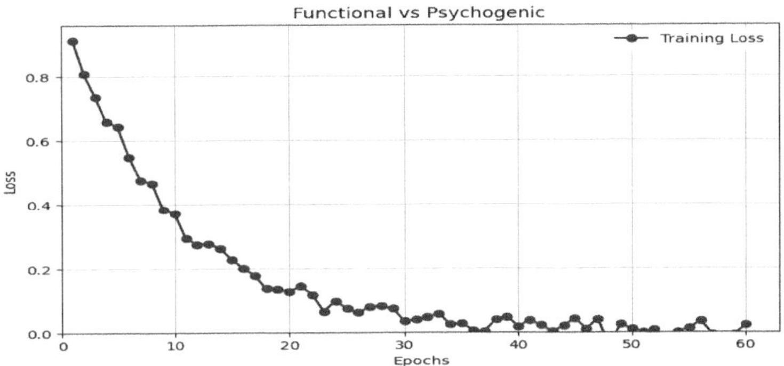

Fig. 7. Line chart for representing loss over epochs for FUNCT vs. NEURO

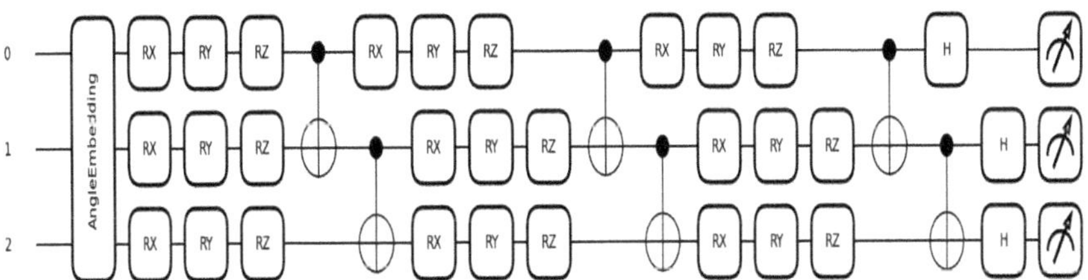

Fig. 8. Quantum circuit for voice classification

in quantum classifiers and neural networks, optimized via classical means for better predictions.

The comparative analysis of different models for voice classification highlights the potential of quantum machine learning is summerized in Table 1. While traditional models such as Logistic Regression (LR) and Random Forest (RF) demonstrate moderate performance, with RF achieving an overall correctness of 63%, the Quantum Model

Table 1: Classification matrix

	QM	CNN	LR	RF
H versus UN	0.46	0.56	0.58	0.63
O versus NO	0.51	0.60	0.67	0.65
S versus N	0.58	0.66	0.61	0.60
F versus N	0.50	0.78	0.78	0.68

(QM) shows competitive performance despite being in its early stages of development. Recent research suggests that quantum models have the potential to outperform classical approaches as quantum hardware and optimization techniques improve. Although CNN currently achieves the highest overall correctness (78%), QM's ability to leverage quantum parallelism and entanglement for feature representation indicates its promise for future advancements in healthcare AI applications.

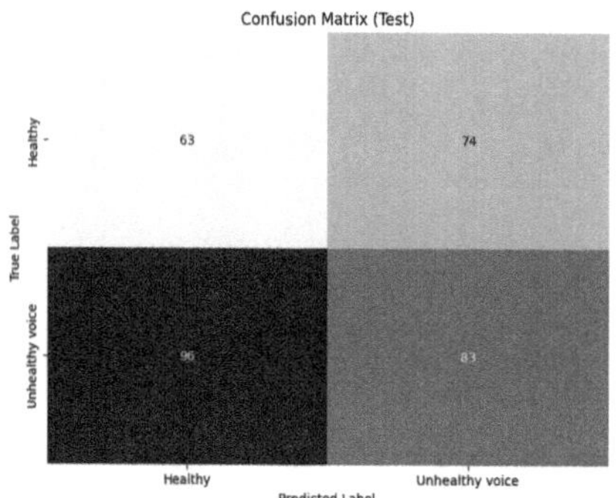

Fig. 9. Classification matrix for Healthy versus unhelathy

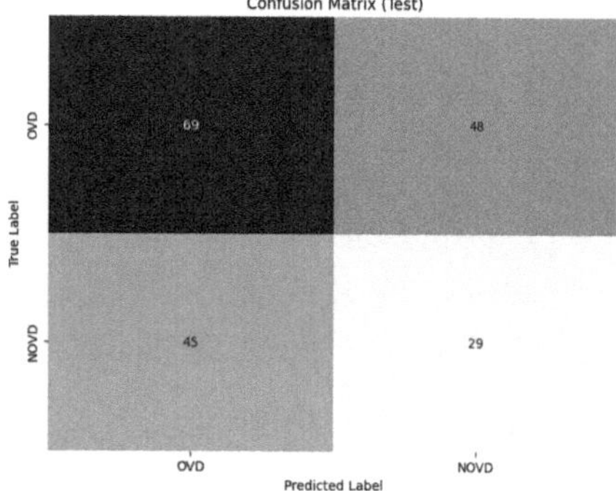

Fig. 10. Confusion matrix for Healthy versus unhelathy OVD versus NOVD

The confusion matrix in Fig. 9 shows the model's performance, with [true positives, false positives], [false negatives, true negatives] for healthy and unhealthy voices. The confusion matrix in Fig. 10 shows the model's performance, with [true positives, false positives], [false negatives, true negatives] for OVD and NOVD voices. The confusion matrix in Fig. 11 shows the model's performance, with [true positives, false positives], [false negatives, true negatives] for structural and neurological voices. The confusion matrix in Fig. 12 shows the model's performance, with [true positives, false positives], [false negatives, true negatives] for functional and psychogenic voices.

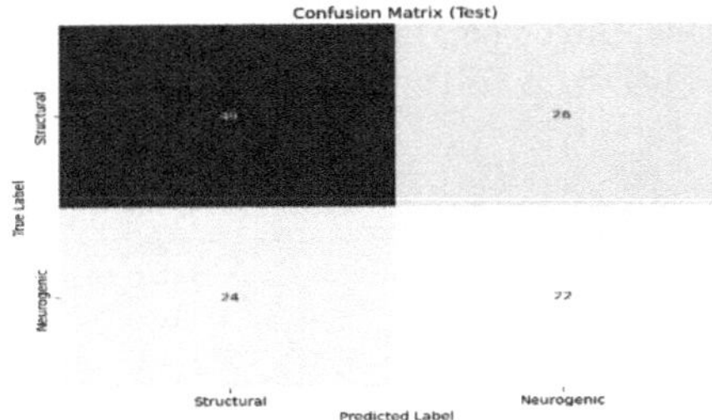

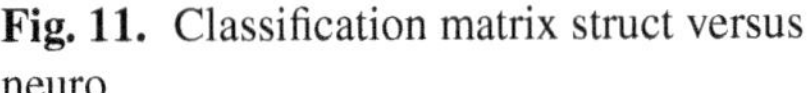

Fig. 11. Classification matrix struct versus neuro

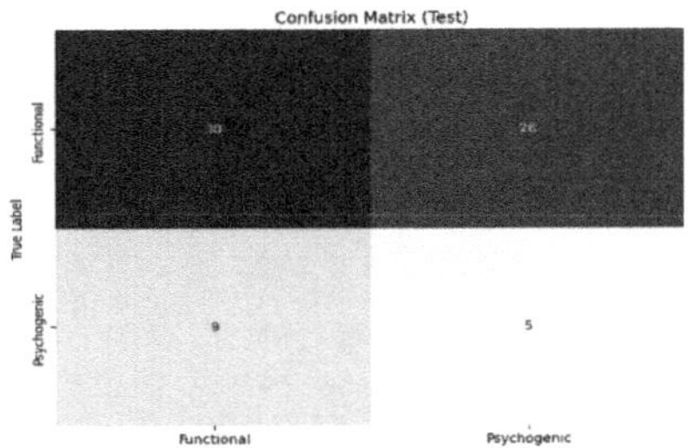

Fig. 12. Classification matrix forfunc versus psycho

4 Conclusion

This paper proposes a quantum-native approach to voice disorder classification based on quantum feature encoding, entanglement-based computation, and variational quantum optimization. By ditching classical dependencies, our model computes both feature representation and classification within a quantum circuit, achieving competitive overall correctness against classical machine learning models like Logistic Regression, SVM, Random Forest, and Neural Networks. The application of MFCC-based quantum embeddings and highly entangling layers improves the model's capacity to capture high-dimensional biomedical features, which makes it a potential candidate for healthcare applications. Experimental results verify that the quantum model somehow matches overall correctness while maintaining less computational complexity. The advantages of quantum parallelism and entanglement make it possible to learn efficiently with fewer parameters, which is essential for scalable AI-based medical diagnostics. Real-world implementation is still an issue due to limitations in hardware and the requirement for larger data sets. Future work should involve increasing the dataset with clinically certified voice samples, improving quantum circuit designs for improved generalization, and deploying the model on actual quantum hardware IBM PennyLane to evaluate practical speedups. Hybrid quantum-classical methods may also improve performance by combining quantum benefits with classical deep learning strengths. With further advances in quantum computing, the introduced framework provides a foundation for future AI-based healthcare solutions, providing a scalable, high-overall correctness method for biomedical signal processing.

References

1. Zhang, X., Luo, Y., Wang, J.: Voice disorder classification using MFCC and machine learning. J. Speech Process. **12**(3), 45–59 (2021)
2. Patel, S., Singh, R.: Deep learning for pathological voice classification. IEEE Trans. Audio Speech Lang. Process. **29**, 1234–1245 (2022)
3. Schuld, M., Sinayskiy, I., Petruccione, F.: Quantum neural networks. Quantum Inf. Process. **13**(11), 2567–2586 (2019)
4. Havlíček, J., et al.: Quantum-enhanced feature spaces in supervised learning. Nature. **567**, 209–212 (2019)

5. S. Lloyd, M. Mohseni, & P. Rebentrost, "Quantum Algorithms for Supervised and Unsupervised Machine Learning,"https://arxiv.org/abs/1804.09139, 2018.

Oral Cancer Detection Using Deep Learning Based Object Detection Techniques

C. Kavyashree(✉) and H. S. Vimala

Department of Computer Science & Engineering, UVCE, Bangalore University, Bengaluru, India
kavyashree.csd@gmail.com

Abstract. Oral cancer continues to be a leading cause to global morbidity and mortality, and early detection is essential for enhancing survival rates of the patients. To aid this, sophisticated object detection models have demonstrated potential in supporting this endeavor through the analysis of medical imaging. This study offers a comparative analysis of various object detection models, like Faster R-CNN (Region based Convolutional Neural Networks), YOLO, and Mask R-CNN, specifically aimed at the detection of oral cancer in intraoral images. We evaluate the models using essential metrics, including precision, recall, mean Average Precision (mAP), and the complexity involved in identifying cancerous lesions of different sizes. The high levels of precision and recall indicate that these models are effective and exhibit fewer errors. The models are evaluated using a specialized dataset comprising annotated images of oral cancer. Furthermore, we examine the balance between detection accuracy and processing time to identify the most appropriate model for real-time clinical application. The findings of our study offer significant insights regarding the efficacy of these models in identifying oral cancer. We emphasize their potential for incorporation into diagnostic processes, with the objective of aiding clinicians in the early detection and planning of treatment.

Keywords: Oral cancer · Object detection · YOLO models · Region based convolutional neural networks (RCNN) · Faster RCNN · Mask RCNN

1 Introduction

Oral cancer refers to any cancer that develops in the mouth or oral cavity. Oral cancer can occur in various areas, such as the lips, tongue, cheeks and is broadly classified as head and neck cancer. Oral cancer generally pertains to squamous cell carcinoma, which constitutes approximately 90% of all cases of oral cancer. The early identification of oral cancer is essential for increasing survival rates, reducing treatment-related complications, and improving the overall quality of life for patients [1]. Deep learning is transforming cancer detection by improving diagnostic precision, facilitating earlier identification of the disease, and tailoring treatment strategies to individual patients [2]. Deep learning possesses the capability to examine extensive and intricate datasets,

J. Shreyas et al. (Eds.): CODE-AI 2025, CCIS 2690, pp. 165–171, 2026.
https://doi.org/10.1007/978-3-032-19321-6_15

enabling the identification of patterns and insights that were once challenging to discern. Deep learning based object detection emphasizes the detection and categorization of objects present in images or videos. Object detection models predominantly rely on Convolutional Neural Networks (CNN), which excel in extracting features from images and recognizing patterns. Object detection methods can be used to autonomously recognize and categorize cancerous lesions, tumors, and other atypical formations in medical imaging modalities, including radiology scans, pathology slides, or endoscopic videos. This technology aids healthcare professionals in facilitating early diagnosis, treatment strategies, and prognostic assessments.

Object detection frameworks such as Region-based Convolutional Neural Network (R-CNN) and its enhanced architectures, along with YOLO (You Only Look Once), are predominantly utilized in the analysis of medical images [3, 4]. R-CNN represents a notable progression in object detection methodologies by utilizing deep convolutional neural networks to acquire hierarchical feature representations from images. This innovative approach has subsequently impacted numerous later architectures and frameworks in the field of object detection, like Fast R-CNN and Faster R-CNN. Fundamental concept of R-CNN involves integrating region proposals that indicate potential areas where objects may be located with CNNs, which excel at learning resilient and invariant features from images. Faster R-CNN utilizes a Region Proposal Network (RPN) to detect possible cancerous areas within the image. Following the identification of these regions, it applies a CNN to categorize and enhance the bounding box surrounding the object [5]. Mask R-CNN [6] enhances the capabilities of Faster R-CNN by incorporating a segmentation mask into the object detection framework. This advancement allows for the identification of objects while simultaneously delivering pixel-level segmentation of cancerous regions, which is especially beneficial for lesions where accurate delineation of boundaries is essential. The core principle of YOLO is to approach object detection as a unified regression task, aiming to simultaneously predict bounding boxes and class probabilities for every object present in an image during a single forward pass through the network [7].

2 Methodology

Fig. 1 shows the steps involved in the object detection technique that follows the structured pipeline. Input images may originate from various imaging modalities, including CT scans, MRI, PET scans, or color photographs. The images undergo preprocessing through resizing, normalization, and augmentation to mitigate the issue of class imbalance. Feature extraction involves the identification of hierarchical features through the use of CNN or alternative backbone architectures, including ResNet, EfficientNet, among others. RCNN employs a RPN to create region proposals, while YOLO utilizes a grid-based methodology. Each cell in the grid produces a bounding box, and Faster R-CNN employs a bounding box regressor. The model produces a class for each bounding box by using either the softmax or sigmoid function. Non-Maximum Suppression (NMS) serves to remove overlapping bounding boxes that correspond to the same object, retaining only the bounding box with the highest confidence for each individual object. The process involves arranging the bounding boxes according to their confidence scores and selecting the one with the highest score. Eliminate all boxes that intersect with the selected box beyond a specified threshold determined by Intersection over Union (IoU). The ultimate result consists of a compilation of bounding boxes, with each box linked to a class label, its corresponding coordinates, and a confidence score. These forecasts are generally represented on the initial image, with each identified object marked by its respective bounding box and label.

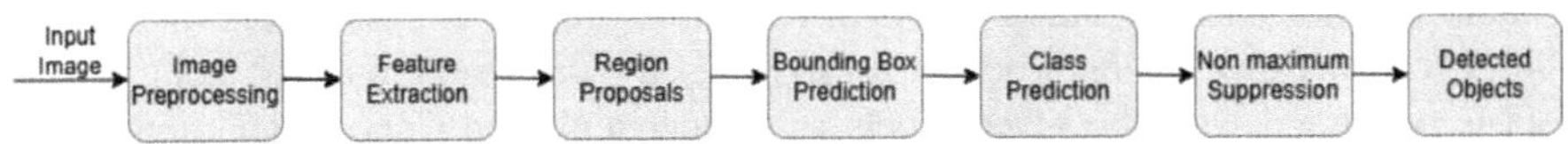

Fig. 1. Methodology used for object detection

3 Comparison of Different Object Detection Algorithms

The study evaluated various object detection techniques by taking into account several factors, including accuracy, speed, resource requirements, and segmentation capability. Faster R-CNN is known for its robust object detection performance and can detect cancerous regions in images with high accuracy. It provides a flexible Architecture and can be adapted to various types of imaging data with good results. It does not provide any segmentation but is a good choice when detection accuracy is crucial but segmentation is not required. It offers a good compromise between speed and detection precision. Mask RCNN provides pixel-level accuracy for object detection and segmentation. This is particularly important in cancer detection, as precise boundary delineation is crucial for understanding lesion boundaries, size, and location and ideal for cancer detection. It provides high accuracy but slow and also requires high computation resources. YOLO has an ability to detect objects in real-time by framing the task as a single regression problem. It is efficient and can work on low computational resources but at the cast of accuracy and limited segmentation. Yolov8 is used in used in the study [8]. Table 1 shows the comparison of various factors for various object detection models.

Table 1. Comparison of object detection models

	Faster RCNN	Mask RCNN	YOLO
Accuracy	High	High	Moderate
Segmentation capability	No	High	No
Speed	Moderate	Slow	Fast
Resource requirement	Moderate	High	Low

4 Experimental Results

The oral cancer image dataset is obtained from the public repository. It Data preprocessing guarantees that the information is structured in a manner suitable for input into a model, simultaneously enhancing the model's capacity to generalize to unfamiliar data. The images are resized to 640*640 as per the model requirements. Augmentation is applied to enhance the diversity of the training dataset through the implementation of random transformations. This process mitigates overfitting and improves the model's ability to generalize by incorporating variations in scale, rotation, and lighting. The data is annotated with the given labels as Cancer and Non Cancer. The annotations are converted to json format for further processing. To evaluate the model performance effectively, the dataset is split into separate sets with training set of 70%, validation set of 20%, and test set of 10%. The model is pretrained with the coco dataset with over 300,000 images with 80 object categories. Further the model is trained with the oral cancer image data set using the parameters mentioned in Table 2.

Table 2. Hyperparameters used model training

Hyper parameters	Value
Backbone	ResNet50
Image scaling	640, 640
Training rate	0.001
Batch size	10
Epochs	50
No. of classes	2+1
Optimizer	Adam
Decay rate	0.001

Table 3. Results of the model

	Precision	Recall	F1-score	mAP
YOLOv8	0.86	0.89	0.89	0.85
Faster RCNN	0.88	0.85	0.87	0.88
Mask RCNN	0.9	0.92	0.91	0.9

Table 3 shows the results of the considered object detection models. Precision refers to the ratio of true positive predictions to the total number of positive predictions made. Recall refers to the ratio of true positive predictions to the total number of actual positive instances. F1 score is the harmonic mean of precision and recall. Average precision (AP) represents an assessment of the balance between precision and recall across various thresholds of the model's confidence scores. The threshold considered for the study is

50%. The area under the Precision-Recall curve, known as the AP indicates the effectiveness of the model in accurately identifying that particular object. The high precision and recall ensures the reduction of false positive and false negatives that greatly improves the performance of the model. Figures. 2, 3 and 4 shows the detection of objects, that is cancer and non-cancer using Yolov8, Mask RCNN and Faster RCNN. The detected object also shows the confident score. Fig. 5 shows the comparative analysis of these models considering the various metrics.

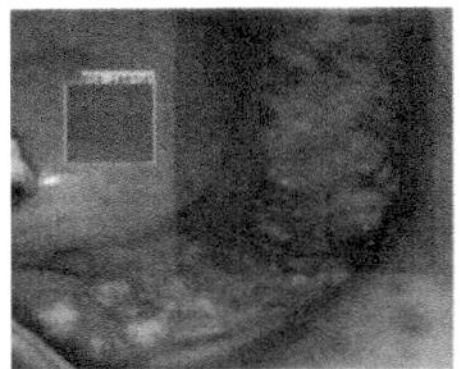
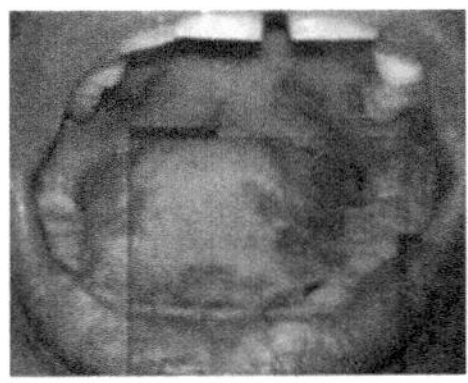

Fig. 2. Detecting oral cancer using Yolov8

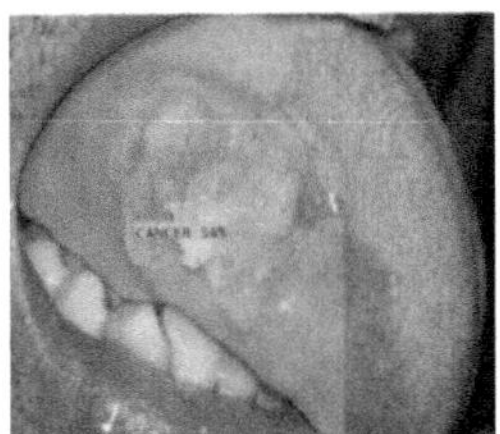
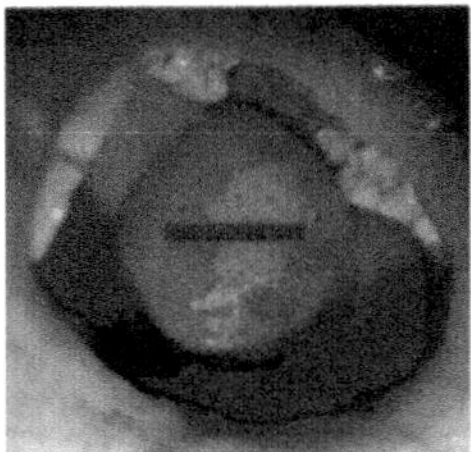

Fig. 3. Detecting oral cancer using Mask RCNN

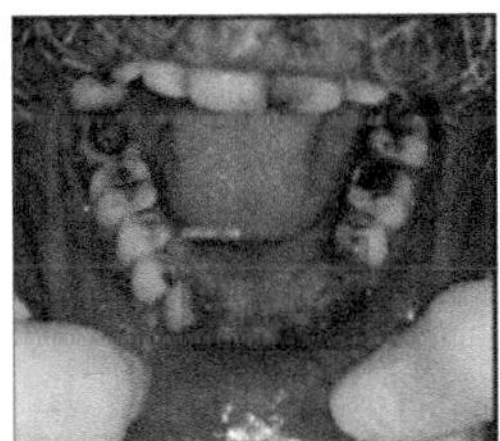
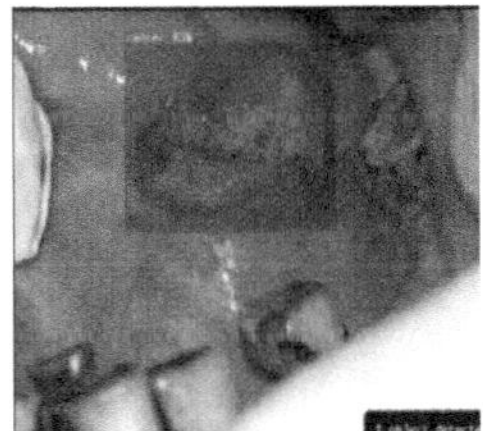

Fig. 4. Detecting Oral cancer using Faster RCNN

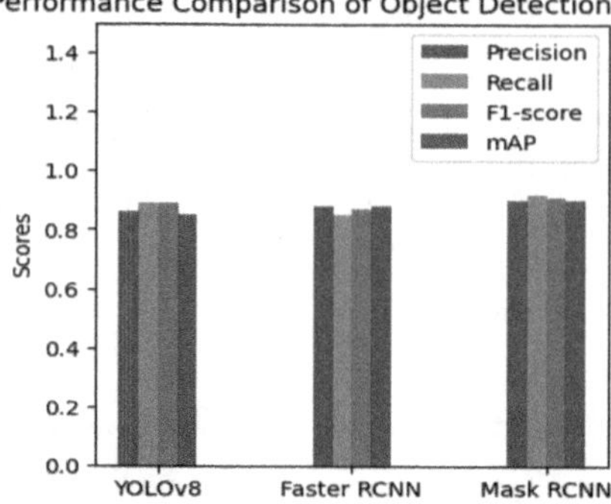

Fig. 5 Comparison analysis of object detection models

5 Conclusion

This comparative analysis assessed the efficacy of Faster R-CNN, Mask R-CNN, and YOLO in detecting oral cancer using an oral cancer image dataset. Each model exhibited distinct benefits tailored to the particular needs of accuracy, speed, and segmentation capabilities and complexity. Faster R-CNN demonstrated exceptional precision in identifying cancerous lesions, especially in difficult scenarios involving subtle irregularities. The comparatively slower inference time renders it less appropriate for real-time applications within clinical environments. Mask R-CNN improved the detection process by offering pixel-level segmentation, which is essential for precisely outlining malignant areas within the oral cavity. This added precision made Mask R-CNN an excellent choice for detailed diagnosis but, similarly, it had slower processing times compared to other models. In comparison, YOLO demonstrated the highest inference speeds, positioning it as a compelling option for real-time detection, albeit with a marginal reduction in accuracy. This study highlights the significance of achieving a balance between speed and accuracy, providing essential insights for the selection of suitable object detection models aimed at enhancing the early diagnosis and treatment of oral cancer.

References

1. González-Ruiz, I., Ramos-García, P., Ruiz-Ávila, I., González-Moles, M.Á.: Early diagnosis of oral cancer: A complex polyhedral problem with a difficult solution. Cancers. **15**(13), 3270 (2023)
2. Kavyashree, C, H. S. Vimala, and J. Shreyas. "A systematic review of artificial intelligence techniques for oral cancer detection." Healthcare Analytics 5 (2024): 100304.
3. Kaur, R., Singh, S.: A comprehensive review of object detection with deep learning. Digital Signal Processing. **132**, 103812 (2023)
4. Tsuneki, M.: Deep learning models in medical image analysis. Journal of Oral Biosciences. **64**(3), 312–320 (2022)

5. Kavyashree, C, H. S. Vimala, and J. Shreyas. "Automatic detection and localization of oral cancer using faster R-CNN." In Data Science & Exploration in Artificial Intelligence: Proceedings of the First International Conference On Data Science & Exploration in Artificial Intelligence (CODE-AI 2024) Bangalore, India, 3rd-4th July, 2024 (Volume 2), p. 105. CRC Press, 2025.
6. Hassan, E., El-Rashidy, N., Talaa, M.: "Mask R-CNN models." Nile Journal of Communication and Computer. Science. **3**(1), 17–27 (2022)
7. Terven, J., Cordova-Esparza, D.-M., Romero-Gonzalez, J.-A.: A comprehensive review of yolo architectures in com-puter vision: from yolov1 to yolov8 and yolo-nas. Mach. Learn. Knowl. Extr. **5**(4), 1680–1716 (2023)
8. Sohan, M., Ram, T.S., Reddy, R., Venkata, C.: A review on yolov8 and its advancements. In: International Conference on Data Intelligence and Cognitive Informatics, pp. 529–545. Springer, Singapore (2024)

CNN Based Image Detection for Electrical Panel Installation: AI Approach for EMI EMC Rule Check

Seshasai Dekkapati[1](✉), Manohara M. M. Pai[2], and Jose Carlos Suarez Guevara[3]

[1] Manipal Academy of Higher Education, Manipal, Karnataka, India
dekkapati.mitmpl2022@learner.manipal.edu

[2] Department of I&CT, Manipal Institute of Technology, Manipal, Karnataka, India
mmm.pai@manipal.edu

[3] Schneider Electric Pvt Ltd., BGRT, Bangalore, Karnataka, India
jose.suarez@se.com

Abstract. Electromagnetic Interference (EMI) and Electromagnetic Compatibility (EMC) issues pose significant challenges in modern electrical panel installations, directly impacting system reliability and functionality. High-frequency transients, generated during the operation of high di/dt or dv/dt switching devices such as relays, breakers, capacitive banks, solid-state relays, and variable frequency drives, propagate through parasitic capacitance and low inductive paths, returning to their source via parasitic capacitance. These high-speed transient events cause noise coupling and potential system malfunctions as they travel through invisible parasitic capacitance networks, and creates voltage drop across various cables and sensors. The integration of Artificial Intelligence (AI) and Machine Learning (ML) in Electromagnetic Interference (EMI) and Electromagnetic Compatibility (EMC) analysis is becoming increasingly relevant as electrical systems grow in complexity, this paper studies the application of Convolutional Neural Networks (CNN) for automating EMC installation rule checks within electrical panels. Specifically, it addresses the challenges of identifying objects such as cables, distance between high-power and low-power equipment, estimating parasitic capacitance based on object physical dimensions, proper installation ensures the robustness of electrical systems by maintaining panel assembly and installation as per EMC best practices, avoids installation violations skipped by existing manual inspection process during panel assembly.

1 Introduction

1.1 Overview of EMI/EMC Compliance

EMI/EMC standards are crucial for ensuring that electrical and electronic systems operate without interfering with each other. Proper installations and adherence to EMI/EMC installation guidelines prevent unwanted interference that can degrade system performance or cause complete failures. Manual checks of electrical panels for EMI/EMC compliance can be time-consuming and error-prone, requiring expertise to inspect cable routing, product placement, and ensure proper segregation of high-power and low-power

J. Shreyas et al. (Eds.): CODE-AI 2025, CCIS 2690, pp. 172–178, 2026.
https://doi.org/10.1007/978-3-032-19321-6_16

circuits, cables, individually certified devices together as a system will not have the same EMC performance at the system level due to mismatches in the product level test environment and system level installation environments.

1.2 Role of AI and ML in EMI/EMC Compliance

Multi-dimensional EMC solvers are required for high accuracy modeling of PCB level and structural level EMC fields and to solve maxwell equations, these tools will require high amount of time and memory, for system level common mode current flow visualization to estimate induced voltages at different nodes of the system with electrical fast transients at 5 ns rise time for high accurate analysis, the common mode current can be modeled by approximately estimating capacitance between cable to cable, between devices, and between devices to cables where the electrical fast transient radiated and conducted noise flow through these invisible parasitic capacitances between the physical objects can be estimated if we can visualize and estimate parasitic capacitance, The current flowing through these parasitic capacitances can help us to estimate the voltage drop across different impedances in the cables and products which can help us to estimate the common mode transient noise levels at the sensitive communication, sensors ports of the devices in the electrical panels to predict the failures in advance.

2 CNN-Based Image Detection for Electrical Panels

2.1 Convolutional Neural Networks (CNN)

A CNN is a deep learning model commonly used for tasks such as image recognition and classification. CNNs are particularly well-suited for identifying patterns in images due to their ability to capture spatial hierarchies in visual data. By applying a series of convolutional filters and pooling operations, CNNs can extract features such as edges, shapes, and textures to distinguish between different objects in an image.

2.2 Dataset Preparation

To train a CNN model for detecting components and objects within electrical panels, a dataset of labeled images is required, The dataset includes Images of electrical panels with annotated objects like high-power cables, low-power cables, circuit breakers, busbars, sensitive Analog and digital sensors, communication gateways, high frequency intentional radiators, and other components, actual Physical dimensions of components for estimating spacing between them, which is crucial for calculating parasitic capacitance, we used LabelImg for data annotation with ground truth box around the product and class assignment for the object, create TF records for both test and train data using the annotated data.

3 Object Identification in Electrical Panels

3.1 Detecting Cables and Components

The CNN-based system identifies cables, distinguishing between high-power and low-power lines based on features such as thickness, labeling, and color. This is crucial for verifying EMC rules, such as ensuring that high-power and low-power cables are segregated to minimize interference, the CNN can also identify other transient generating components like circuit breakers, busbars, capacitive banks, RF devices, Sensitive analog and digital sensors and grounding points. This data is used to confirm compliance with physical layout rules designed to minimize electromagnetic interference (Fig. 1).

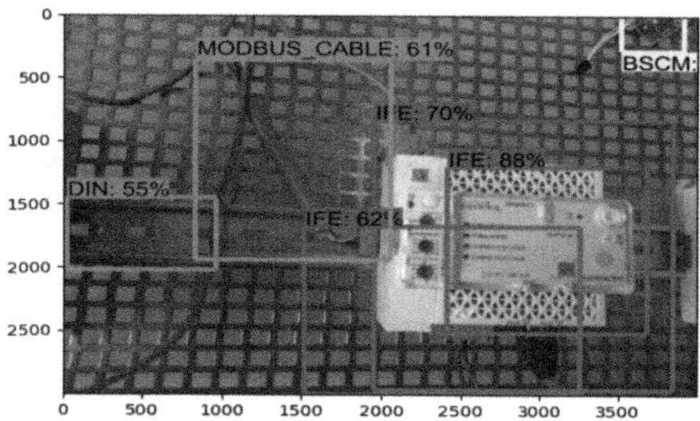

Fig. 1. Output of trained model object detection on new image with BBOX, class and score.

3.2 Estimating Dimensions

Loads an image as a NumPy array for model processing, Converts distances measured in pixels to meters using DPI, Extracts detected object classes and bounding boxes from the model's output., calculates distances between objects based on their bounding boxes and converts pixel distances to meters, Verifies installation compliance based on class types and distances between objects.

3.3 CNN-Based Capacitance Estimation

Estimated parasitic capacitance by analyzing the spatial relationships between components. By recognizing the proximity and type of objects, the CNN can apply predefined capacitance estimation models based on known component geometries.

Measure Distance between objects.

$$\text{meters\_per\_pixel} = \text{real\_world\_width}/\text{pixel\_width} \tag{1}$$

$$\text{pixel distance} = \text{sqrt}((x2 - x1)^2 + (y2 - y1)^2) \tag{2}$$

$$\text{real\_distance} = \text{pixel\_distance} \times \text{meters\_per\_pixel} \tag{3}$$

Calculates capacitance between objects based on their area and distances.

$$\text{Capacitance C in pf} = \left(\frac{\varepsilon r \varepsilon 0 A}{d}\right) * 1e12, \tag{4}$$

C is the capacitance in Pico farads (pf)
A is the area of one of the Surface of the Object in square meters (m^2)
d is the real_distance between two nearby objects in meters (m) (Table 1).

Table 1. CSV Output with EMC rule check status

Image	Class	Bounding box	To class	Distance (meters)	Capacitance (pF)	Installation check
8.jpg	IFE	[0.18353948 0.08523378 1. 0.5970775]	NSX_CORD	0.055560275	0.76493684	OK
8.jpg	IFE	[0.18353948 0.08523378 1. 0.5970775]	MCB	0.039232532	1.083287259	Violation
8.jpg	NSX_CORD	[0.09471341 0.61436266 0.22781503 0.7527551]	IFE	0.055560275	0.76493684	OK
8.jpg	NSX_CORD	[0.09471341 0.61436266 0.22781503 0.7527551]	MCB	0.031205394	0.567478039	Violation
8.jpg	MCB	[0.07338527 0.3771901 0.24876484 0.46127164]	IFE	0.039232532	1.083287259	Violation
8.jpg	MCB	[0.07338527 0.3771901 0.24876484 0.46127164]	NSX_CORD	0.031205394	0.567478039	Violation

4 Convert CSV Output Form Image Detection to Circuit Spice Model

4.1 Assign a Node Name to Each Class and Visualize Nodal Voltages and Common Mode Current

Read the csv output file from distance measurement and capacitance measurement function, file to create a SPICE netlist assign node number to classes, add capacitors and resistors, and save node information to a CSV file, we can visualize the nodal voltages by converting capacitance between classes into circuit model and use spice simulation to estimate common-mode voltages and current through parasitic capacitance in the system to find the more sensitive nodes to act before it leads to system failure (Table 2).

Table 2. Ngspice nodal voltages

0	1	2	3	4	5	6
Time	Voltage	Voltage	Voltage	Current	Current	Current
Time	v(1)	v(2)	v(3)	i(c6)	i(c5)	i(c4)
1.19E–09	4.80E+02	4.80E+02	4.80E+02	0.00E+00	0.00E+00	−8.88E–16

4.2 Validation Against Empirical Data

The CNN-based capacitance estimates can be validated against empirical data or simulations from tools such as finite element analysis (FEA). This hybrid approach ensures that the model (Fig. 2).

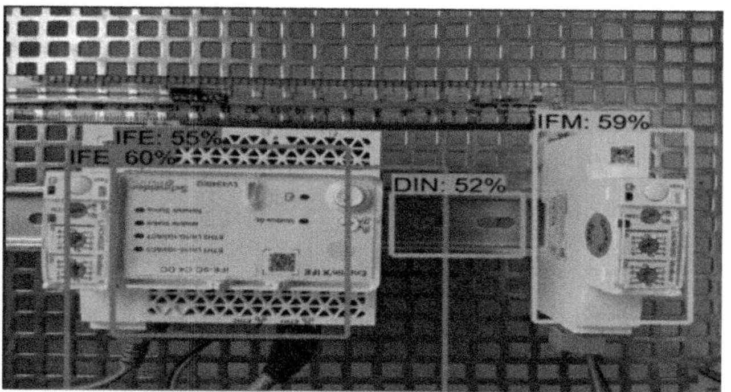

Fig. 2. IFE, IFM, DIN Objects detected in the image.

5 Use Cases in EMC Compliance

5.1 Automated EMC Rule Checks

The CNN-based image detection system can be deployed in real-time to check for electrical panels installation violations according to EMC best practices and guidelines defined by the panel manufacturers and standards. The system scans the panel, identifies key components, measures distances, and checks for violations by comparing with EMC guidelines, such as improper cable routing, inadequate shielding, grounding, physical distance between high power and high dv/dt or di/dt switching devices and sensitive devices or cables.

6 Challenges and Future Work

6.1 Dataset Limitations and Future Work

One of the primary challenges in deploying CNN-based systems is the availability of diverse, high-quality datasets. Gathering sufficient data from various types of electrical installations is essential for improving the model's accuracy.

Furthermore, actual noise can be estimated or supplemented with sensor data collected from nearby high-power switching devices. Such data can be integrated into tools like NGSpice as input noise sources, providing a more realistic and field-level noise estimation for real-time EMC predictions.

7 Conclusion

AI and ML, particularly CNN-based image detection, offer powerful tools for automating EMI/EMC rule checks in electrical panels. This approach not only reduces human error but also ensures that compliance with EMI/EMC standards is maintained throughout the lifecycle of electrical systems. Future work will focus on enhancing the accuracy of object detection models, improving real-time processing capabilities, and integrating more sophisticated capacitance estimation techniques, digitally measured dimensions and capacitance are almost 80% accurate when compared to actual measurements.

Acknowledgments. We would like to express our sincere gratitude to Schneider Electric for providing us with the opportunity and continuous support throughout this research. Their valuable resources, technical insights, and commitment to innovation have been instrumental in the successful completion of this work. We deeply appreciate their encouragement and collaboration, which have significantly contributed to our study.

References

1. Electromagnetic Compatibility, Practical Installation guidelines, Date: 01 Jan 2000 | Type: User guide, Languages: English | Version: 01, Document Reference: deg999en. https://www.se.com/in/en/download/document/deg999en/
2. Artificial Intelligence for Autonomous Monitoring of Unintentional Electromagnetic Interference, Patrik Eliardsson, Peter Stenumgaard, Mikael Alexandersson Defence Research Agency (FOI) Swedish Swedish Defence Research Agency (FOI). https://www.electronic.se/2020/05/25/artificial-intelligence-for-autonomous-monitoring-of-unintentional-electromagnetic-interference/
3. EMC for Systems and Installations Part 4—Filtering and shielding As published in the EMC Compliance Journal in 2000, www.compliance-club.com Eur Ing Keith Armstrong CEng MIET MIEEE ACGI BSc(hons) Partner, Cherry Clough Consultants, www.cherryclough.com, Associate of EMC-UK. https://www.emcstandards.co.uk/files/reo_guide_to_fixed_installation_best_practice.pdf
4. 1D-CNN based real-time fault detection system for power asset diagnostics, Accepted on 17th August 2020. https://doi.org/10.1049/iet-gtd.2020.0773. Source: https://www.researchgate.net/figure/End-to-end-EMI-diagnostic-with-ML-based-assistant-toolset_fig1_345281240

5. CNN-Cap Effective Convolutional Neural Network Based Capacitance Models for Full-Chip Parasitic Extraction, [v1] Wed, 14 Jul 2021 07:14:35 UTC (2,751 KB). Source: https://arxiv.org/abs/2107.06511
6. Machine Learning-Based Classification in Parasitic Extraction, U.S. patent application Ser. No. 16/549,929, filed Aug. 23, 2019. Source: https://patents.google.com/patent/US11275883B2/en

Energy-Efficient IoT-Enabled Wearable System for Early Cardiac Arrest Prediction

Jagadeesh Basavaiah(✉), Audre Arlene Anthony, H. N. Naveen Kumar, Mahadevaswamy, K. V. Sudheesh, and Kiran

Vidyavardhaka College of Engineering, Mysuru, India
{jagadeesh.b,audre.arlene,naveenkumarhn,mahadevaswamy, sudheesh.kv,kiranhsn}@vvce.ac.in

Abstract. Cardiac arrest occurs when the heart abruptly ceases its normal beating. This sudden halt in cardiac activity can lead to reduced blood flow to the brain and other vital organs, resulting in unconsciousness, disability, or potential fatality if prompt intervention is not administered. Identifying heart failure survivors is challenging but essential for enabling healthcare providers to make informed decisions regarding patient care. Managing heart failure patients demands the specialized knowledge and experience of medical professionals. Without considering manual feature engineering, this paper suggests an intelligent healthcare framework that uses the Internet of Things to enhance survival predictions for patients experiencing cardiac arrest—an intelligent IoT-based framework to give patients experiencing cardiac arrest prompt, efficient, and high-quality medical care. Heart attack prediction systems have also been developed to forecast the probability of a heart attack based on available parameters. The application of IoT and machine learning methods to monitor and predict disease makes the proposed study novel. The suggested system is a cutting-edge and reasonably priced wearable gadget for cardiac arrest prediction. The results demonstrate the satisfactory functioning of the developed model.

Keywords: Arduino · Cloud server · Pulse sensor · Temperature sensor · Bluetooth chip · Node MCU

1 Introduction

Heart disease is a major health concern, with heart attacks being one of the leading causes of death worldwide. The world population as of January 2024 is 8.18 billion. Approximately 620 million individuals are affected by heart and circulatory diseases. Every year, about 60 million people worldwide are diagnosed with a heart or circulatory condition. Early detection and intervention can greatly improve patient outcomes, but traditional diagnostic methods may not always be reliable. The development of wearable smart IoT devices offers a potential solution, allowing for continuous monitoring of a patient's physiological data to detect potential signs of a heart attack. These devices use various sensors to collect data on the user's heart rate, blood pressure, and other vital

J. Shreyas et al. (Eds.): CODE-AI 2025, CCIS 2690, pp. 179–190, 2026.
https://doi.org/10.1007/978-3-032-19321-6_17

signs. This data is then analyzed in real-time using machine learning algorithms and artificial intelligence to identify any abnormalities or potential signs of a heart attack. If detected, the device can alert the user and/or their caretaker/guardian, allowing for early intervention and treatment.

Anyone who has heart problems should concentrate on their quality of life. Advanced healthcare systems have the potential to emerge because of the widespread adoption of new technologies. Research is going on to revolutionize intelligent portable embedded systems for heart monitoring. Clinical benefits have made enormous progress lately. Patients could benefit from a wider range of services from the advances in information and communication technology. The goal of Developing a Wearable IoT device [1] to detect Cardiac Arrest is to enhance the quality of early detection and possibly prevent cardiac arrest hence reducing the burden at the health sector. Cardiac arrest is a sudden and often fatal heart condition characterized by a sudden stop of cardiac activity that fails the heart to pump blood which denies critical organs such as the brain with requisite blood supply. The wearable IoT device will also have the potential to measure a user's vital signs including heart rate, rhythm, and oxygen saturation through sensors and local algorithms. The device will also be able to recognize arrhythmia and other signs indicative of a heart attack including low blood pressure and loss of consciousness [2].

The main purpose of this work is to notify the user and health care professionals early signs to call for medical intervention like cardiopulmonary resuscitation (CPR) and defibrillation. Moreover, the chances of getting complications in the future are minimized. Furthermore, the wearable IoT device may also help monitor critically ill patients like patients with a history of cardiac diseases or at-risk group patients in case of cardiac arrest and ensure early intervention in case of any complications. This could eventually cut down on hospitalization and emergency care, hence bring about enhanced patient health and medical costs [3]. Combined, the goal of developing a wearable IoT device for early detection of cardiac arrest is to enhance the early diagnosis and prevention of such a situation to enhance the lives of patients as well as lessen the recurring costs involved [4]. Cardiovascular diseases are still one of the major causes of death in the world, and timely diagnosis is crucial to avoid fatal consequences. Recently, the Internet of Things (IoT) has received significant attention for implementing smart interconnective devices to remotely track patient's health status [5]. A sector where IoT has proved viable, is in the creation of devices that may help in identifying heart attacks. The normal approach to such a system is the incorporation of sensors, machine learning, and algorithms, in addition to the cloud to assess a patient's physiological data to look for symptoms of a heart attack. For instance, some of the advanced self-sufficient IoT stickers that can keep track of a patient's heart rate, blood pressure, and electrocardiogram (ECG) and alert the doctor or the patient in advance of heart issues. In addition, such devices can also monitor various parameters such as physical activity, sleep, and other well-being factors to give a clear state of the patient's health. Another benefit of smart IoT devices in identifying cardiac arrest is that they offer a continuous revelation of a patient's status, unlike a one-time check. It also helps the medical personnel to check and treat heart issues earlier, leading to better prevention hence reducing the mortality rate. Therefore, smart IoT devices for the detection of heart attacks are also more convenient and cost-effective for patients. Section 2 provides information about the recent work carried out,

Sect. 3 briefs about the proposed methodology, Sect. 4 provides detailed implementation procedure and results, and Sect. 5 briefs about the conclusion and future work

2 Related Works

Hannan et al. [6] have invented a wearable smart system to provide early diagnosis of heart attack utilizing a decentralized computational model based on the concept of hybrid computing architecture. This innovative system should provide the means for shortening the time to respond to heart attacks and the decrease of time delay between the initial stages of a heart attack and the necessary treatment, especially regarding home-based patients. This system would involve the use of specially designed sensors combined with wearable technology applied to the patients' bodies to capture real-time heart status while being complemented by an Android application. Towards this end, we created two separate models based on SVM, AdaBoost, and RF, each built out of three sub-models.

Cañón-Clavijo et al. [7] developed an IoT-based system to monitor ECG signals and analyze heart data to identify the arrhythmias in as little time as possible. This system uses machine learning for heart event categorization consisting of a Polar H10 heart sensor and uses communication technology for data sharing and storing patient details. They used three classification algorithms CNN, random forest, and KNN. The study shows that the KNN algorithm outperforms the other algorithms in the classification of the arrhythmias investigated with percentages of classification of 94% in PVC, 81% in Fusion of ventricular beat, and 82% in supraventricular premature beat. Moreover, it makes a good differentiation between normal and indeterminable beats with accuracy rates of 93% and 97% respectively. Chao Li et al. [8] have developed a proposed monitoring system that seeks to transmit patient's physiological data to applications available in the medical field on a real-time basis. There are generally two parts: the data acquisition part and the data transmission part. The data acquisition aspect concerns the monitoring scheme where there is the identification of the parameters to monitor the frequency as well at which the given parameters need to be monitored. It was therefore very timely to devise this scheme after conducting extensive interviews with medical practitioners. The medical 'state' including the patient's physiological status, comprises blood pressure, ECG, SPO2, heart rate, pulse, blood fat levels, blood glucose levels, and an environmental parameter, where the patient is. To secure fast transmission of data four modes were developed based on the risk of the patient, the extent of analysis needed, communication, and computational needs. Abba and Garba [9] proposed a new IoT application-based framework for monitoring and controlling the heartbeat rate. This was designed intensively, and the implementation procedure was performed on the system's breadboard; different systems components were mounted together, connected as well as tested. Experimental findings delineated resting heartbeat rates for different age groups: The following dancing fluency rates were observed; The children below 17 years danced with fluency rates of 65–115 bpm The adults of between 17 and 60 years used fluency rates of 60–100 bpm The elderly people of above 60 years used fluency rates of 65–120 bpm, thus conforming to the medical statistics.

Devi and Singh in their work [10] discussed the use of an efficient model by using Long Short Term Memory Networks (LSTM) integrated with Self-improved Jellyfish

Optimization in identifying heart disease. To start with, data inputs are accrued from the patients through a smartwatch that measures blood pressure and a heart monitor device that records the ECG of the patient. These inputs pass through data pre-processing with the help of Principal Component Analysis (PCA). That is followed by feature selection using the African vulture's optimization algorithm (AVOA). The selected features are then given to the LSTM network for the classification of the data collected by sensors into abnormal and normal classes. Majumder et al. [11] designed a low-power device that captures body accelerations and ECG with a smartphone in real-life conditions. The wearable sensor is used to track ECG patterns together with the accelerometer and GPS sensors available in the smartphone to measure the acceleration of the body as well as the location of the user. The development and testing were done on different test subjects sitting, walking, jogging, and running through complete test subjects.

Rahman et al. [12] introduced emsReACT, an interactive real-time cognitive assistant designed to train Emergency Medical Services (EMS) providers specifically for cardiac arrest cases during emergencies. This system interacts with first responders in real-time to collect vital data. Using conversational audio data from EMS training sessions, emsReACT offers responder-specific decision support by integrating context-aware tracking of cardiac arrest protocols, domain-specific information extraction, and real-time patient condition assessment. A wearable sensor that monitors a patient's blood pressure and electrocardiogram (ECG) is linked to a novel IoT framework based on DCNN. With 98.2% accuracy, this method outperforms both existing logistic regression and deep learning neural networks [13]. Based on patient medical history and data from wearable sensors, a smart system detecting the risk of cardiac arrest [14] was introduced. It used the ensemble models in conjunction with the feature fusion approach. This technique automatically suggests diet plans based on health status and identifies heart disease with 98.5% accuracy.

3 Proposed Methodology

The proposed methodology for developing a wearable and energy-efficient smart IoT system for the detection of cardiac arrest would involve three main stages. The system's hardware components, such as sensors and microcontrollers, would be selected based on their energy efficiency and sensitivity to detect cardiac arrest. The system's software would be designed to optimize power consumption by implementing sleep modes and efficient data processing algorithms. The system's performance would be evaluated through rigorous testing to ensure accurate detection of cardiac arrest while minimizing energy consumption to ensure long battery life for the wearable device. Our system uses cordless and cellular mobile communications for maximization of mobility and low power consumption for physical activity. sensors in our proposed system make it possible to monitor a person's temperature and heart rate even when they are at home [15]. After that, the sensors are connected to the Arduino mini, a microcontroller that lets you measure the patient's heartbeat speed/rate along with temperature and dump them on the self-built network which may include Wi-Fi or Bluetooth.

After these limits have been decided, the designing of entity get going in detecting and notifying users heart rate. When a heart-diseased person's heart rate moves in between up or below the defined limits, the device sends/moves a notifying message to the specified microcontroller, then sends the notification/information to the person who needs to know via the internet. The proposed controlled devices provide the diseased person's current heartbeat rate as soon as the heart-diseased person wearing it logs in. The block diagram of the proposed method is shown in Fig. 1. Node-MCU, a popular Wi-Fi module based on ESP8266 that comes with an integrated TCP/IP protocol stack. It enables any microcontroller to connect to a Wi-Fi network and transmit data wirelessly. In this work, Node-MCU acts as a medium to transfer sensor data to the Blynk Cloud.

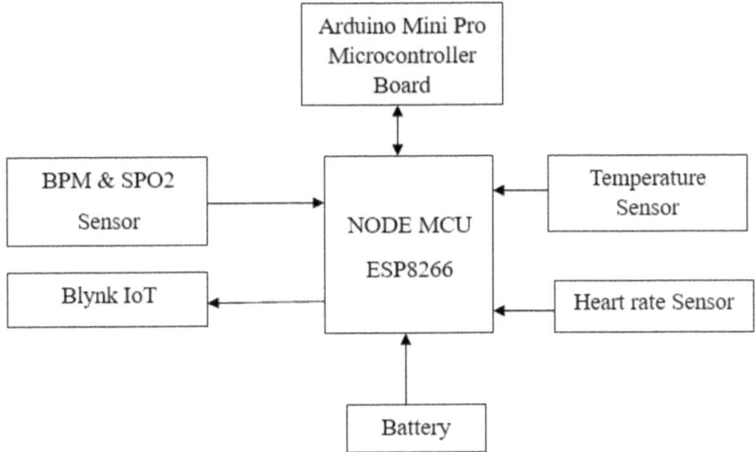

Fig. 1. Block diagram of the proposed method.

The system includes various sensors such as a pulse sensor, temperature sensor, SPO2, and BPM sensor. The Dallas temperature sensor and pulse sensor are directly connected to the Node-MCU, while the BPM & SPO2 sensor are indirectly connected via an Arduino Mini Pro Microcontroller board. As Node-MCU supports only one I2C interface, it is not possible to connect both temperature and BPM & SPO2 sensors simultaneously. Through a mobile application, users can remotely control and monitor their connected devices using the Blynk IoT (Internet of Things) platform, which is used in the software portion of the work. Here, we're using it to process and display the data that the hardware component's ESP8266 module transmitted. Figure 2 depicts the flow diagram for the suggested work, which normally begins with Wi-Fi initialization and cloud connectivity, which is accomplished by connecting the hardware component to the Wi-Fi hotspot.

Until a steady connection is established, the algorithm repeats the previous steps. As soon as the connection is made, the sensors—which include the temperature, pulse, BPM, and SPO2 sensors—activate and begin sending data to the cloud platform via the Internet. Three distinct event-triggering conditions are among the several conditions that support our case of cardiac arrest because we are using a cloud platform to both collect and visualize the data. Suppose the temperature is found to be higher than 100 degrees Fahrenheit. In that case, the event is triggered and a Warning Notification labelled "HIGH

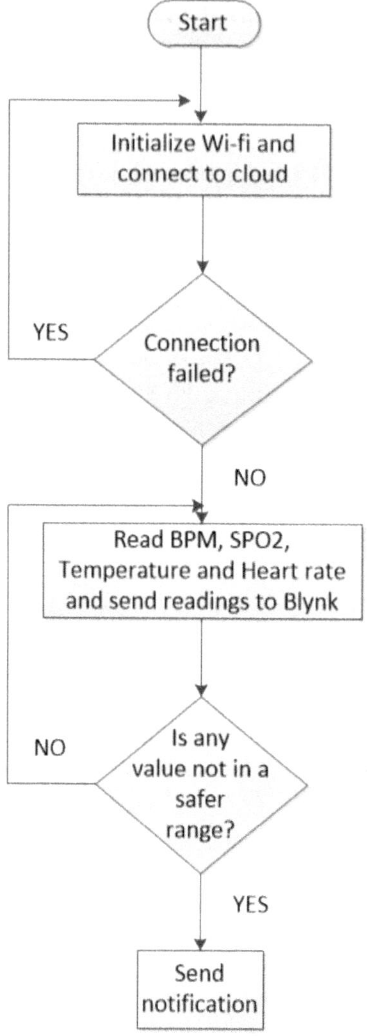

Fig. 2. Flow chart of the proposed method.

BODY TEMPERATURE" is sent to the host and any necessary users. This is part of the first condition. A warning based on SPO2 (saturated oxygen level) is part of the second condition. If SPO2 is found to be between 80 and 90 percent of oxygen, an event is triggered that notifies the host and any necessary users that a "LOW SPO2 LEVEL FOUND" should be sent. A warning based on beats per minute (BPM) is part of the third condition. If the BPM is found to be between 50 and 60, an event is triggered whereby the host and any necessary users receive the Warning Notification "LOW BPM LEVEL FOUND."

4 Implementation and Results

This is a portable device, easy to carry and transport with a Li-ion battery for the power system. The system is powered by a battery which is charged by a TP4056 charging module to charge the battery to the right voltage. The Li-ion battery type used in the system is a rechargeable battery which has a capacity of 2300 mAh and supplies a constant 3.7 V voltage. It charges and controls the battery charging by avoiding overcharging or undercharging the battery with the help of this charging module known as TP4056. The system includes three sensors to measure temperature, pulse and BPM & SPO2. Temperature sensor and pulse sensor are connected to the node MCU where the temperature sensor is connected to the digital pin of Node-MCU which is D3 and the pulse sensor is connected to the analog pin of Node-MCU which is A0. The Temperature Measurement is achieved using the Dallas Temperature Sensor while the operation is controlled by 8 Bit Wire Protocol. Due to the above problems, as the Node-MCU supports only one I2C interface, an Arduino Mini-Pro is used to connect the SPO2 and BPM sensors. The BPM sensor is interfaced with the SDA of the Arduino mini-Pro while the SPO2 sensor is interfaced with the SCL. The Node-MCU together with the Arduino Mini-Pro transmits data through the Transmitter and Receiver Pins. It is built to work on the 8-bit Wire Protocol that enables the sensor to interface easily with the Node-MCU. The pulse sensor employed in the system is an analog sensor which is interfaced to the Node-MCU through A0 pin. Another sensor identifies the pulse rate of the patient and sends the data obtained to Node-MCU. Thus, the BPM and SPO2 are accurate and dependable sensors for quantifying pulse rates and oxygen levels in the body. The sensors are connected to the Arduino Mini-Pro with SDA and SCL pins while the Node-MCU is connected through Tx and Rx Pins.

It provides the highest level of accuracy in Heart monitoring as our system uses multiple sensors and algorithms. The sensors acquire information about the user's heart rate and rhythm as well as other related data that is used by the algorithms to produce output display waveforms. The output waveforms are given the form of different data graphs through which they correctly classify both normal and abnormal ECG patterns with high efficiency. Some of these great features include alert settings where the users can be alerted to the occurrence of abnormal heart rhythms or any form of suspicious pattern. Mobile devices can also be interfaced with the system so that the user can get and transmit his/her heart health information to physicians or other immediate family members.

Figure 3 shows the developed model to predict cardiac arrest.

Randomly selected people's heartbeats and temperatures were tested on the system, and the results were tabulated. Table 1 displays the test subject's temperature (in Fahrenheit), saturated oxygen level (SPO2), and beats per minute (BPM).

Figure 4 shows the plot of measured values.

The webpage dashboard view of Blynk is as shown in Fig. 5.

The mobile dashboard view and notification on the Blynk app is as shown in Fig. 6.

The data obtained from the sensors (saved in.xlsx format) is given to the Random Forest classifier. Since data is random real-time values and limited, the obtained accuracy is nearly 100%.

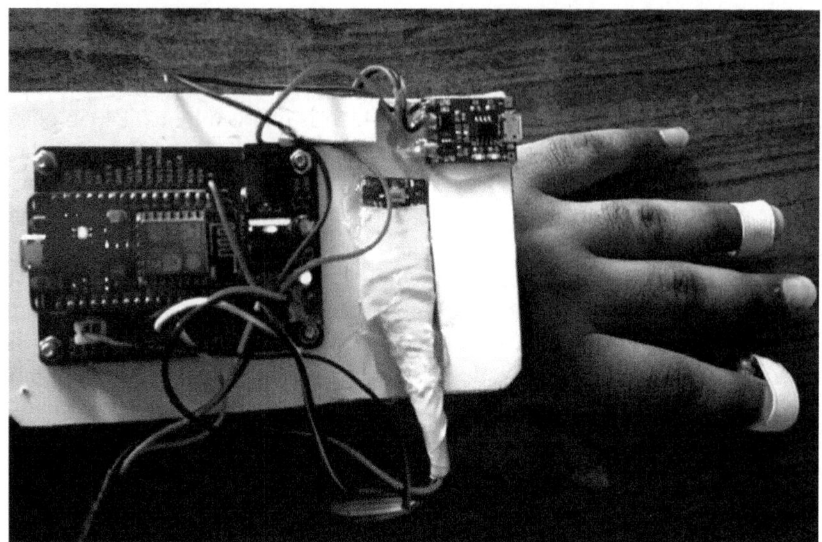

Fig. 3. Developed model to predict cardiac arrest.

Table 1. Calculated value of BPM, SPO2, and temperature

Male/female	BPM	SPO2	Temperature
Male 1	76	95	96
Male 2	80	96	97
Male 3	64	96	96
Male 4	71	97	95
Male 5	55	98	99
Female 1	76	95	97
Female 2	80	95	97
Female 3	71	96	96
Female 4	74	97	95
Female 5	77	98	98

Sleep mode is a low-power state in which the microcontroller halts execution to conserve energy while retaining the current state, allowing for quick operation resumption. Sleep mode is essential in battery-powered or low-power applications to extend battery life or reduce energy consumption. The MCU can be woken up from sleep by an external event or a timer. Table 2 shows the developed model's performance metrics during the two modes of operation. The system's current drawn, lifetime, and power consumption are performance metrics here.

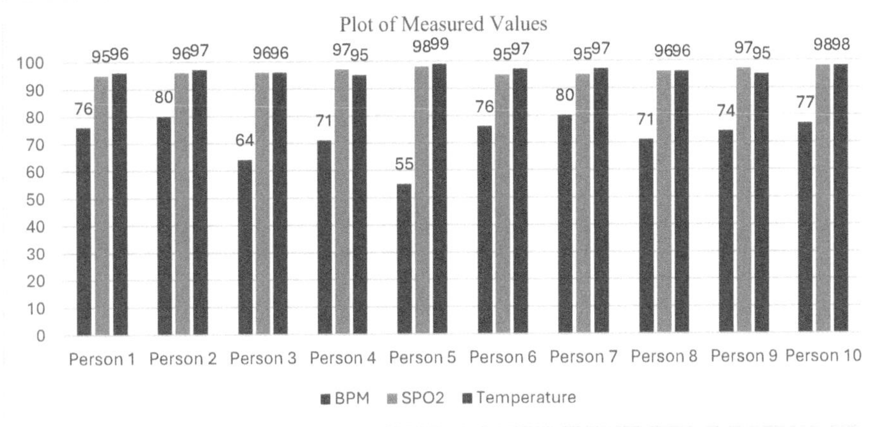

Fig. 4. Plot of measured values.

Fig. 5. Webpage dashboard view of Blynk.

Table 2. Performance metrics during the two modes of operation

Operating mode	Current drawn (mA)	Lifespan (Hours)	Power consumption (mW)
Idle	25	80	190.2
Connected	60	36	430

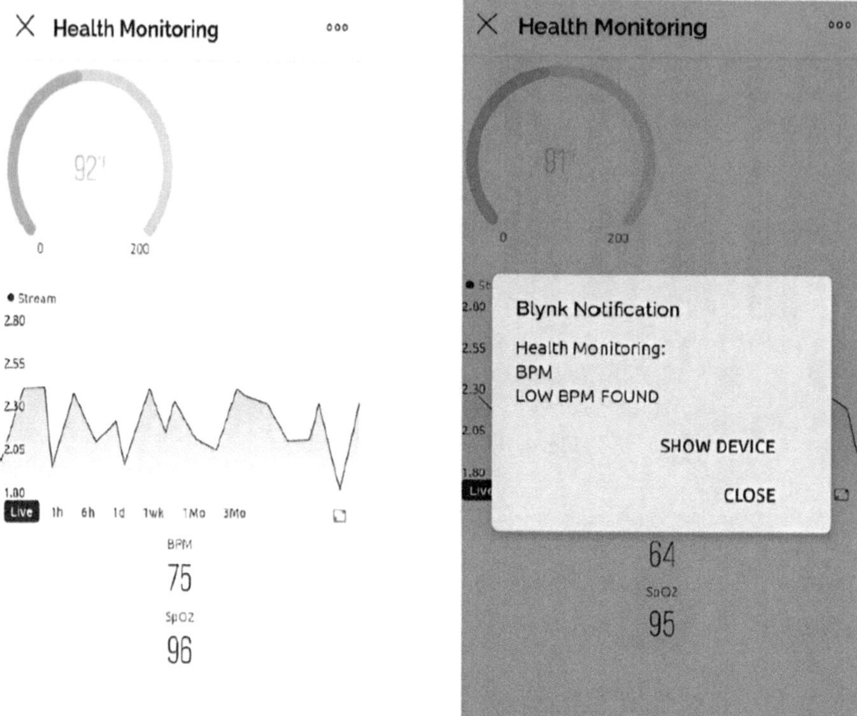

Fig. 6. Mobile dashboard view and notification on the Blynk.

5 Conclusion and Future Work

An energy-efficient heart rate monitoring system using Blynk IoT can be enormously helpful for the person who wants to monitor their heart's health. It is very simple to install the system with various sensors and hardware, where you can monitor your heart health at home. The Blynk IoT platform also has other features, including different alerts and notifications so that early intervention and prevention of any heart problems may be realized. Furthermore, the data fed into the system can be collected and used for the improvement of the overall health of the patient as well as the assistance of the healthcare provider in making decisions in the framework of the treatment process. The efficient heart rate monitoring system using Blynk IoT also has the potential to expand to other fields, especially the sports and fitness industry, where timely heart rate monitoring can benefit athletes to perform better and avoid possible injuries on the field. To sum up, proper heart rate control using IoT with the Blynk app can play a significant role in improving people's experience of monitoring their health results and can change the approach to heart health in different fields.

Machine learning algorithms can be added to the IoT system to combine big amounts of data from various sources, including databases with patient records and environment monitoring devices. This would enable the system to analyze the probability of a heart attack occurrence due to identified patterns. To make certain the recommended IoT systems work apace and precise, they can be possibly affirmed through clinical trial studies with a large cohort of patients. It would assist in sharpening the system to provide better patient care to the people by increasing the percentage of success rate as intended. There are many ways to improve the energy of IoT to produce even better results; for instance, low-power sensors and IoT devices can be used along with intelligent energy management systems. This, in turn, would make it quite possible for a system to run longer without having to be recharged or batteries replaced constantly.

References

1. Sethi, P., Sarangi, S.R.: Internet of Things: architectures, protocols, and applications. J. Electr. Comput. Eng. 1–25, 9324035 (2017). https://doi.org/10.1155/2017/9324035
2. Zhang, Y., Liu, H., Su, X., Jiang, P., Wei, D.: Remote mobile health monitoring system based on smart phone and browser/server structure. J. Healthc. Eng. **6**(2), 717–738, 590401 (2015). https://doi.org/10.1260/2040-2295.6.4.717
3. Rajkumar, G., Gayathri Devi, T., Srinivasan, A.: Heart disease prediction using IOT based framework and improved deep learning approach: medical application. Med. Eng. Phys. **111**, 103937 (2023). https://doi.org/10.1016/j.medengphy.2022.103937
4. Sarmah, S.S.: An efficient IoT-based patient monitoring and heart disease prediction system using deep learning modified neural network. IEEE Access **8**, 135784–135797 (2020). https://doi.org/10.1109/ACCESS.2020.3007561
5. Nancy, A.A., Ravindran, D., Raj Vincent, P.M.D., Srinivasan, K., Gutierrez, R.D.: IoT-cloud-based smart healthcare monitoring system for heart disease prediction via deep learning. Electronics **11**(15), 2292 (2022). https://doi.org/10.3390/electronics11152292
6. Hannan, A., Cheema, S.M., Pires, I.M.: Machine learning-based Smart wearable system for cardiac arrest monitoring using hybrid computing. Biomed. Signal Process. Control **87**, 105519 (2024). https://doi.org/10.1016/j.bspc.2023.105519
7. Cañón-Clavijo, R.E., Montenegro-Marin, C.E., Gaona-Garcia, P.A., Ortiz-Guzmán, J.: IOT based system for heart monitoring and arrhythmia detection using machine learning. J. Healthc. Eng. **2023**, 1–13 (2023). https://doi.org/10.1155/2023/6401673
8. Li, C., Hu, X., Zhang, L.: The IOT-based heart disease monitoring system for pervasive healthcare service. Procedia Comput. Sci. **112**, 2328–2334 (2017). https://doi.org/10.1016/j.procs.2017.08.265
9. Abba, S., Garba, A.M.: An IoT-based smart framework for a human heartbeat rate monitoring and control system. In: Proceedings, vol. 42, p. 36 (2020). https://doi.org/10.3390/ecsa-6-06543
10. Devi, N.G.S., Singh, N.S.: Enhancing heart disease detection in IoT: optimizing long short-term memory with enhanced jellyfish optimization. Multimed. Tools Appl. (2024). https://doi.org/10.1007/s11042-024-18503-6
11. Majumder, A.J., Elsaadany, M., Izaguirre, J.A., Ucci, D.R.: A real-time cardiac monitoring using a multisensory smart IoT system. In: 2019 IEEE 43rd Annual Computer Software and Applications Conference (COMPSAC), Milwaukee, WI, USA, pp. 281–287 (2019). https://doi.org/10.1109/COMPSAC.2019.10220
12. Rahman, M.A., et al.: emsReACT: a real-time interactive cognitive assistant for cardiac arrest training in emergency medical services. In: 2023 19th International Conference on Distributed Computing in Smart Systems and the Internet of Things (DCOSS-IoT), Pafos, Cyprus, pp. 120–128 (2023). https://doi.org/10.1109/DCOSS-IoT58021.2023.00031

13. Khan, M.A.: An IoT framework for heart disease prediction based on MDCNN classifier. IEEE Access **8**, 34717–34727 (2020)
14. Ali, F., et al.: A smart healthcare monitoring system for heart disease prediction based on ensemble deep learning and feature fusion. Inf. Fusion **63**, 208–222 (2020)
15. Basavaiah, J., Anthony, A.A., Mahadevaswamy, S., et al.: An efficient approach of epilepsy seizure alert system using IoT and machine learning. J. Reliable Intell Environ. (2024). https://doi.org/10.1007/s40860-024-00228-w

Leveraging Fintech Innovations in Advancing SDG's—Adoption of Hybrid Model Approach to Forecasting Indices

Asfarin Qureshi[1](✉), Gouher Ahmed[1], Jennifer Dafoodils[2,3], and Saisanjeev Reddy Pereddy[4]

[1] Horizon University College, Ajman, UAE
caasfarin@gmail.com
[2] Victorian Institute of Technology, Melbourne, Australia
[3] Polytechnic Institute, Sydney, Australia
[4] Manipal Academy of Higher Education, Dubai, UAE

Abstract. This study focuses on ESG criteria and gives a deep understanding of how Fintech innovations can enhance transparency, accountability and accessibility within the ESG frameworks. Additionally various ESG indices have also been taken into consideration for the connectedness graph analysis, to show how they respond to global events. Along with this, forecasting with a hybrid LSTM-CNN-Transformer with Added Attention mechanism Architecture has also been done.

Keywords: Financial inclusion · Fintech · ESG · Sustainability · SGDs

1 Introduction

Fintech which refers technological innovations in the financial sector and ESG focused on environmental, social and governance metrics. ESG as a result, serves a very important function by promoting responsible investment, providing stable and sustainable long-term returns and better risk management [2]

ESG is calculated in the following way:

- Data Collection: ESG data is collected through publicly available domains rating agencies like MSCI and S&P Global.
- Metrics are then defined and weighting is done based on industry materiality and relevance.
- Scores are then normalised based on company size, benchmarked against competitors and regional standards and then a final score is then given [3].

MSCI and S&P Global are the two main ESG rating agencies with slightly different methods and varying criteria. Issues arise when there are inconsistencies between the two and limited data verification which makes it less transparent and more prone to cases of Greenwashing.

With all of these points in mind, this work aims to achieve three major objectives:

J. Shreyas et al. (Eds.): CODE-AI 2025, CCIS 2690, pp. 191–205, 2026.
https://doi.org/10.1007/978-3-032-19321-6_18

- Analyse the current trends of ESG indices, their connectedness and their need in an ever-changing global order.
- Forecast ESG indices using a novel hybrid model combining CNNs, LSTM and Transformer Architecture.

2 Literature Review

This research paper has referred to various recent works, which have inspired the author to delve deeper into the world of Fintech and Financial inclusion, specifically highlighting the use and relevance of ESG. Fintech has transformed financial institutions fostering inclusivity and openness. Ding et al. [4] and Zhu et al. [5] proved that Fintech is essential in the long run when we are working towards sustainability (by reducing equity costs) and governance. This was also supported by Dicuonzo et al. [6] which used hypothesis testing and corpus processing tools to establish how ESG is helpful and advantageous when tested in banks. This was further supported by Galeone et al. [7] who took the case study of BNP Paribas and saw a net positive correlation for Fintech to further the goals of financial inclusion using ESG.

Karim and Lucey [8] used Random Forest to predict the ESG goals based on Fintech's complex relationship. Many have used technology to predict investor sentiment which was earlier swayed by bears and bulls. Zhang and Wang [9] used CNN and LSTM based on data from the Shanghai stock exchange and investor sentiment, performing significantly better than traditional models. Ouyang et al. [10] used CNN to identify sentiment patterns and achieved a 45.4% accuracy on social media dataset. Chen et al. [11] used a hybrid BiCUDNNLSTM and CNN model with 93.84% accuracy and a mean-squared error of 0.6% for predicting stock prices.

Several studies have used these Fintech prices and data to measure how interconnected we have become in the globalized world order. Mensi et al. [12] used the Time-Varying Parameter Varying Autoregression Model (TVP-VAR) to examine interconnections in green and non-green bonds, USDX, precious metals and stocks. The study showed that USDX and green bonds are effective for risk management. Naysary et al. [13] studied Fintech and ESG markets and proved that they have a direct relationship, each affecting the other to a certain extent through conditionally and dynamically correlating, with the GARCH model, providing bi-directional insights that are stronger at lower frequencies.

While there is already research present in the field of Fintech and ESG to increase inclusivity and sustainability, gaps persist in terms of:

- Deep understanding of ESG metrics and indices, its connectedness in terms of world events and exploring network based models
- A novel architecture to predict ESG indices based on combination of two-deep learning architectures
- AI-based ESG Analytics work

3 Methodology

This study, focuses on the effectiveness and connectedness of various ESG indices.

3.1 Connectedness Analysis of ESG Indices

The connectedness analysis would show the behavior of ESG indices in relation to each other. This work will also entail its behavior in relation to world events. Based on common logic, the demand for ESG indices decreases during times of war, as oil-based commodities play in a much larger factor and experience much higher growth. The situations which we will be considering are:

- Announcement of UN Sustainability Development Goals
- Brexit
- 2016 US Election
- 2020 Covid-19 Pandemic
- 2020 US Election
- 2022 Russia's invasion of Ukraine
- 2023 Hamas-Israel War
- 2024 Iran-Israel Conflict
- 2024 US Elections

All of these global events, significantly impacted markets and companies all over the world. The indices that will be covered in this study at the moment include:

- FTSE4Good Global 100 Index: shows a broad view across multiple industries globally.
- MSCI All Country World Equity Index: Focuses on global companies with high ESG indices.
- S&P Global Clean Energy Index: It is an index that analyses the top 30 largest companies on the stock market based on their clean-energy use.
- FTSE Environmental Opportunities USA Index: Involved US companies involved in clean energy and waste management sectors.
- iShares MSCI ACWI Low Carbon Target ETF: The main goal of this index is to analyse and monitor corporates with their carbon footprint.
- Dow Jones Sustainability World Index: Tracks companies in various sectors that are leading in sustainability.
- S&P 500 ESG Index: An ESG-friendly version of the S&P 500.
- NASDAQ Clean Energy Index: Involves companies involved in clean energy technologies.
- SNPE ETF: Involves US Stocks that have high ESG ratings.

The dataset would contain day-to-day historical data of these indices, consisting of the following dependent variables:

1. Date
2. Closing Price
3. Opening Price
4. High
5. Low
6. Volume Traded
7. Change %

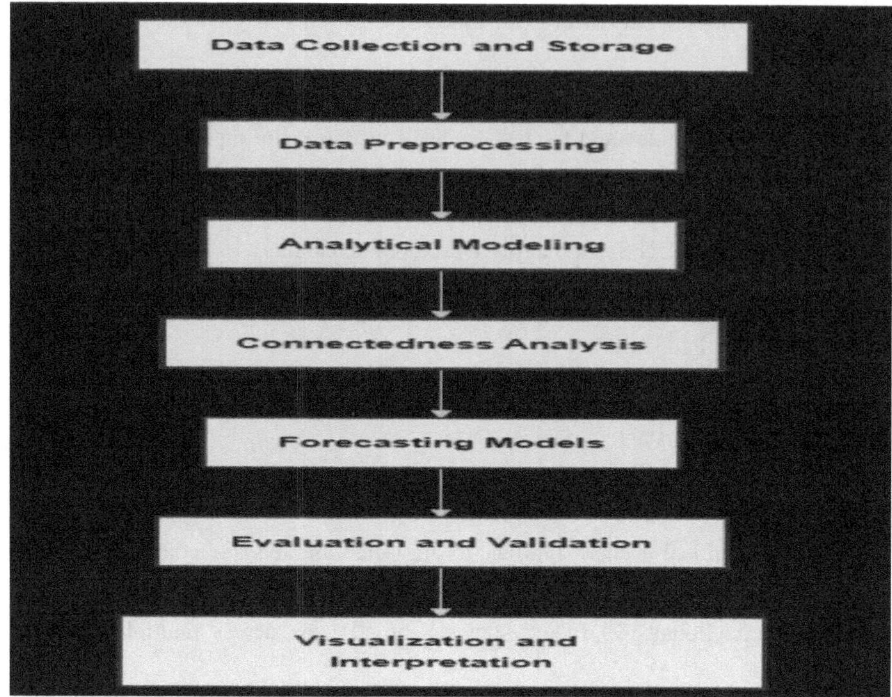

Fig. 1. Flow of work done

The work was done as follows (Fig. 1).

Tools used for this are:

- Python 3.10.0 or higher
- Python Libraries: Pandas, NumPy, Matplotlib, Seaborn, Networkx
- Jupyter Notebook

3.2 Price Trends of Various ESG Indices

10 year period data was taken of day-to-day prices of above-mentioned ESG indices. The mean price is plotted over years. Here, FTSE4Good shows steep upward trend especially after 2019, showing growing positive investor sentiment. Dow Jones shows steady improvement showing consistent performance. SNPE ETF shows low volatility whereas ICLN and S&P ETF shows moderate increases signifying that the market is lowly adapting to green initiatives.

3.3 Analysis of ESG Asset Price During Covid-19 Pandemic

All assets show major dip during this period, but FTSE4Good shows the biggest dip, slowly recovering in 2021 and stabilizing to a lower level in 2022. MSCI All Country World Equity has a more gradual rise, with lower volatility, portraying itself as a low-risk investment option. S&P Global shows significant rise during 2021 driven by policy shifts in renewable and sustainable energy and stabilizes afterwards. ICLN and NASDAQ Clean Edge Green Energy show gains but with high market volatility. The rise can also be attributed to other positive policy shifts by various governments

3.4 Network and Centrality Analysis

Statistically significant correlations have been given edges (threshold of 0.2), while the edge thickness gives the strength of statistic correlations. Louvain Algorithm has been used to detect communities and group them based on color. Here, in Fig. 2. Strong correlated clusters in purple show high interdependence on one another while yellow nodes form smaller, connected groups possibly on certain niches like clean energy. ACWI ETF acts like a hub linking multiple indices and acting as a benchmark for global ESG trends.

t statistic

$$t = \frac{r \cdot \sqrt{n-2}}{\sqrt{1-r^2}}$$

where r is the correlation coefficient and n is the number of data points

Modularity, Q

$$Q = \frac{1}{2m} \sum_{ij} \left[A_{ij} - \frac{k_i k_j}{2m} \right] \delta(C_i, C_j)$$

where A_{ij} is the adjacency matrix (1 if connected else 0), $k_i k_j$ are the degree of nodes i and j.

m is the total number of edges and $\delta(C_i, C_j)$ is 1 if nodes i and j belong to the same community else 0.

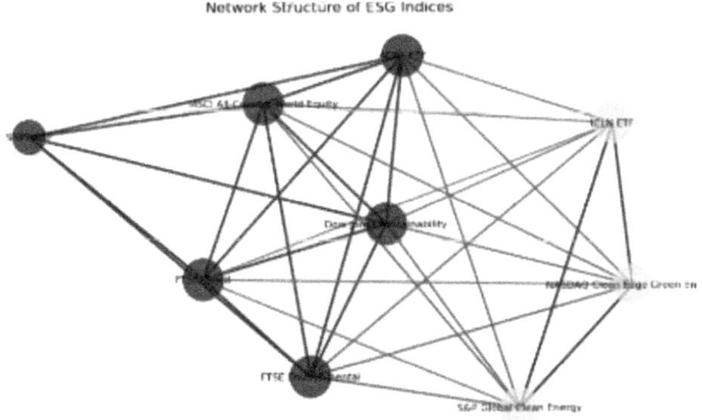

Fig. 2. Centrality study

- Various Centrality metrics have been compared such as (Fig. 3).

- Degree Centrality (Blue): It is the measure of the direct connections a node has, with ACWI ETF as the highest.

  Degree centrality of node v,

$$C_d(v) = \frac{d(v)}{N-1}$$

where d(v) is number of connections/edges for node v and N is total number of nodes

- Betweenness Centrality (Red): It is the measure of how many times a node acts as a bridge, ACWI ETF being the highest here again.

  Betweenness centrality of node v,

$$C_B(v) = \sum_{s \neq v \neq t} \frac{\sigma_{st}(v)}{\sigma_{st}}$$

where σ_{st} is total number of shortest paths from s to t and $\sigma_{st}(v)$ is number of shortest paths from s to t through v

- EigenVector Centrality (Green): This is the measure of influence of of connections to nodes, with MSCI World Equity dominating.

  Eigenvector centrality of node v,

$$C_E(v) = \frac{1}{\lambda} \sum_{u \in N(v)} A_{vu} C_E(u)$$

where λ is largest eigenvalue of the adjacency matrix and A_{vu} is the adjacency matrix value (1 if v and u are connected else 0)

- Closeness Centrality: It is the measure of how quickly the information can spread across a network from each node, here again ACWI ETF being the highest. Through this it shows ACWI ETF and MSCI World Equity are critical structurally, acting as an information hub and as a trendsetter. SNPE ETF functions as an outlier providing diversification option

  Closeness centrality of node v,

$$C_c(v) = \frac{N-1}{\sum_u d(v, u)}$$

where d(v, u) is the shortest path distance between nodes v and u and N is the total number of nodes

High Clustering, means dense local connectivity showing that these indices move together.

Clustering coefficient of node v,

$$C(v) = \frac{2T(v)}{d(v)(d(v)-1)}$$

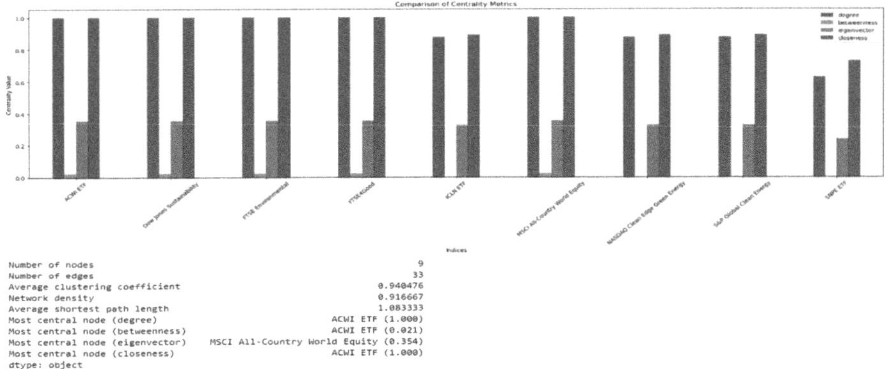

Fig. 3. Centrality metrics comparison

where T(v) is number of triangles through node v and d(v) is degree of node v

Network density,

$$D = \frac{2E}{N(N-1)}$$

where E is number of edges and N is number of nodes.

A low shortest path length of 1.03 also indicates that information flow happens quickly over the network.

3.5 ESG Indices-Based Descriptive Statistics Test For Normality

Table: Descriptive Statistics of ESG Indices (Table 1).

Looking at Table 1, we will mostly look at the different performance metrics, such as Skewness, which measures the asymmetry of the distribution.

Values greater than 0 have skewness towards to the right, while those below 0 have skewness towards the left. Values closer to 0 are considered to be symmetric.

Skewness,

$$S = \frac{\frac{1}{n}\sum_i (X_i - X)^3}{\left(\frac{1}{n}\sum_i (X_i - X)^2\right)^{\frac{3}{2}}} - 3$$

where X_i is an element and $\overline{X}$ ithe mean.

FTSE4Good with 0.68 is right skewed whereas Dow Jones Sustainability with – 0.004 is almost symmetrical. Most are right skewed based on this table.

Kurtosis deals with tails or peakedness, where values below 0 have light tails and flat peaks like S&P Global Clean Energy (−0.41) whereas those above have heavy tails and sharp peaks. 0 follows normal distribution like in the case of ICLN ETF (−0.03).

Kurtosis,

$$K = \frac{\frac{1}{n}\sum_i (X_i - X)^4}{\left(\frac{1}{n}\sum_i (X_i - X)^2\right)^2} - 3$$

Table 1. ESG indices-based descriptive statistics table for normality

Index	Count	Mean	Std	Min	25%	50%	75%	Max	Skewness	Kurtosis	JB Test	Probability
ACWI ETF	2517	79.419984	17.742230	49.5200	63.7600	74.3700	93.9100	121.62	0.406146	−0.918126	157.628720	5.91e−35
Dow Jones Sustainability	2230	2704.958148	944.732118	1278.9100	1609.2250	2801.6350	3517.7025	4506.79	−0.004907	−1.337923	166.256709	7.90e−37
FTSE Environmental	2875	448.537649	155.142013	233.9600	317.5600	376.0200	596.3250	808.07	0.417715	−1.235994	266.505288	1.35e−58
FTSE4Good	2609	8296.242442	2637.519921	4768.7600	6048.7200	7225.6200	10352.3900	14871.19	0.683860	−0.579711	239.805008	8.45e−53
ICLN ETF	2516	13.798247	5.462301	7.7600	9.2300	11.5650	18.6225	33.41	0.944631	−0.034150	373.875747	6.51e−82
MSCI All-Country World Equity	2610	563.329536	126.732323	353.3500	451.2850	526.4600	666.7425	863.10	0.415758	−0.920564	167.371490	4.53e−37
NASDAQ Clean Edge Green Energy	2517	437.331383	248.295586	165.7000	242.5000	289.2000	647.1800	1152.37	0.840108	−0.572073	330.194532	1.99e−72
S&P Global Clean Energy	2589	1120.961409	478.072406	575.5436	705.5701	895.9414	1565.7500	2720.79	0.810347	−0.413474	301.599554	3.22e−66
SNPE ETF	1392	36.686185	7.607771	19.5100	30.4050	36.9000	40.9700	54.69	0.225587	−0.482863	25.448588	2.98e−06

Jarque Bera Test for Normality, tests whether the data follows normal distribution or not.

$$JB = \frac{n}{6}\left(S^2 + \frac{(K-3)^2}{4}\right)$$

If the probability value is greater than 0.05, it is following normal distribution else it is not. In this case, none of the indices are following normal distribution, making it more complex to analyses and predict. FTSE4Good has the standard deviation (2637) showing that it is highly volatile while SNPE ETF is very stable with low standard deviation of 7.6.

3.6 Log Returns and Frequency Analysis with Major Events

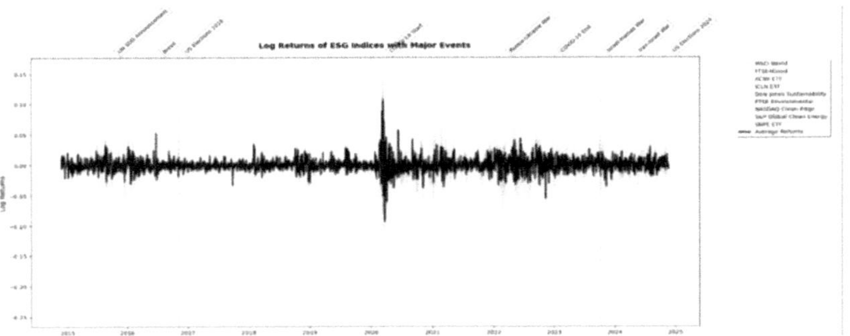

Fig. 4. Overall log returns and frequency components of returns

Looking at Fig. 4 we can see the sharpest volatility spikes during the 2020–22 Pandemic, and followed by the Russian Invasion of Ukraine in early 2022. These log returns are calculated as:

$$Log\ Returns = ln\left(\frac{Price\ at\ time\ t}{Price\ at\ time\ t-1}\right)$$

This is helpful as it provides stationarity in time-series

The bottom graph decomposes returns into low, medium and high frequency components. Low frequency component attributes to long term growth and a recent upward trend is seen in that aspect aligning with the sustainability goals. Medium frequency is flat, showing policy driven shifts, signifying ESG is less prone to market shocks and the high frequency component has been showing a downturn in short-term fluctuations, confirming market maturity (becoming less speculative and more institutionally driven) and showing that ESG investments are stable. They have been calculated using the Butterworth Filter which are given by two filter functions:

$$\frac{1}{1+\left(\frac{s}{cut\text{-}off\ frequency}\right)^{2n}}$$

(i) Transfer function = where n is the filter order

3.7 Forecasting of ESG Indices

Now that, we have understood the crux of the ESG indices, it is only natural that we predict indices' prices for 1 month after taking 10 years data using a novel architecture that involves a combination of various deep learning algorithms.

3.8 Model Architecture

This model leverages Convolutional Neural Networks, Squeeze-Excitation block, Long Short Term Memory (LSTM) and Attention mechanisms. We will look at each part one-by-one and its contributions to the model. The whole idea is to combine the strengths of the individual models to create a novel hybrid architecture that is stronger than its individual parts.

(i) CNN: It captures pattern variations across different time scales and extracts local features such as trends using multiple filters.
(ii) Squeeze-Excitation Block: It assigns weights dynamically and recalibrates based on their importance.
(iii) LSTM: It has been selected due to its ability to learn temporal dependencies effectively and handling sequential dependencies.
(iv) Multi-Head Attention Mechanisms: It is an important part of any transformer model and captures dependencies, between time steps even if they are far apart.

Technical indicators used here are volatility, trend strength and rate of change. An Adam Optimiser with learning rate decay using ReduceLRonPlateau has been used. The loss function (Huber Loss) handles loss effectively, by combining Mean Absolute Error and Mean Squared Error (Fig. 5).

$$L(y, \hat{y}) = 0.5 \times (y - \hat{y})^2 \, for\ small\ errors$$

$$L(y, \hat{y}) = \delta|y - \hat{y}| - 0.5\delta \, for\ large\ errors$$

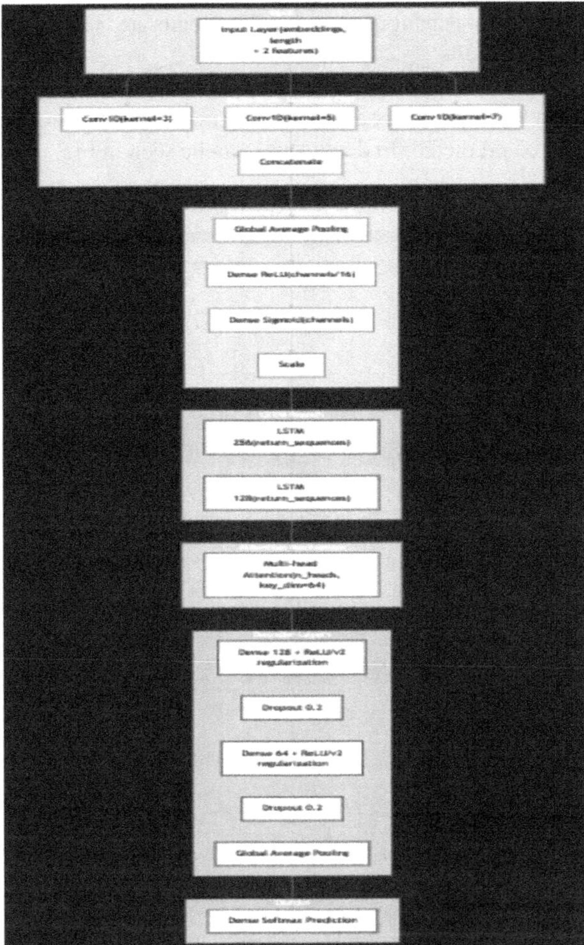

Fig. 5. Hybrid model architecture

3.9 AI Predictions

A Random Forest Regressor has been used to predict the ESG metrics. It checks based on the time-stamp the data has been added to give all stakeholders various insights into the SMEs, which are often ignored. It also gives insights based on region-based ESG data, possibly showing where investors can look into investing sustainably.

Rolling Averages are calculated with smooth trends for numeric features and these trends are predicted for the next six months

Steps:

(i) Data is first filtered based on company or location or both combined.
(ii) Model is then trained with scaled data and rolling averages. At least two data entries for the same entity must be required to, so some prediction can be done. However,

the algorithm works when more and more data points are recorded over long periods of time for better results.

(iii) Confidence intervals are also predicted, giving a range of values.

(iv) Trends are also assessed if they are increasing or decreasing, and the complete ESG score can also be predicted. In depth steps can be seen in Fig. 6.

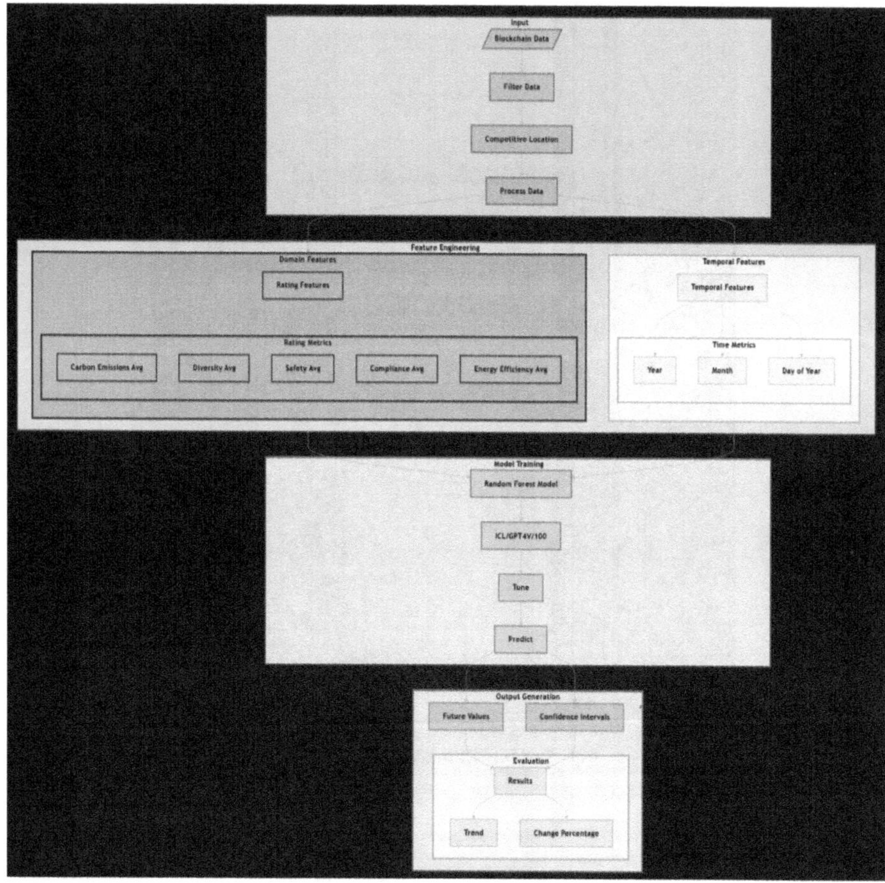

Fig. 6. Model architecture for random forest regressor

4 Results

4.1 ESG Indices Forecasting Results

Table: Model Prediction Results by Index (Table 2).

Table 2. Summary of deviation of predictions from actual change

Index	Last actual price	First predicted price	Last predicted price	Prediction change %
MSCI	426.75	427.3300781	424.7242126	−0.60980157
FTSE4Good	10.43	10.28195763	10.01839447	−2.563355863
SP	768.3205	752.7130127	755.3108521	0.345130125
FTSEEnv	277.9	277.9108582	280.099762	0.78762807
ACWI	60.56	60.74846268	60.87294388	0.204912503
DowJones	1558.43	1553.037476	1558.004639	0.319835357
SNPE	25.09	25.07217789	24.95304489	−0.475160126
NASDAQ	238.6	236.8478394	242.9608765	2.580997534
ICLN	10.43	10.28195763	10.01839447	−2.563355863

Fig. 7. Predicted prices for the month of December

Through Fig. 7 and Table 2, we can see that the prediction % is very minimal indicating the model is performing well. The model maybe underestimating short-term shocks and maybe better in predicting trends than short term trading.

ACWI and DowJones predictions closely follow actual prices, displaying model's robustness for low volatility indices. FTSE4Good and ICLN show downward biases and may need further investigation if it is sector specific.

5 Conclusion

This study provides a multifaceted approach to ESG indices analysis and hybrid-model forecasting based solution for greater transparency and thus address the key challenges of risk management, predictive forecasting and data transparency.

In the first part which incorporates the ESG indices analysis we graphed several plots, to see and investigate their behavior in a convoluted, global and ever-changing market that is marked by various global and geopolitical scenarios. Not only that, we plotted various centrality and spillover graphs to understand their interconnectedness amongst each other. The study saw that central indices such as FTSE4Good and MSCI acted as risk anchors while ICLN and NASDAQ showed greater volatility due to sectoral risks with regards to investment in renewable energy.

A hybrid deep learning architecture was also curated combining the strengths of CNN, LSTM and Self Adaptive Attention to forecast ESG index prices which reflected smooth trends and slowly matched the actual prices. It also gave more insights into which indices are more volatile than the other based on how close the percentage changes are. NASDAQ and ICLN promise positive growth whereas FTSE4Good and MSCI show stable returns.

This study bridges AI and Fintech to advance Financial Inclusion and work towards UN Sustainability Goals for No Poverty. By leveraging ESG index analysis and deep learning forecasting models system contribute to a resilient and inclusive financial ecosystem.

6 Limitations and Future Work

1. Use of Random Forest Regression: Random Forests have been used to it being lightweight and quicker than deep learning models but have also been proven to perform better than other machine learning models [8].
2. Use of Static Graphs: The Network Analysis used static graphs, but dynamic graphs capture temporal relationships better.
3. Data Limitations and Market Bias: The forecasting model relies completely on historical data and no sentiment analysis based on social media or newspaper articles, which may be skewed due to COVID19 Pandemic. The analysis also mostly focused on equity indices and not on other commodities or fixed income assets.
4. Assumptions in Network Analysis: The TVP-VAR model assumes a linear relationship and over-simplifies non-linear dependencies observed during market shocks. Also stationarity is assumed in fixed windows, ignoring dynamic shifts during crises.

Looking at these limitations, future research can be done on the use of Dynamic Graph based Neural Networks for the connectedness study. Sentiment Analysis and use of SHAP values can also be incorporated into forecasting models to predict the prices with greater accuracy.

References

1. The 17 Goals | Sustainable development. (n.d.). The United Nations. https://sdgs.un.org/goals. Accessed 2 October 2024

2. Pagano, M.S., Sinclair, G., Yang, T.: Understanding ESG ratings and ESG indexes. In: Research Handbook of Finance and Sustainability. Edward Elgar Publishing (2018). https://doi.org/10.4337/9781786432636.00027
3. Kotsantonis, S., Serafeim, G.: Four things no one will tell you about ESG data. J. Appl. Corp. Finance **31**(2), 50–58 (2019). https://doi.org/10.1111/jacf.12346
4. Ding, J., Li, L., Zhao, J.: How does Fintech prompt corporations toward ESG sustainable development? evidence from China. Energy Econ. **101**, 23–34 (2024)
5. Zhu, M., Chen, Z., Shen, W., Li, C.: The impact of Fintech on corporate ESG performance—evidence from China. Finance Res. Lett. **47**, 203–215 (2024)
6. Dicuonzo, G., Palmaccio, M., Shini, M.: ESG, governance variables, and Fintech: an empirical analysis. J. Bank. Finance **130**, 67–82 (2024)
7. Galeone, G., Ranaldo, S., Fusco, A.: ESG and Fintech: are they connected? Appl. Econ. **56**, 1456–1465 (2024)
8. Karim, S., Lucey, B.M.: What do advanced machine learning techniques reveal about the ESG-Fintech nexus? SSRN (2024)
9. Zhang, Y., Wang, X.: A hybrid model integrating deep learning with investor sentiment analysis for stock price prediction in Fintech-enabled ESG markets. J. Financ. Markets **22**, 245–260 (2024)
10. Ouyang, X., Zhou, P., Li, C.H., Liu, L.: Sentiment analysis using convolutional neural network. In: 2015 IEEE International Conference on Computer and Information Technology; Ubiquitous Computing and Communications; Dependable, Autonomic and Secure Computing; Pervasive Intelligence and Computing, pp. 2359–2364. IEEE, Liverpool, UK (2015). https://doi.org/10.1109/CIT/IUCC/DASC/PICOM.2015.349
11. Chen, T., et al.: BiCuDNNLSTM-1dCNN—a hybrid deep learning-based predictive model for stock price prediction. IEEE Access **12**, 14475–14483 (2024)
12. Mensi, W., El Khoury, R., Al-Kharusi, S., Kang, S.H.: Extreme dynamic connectedness and hedging strategy across commodity, bond, currency, and stock markets: evidence from Asian Pacific, Canada, Mexico, and US countries. Int. Rev. Econ. Finance **96**, 103533 (2024). https://doi.org/10.1016/j.iref.2024.103533
13. Naysary, et al.: Financial technology and ESG market: a wavelet coherence and DCC-GARCH nalysis. J. Financ. Intermed. **59** (2024)

Performance Evaluation of Quantum Machine Learning Models for Breast Cancer Classification: A Comparative Study

V. Bhuvaneshwari and P. Shanmugavadivu(✉)

Department of Computer Science and Applications, The Gandhigram Rural Institute, Dindigul, India

bhuvaneshwari.cs@bhc.edu.in, psvadivu@ruraluniv.ac.in

Abstract. The breast cancer is widely prevalent, with millions of new cases and deaths reported annually. Globally, one in every 70 women is affected with breast cancer. It is apparent from the statistics that the number of breast cancer subjects are exponentially rising day by day. Early detection of the disease plays a crucial role to decrease the mortality rate and increase the survival chance. Quantum machine learning is an upcoming field in the era of big data, which helps to process the complex data more efficiently. This study aims to explore the QML models to accurately predict the breast cancer subjects and diagnose the disease in the early stage. Additionally, QML-based models aid in the classification of breast cancer as either benign or malignant. This study presents the comparative analysis of three QML models, namely QRF, QkNN and QSVM, for breast cancer classification using the Breast Cancer Wisconsin dataset. Each model is evaluated based on key performance metrics and the models achieved the accuracy of 97%, 94% and 90%, respectively. Hence, the results showed that the QRF outperformed QSVM and QkNN in classification accuracy, demonstrating superior ability to distinguish between malignant and benign tumors. This study highlights the potential of QML in medical diagnostics while emphasizing the need for further research in hybrid quantum-classical approaches, improved feature mapping techniques, and optimized quantum circuits.

Keywords: Breast cancer · Quantum machine learning · Quantum support vector machine · Quantum k-nearest neighbour · Quantum random forest

1 Introduction

Breast cancer is a disease in which breast cells grow abnormally, forming malignant tumours. If it is not treated at the early stage, it will lead to a process which we call it as metastasis. In India, the mortality rate is approximately 28.2% of all cancer types, making it the most common type of cancer. Therefore, early detection of the disease is the need of the hour to drastically reduce the mortality rate. The combination of machine learning with quantum computing will lead to QML [1]. QML algorithms are used to solve difficult computational problems that are not solved by classical ML algorithms [2].

J. Shreyas et al. (Eds.): CODE-AI 2025, CCIS 2690, pp. 206–219, 2026.
https://doi.org/10.1007/978-3-032-19321-6_19

As opposed to classical computers that store data in 0's and 1's (bits), quantum computers deal with quantum states $|0\rangle$ and $|1\rangle$, which is called as ket notation. The smallest informational unit is called qubit. Existence of qubits in multiple states, which we call superposition, enables parallel processing. Quantum feature that permits qubits to be interconnected is called quantum entanglement. A quantum circuit is a model that uses qubits and quantum gates to carry out quantum operations. Quantum logic gates are represented by unitary matrices, which form building blocks of quantum circuit (QC).

There is a huge volume of data and the size of the data stored is increasing rapidly at a rate of 163 zettabytes, which is ten times the data generated in 2016. Though ML algorithms are extensively used for processing tasks, it suffers in terms of computational complexity while handling large volume of data [3]. Therefore, companies like IBM, Google, Microsoft, etc. are involved in building quantum computers, with IBM as a global leader developing both quantum hardware and software [4]. Quantum computing has the ability to process high-dimensional data effectively and it outperforms conventional ML algorithms.

The objectives of this study are as follows:

- To perform exploratory data analysis to examine and visualize the dataset to acquire a deep understanding of relationship between variables and to find the outliers.
- To apply QML models, namely QSVM, QkNN and QRF, to obtain the good accuracy.
- To analyse the QML model's performance against various metrics for classification and prediction of breast cancer tumours.

QML models for classification leverage quantum algorithms to categorize data points by changing classical data (0's and 1's) into quantum states using quantum feature maps. These states are then processed through a quantum classifier circuit, and the results are used to make predictions. Quantum computing allows for parallel computation, which significantly reduces training time and accelerates classification tasks. Hence, in this study, QML models, namely QSVM, QkNN and QRF, are implemented for classifying tumors into benign and malignant. These models produced better results in lesser time.

The remainder of this study is structured as follows. The relevant work is presented in Sect. 2. The suggested methodology is explained in Sect. 3. The findings and discussion are presented in Sect. 4, and the conclusion is presented in Sect. 5.

2 Related Work

Vashisth et al. used the linear, nonlinear and QSVM classifiers to perform the classification task of breast cancer tumours as benign or malignant [5]. This study also highlighted the comparative analysis of three SVM classifiers and found the accuracy of linear, nonlinear and QSVM variants to be 89%, 95% and 92%, respectively. They also applied hyperparameter tuning by making use of GridSearchCV technique by the Scikit-learn library to improve the performance of the SVM classifiers. They found that nonlinear SVM model produced the notable results with 95% accuracy, but its worst-case time complexity was quite high when compared to other SVM models.

Kumar et al. used QSVM for brain tumour classification into benign and malignant [4]. They compared the accuracy obtained from QSVM models with its classical equivalent. Kernel-based QSVM and variational QSVM models produced 95% accuracy, whereas its counterpart, the classical SVM gained 93% accuracy. In comparison with SVM model, they found that a 5-qubit quantum processor required 24.19% less time to execute QSVM, which is kernel-based.

Nan Jiang et al. Showed the quantum computer's ability to do parallel computing to enhance the image classification efficiency. The datasets used are Graz-10 and Caltech-101 [6]. Using QkNN, feature vectors of images are extracted and then fed as an input to the system. For Graz-10 and Caltech-101 datasets, the study showed the accuracy of 83.1% and 78%, respectively.

Yu-Ling Chen et al. Proposed a QkNN algorithm especially for image classification. They considered three datasets, namely MNIST, Fashion-MNIST and CIFAR-10 [7]. Extracted features are mapped into quantum states and the Hamming distance between each feature is calculated. This study shows that proposed QkNN could reduce the time complexity and has increased the classification accuracy.

Sridevi et al. used quantum kernel (QK)-SVM to diagnose breast cancer [8]. The datasets comprise of 106 normal, 87 benign, and 87 malignant cases, where expert radiologists are involved in marking lesion areas. The proposed QKSVM model outperformed previous QKSVM variants with a testing accuracy of approximately 84.33%. The real-time quantum processing unit (QPU) "Aspen-M-3" and quantum Statevector simulator "SV1" are used to implement the proposed model.

Elsedimy et al. proposed QPSO-SVM model, which incorporates quantum particle swarm optimization (QPSO) for SVM parameter optimization [9]. By avoiding local minima throughout the optimization process, this model seeks to improve the accuracy of cardiac disease prediction. To fine-tune the parameters, a self-adaptive threshold technique is introduced, which helps to optimize the decision threshold for classification. QPSO-SVM outperformed other traditional models, with an accuracy of 96.31%.

Anas Bilal et al. suggested IQI-BGWO-SVM model optimizes SVM parameters and enhances feature selection by combining quantum computing concepts with metaheuristic algorithms inspired by nature [10]. The findings showed that the optimized model outperformed conventional models on the MIAS dataset, with 99.25%, 98.96%, and 100% of accuracy, sensitivity and specificity, respectively.

With an emphasis on the Canberra distance metric, this study compares the accuracy of conventional and QkNN algorithms across multiple datasets. Prasanna et al. compared the accuracy and computing efficiency of QkNN against classical kNN using a variety of distance measures [11]. Experimental findings showed that while classical kNN outperforms QkNN in overall accuracy, QkNN achieves competitive performance in datasets like breast cancer (93.85%) and iris (93.33%) for $k = 5$.

3 Materials and Methods

3.1 Dataset

Of 569 instances and 30 features in WDBC, 10 are real-valued features that are calculated using fine needle aspiration (FNA) images of breast cancer masses [12]. These ten characteristics include symmetry, fractal dimension, concavity, concave points, smoothness, compactness, texture, perimeter, area and radius. This dataset is mainly used to train and test various ML and DL algorithms.

3.2 Exploratory Data Analysis (EDA) of WDBC Dataset

Exploratory data analysis is conducted in this study. The dataset is scrutinized for any NA entries [13]. Since there are no missing values, the isnull() method returns 0 for all columns. Benign tumors will have lower values for radius, compactness and texture variation but malignant tumors will have larger mean radius, higher compactness, and rougher texture, indicating irregular cell growth. Additionally, the 18 unwanted features are eliminated from the dataset and only 12 features are selected because of their high correlation with malignancy and importance in breast cancer diagnosis. Finally, by applying data splitting, training set is 80% (n = 455) and testing set is 20% (n = 114), respectively.

To examine the features and their correlation in WDBC, feature correlation matrix is computed and the matrix obtained is used to generate a heat map, in which each cell represents the correction between two features. Histograms illustrating the distribution of selected features from the WDBC dataset are shown in Fig. 2. Table 1 presents the selected diagnostic features, their value ranges, and corresponding clinical inferences used for breast cancer analysis.

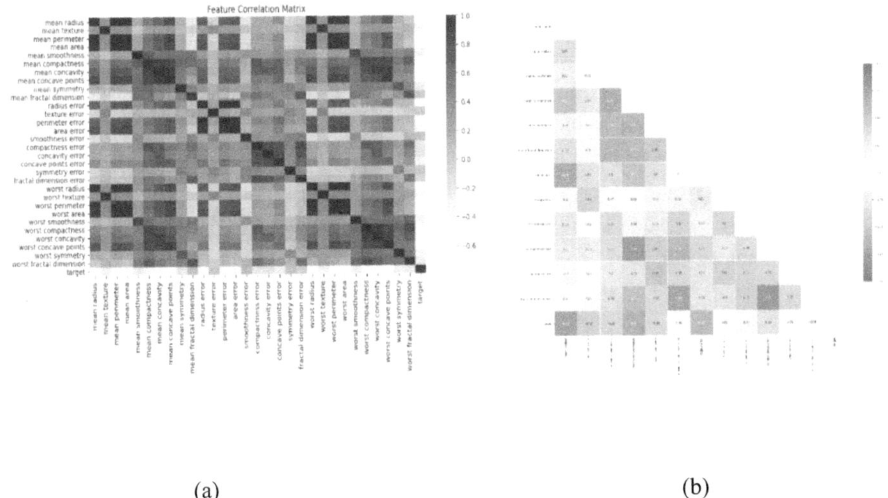

(a) (b)

Fig. 1. Feature correction matrix of WDBC dataset. **a** Heat map and **b** New correlation matrix

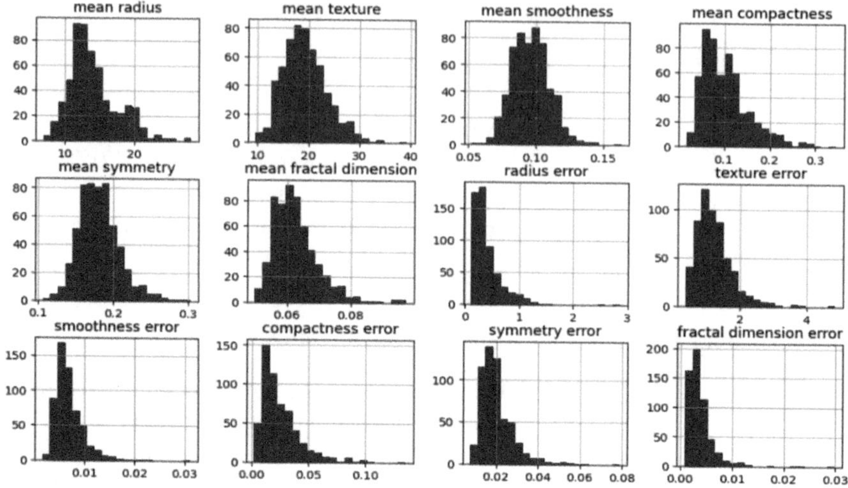

Fig. 2. Histograms for input features selected from WDBC dataset

Table 1. Selected features and their inference

S. No.	Feature	Values	Inference
1	Mean radius	10–25	Larger radius = Higher chance of malignancy
2	Mean texture	12–30	Higher texture variation → More likely to be cancerous
3	Mean smoothness	0.08–0.12	Higher smoothness → More likely benign
4	Mean compactness	0.06–0.2	Higher compactness → Higher malignancy risk
5	Mean symmetry	0.15–0.25	Asymmetry is a strong cancer indicator
6	Mean fractal dimension	0.05–0.08	Higher complexity = More aggressive cancer
7	Radius error	0.2–1.0	High variation → Higher malignancy risk
8	Texture error	0.5–3.0	Greater texture error → More irregular & potentially cancerous
9	Smoothness error	0.003–0.015	Higher error → Higher risk
10	Compactness error	0.01–0.06	High fluctuation suggests malignancy
11	Symmetry error	0.01–0.05	More asymmetric = More aggressive

(continued)

Table 1. (*continued*)

S. No.	Feature	Values	Inference
12	Fractal dimension error	0.002–0.01	Greater error → Less regularity → Higher cancer probability

3.3 Experimental Quantum Machine Learning Models

Quantum Support Vector Machine

SVM being supervised learning model was developed in 1960s. The QSVM is especially used for classification purpose, which is quantum-enhanced version [5]. Finding a hyperplane to map data into a high-dimensional Hilbert space might not be most optimal approach. With the help of quantum properties, QSVM could able to process the complex patterns more effectively than classical SVMs. The QSVM can be implemented using VQC and quantum kernel methods.

Quantum k-Nearest Neighbour

The classical kNN is a supervised/unsupervised learning method and is used to perform both classification and regression. QkNN mimics the classical kNN which computes distance between the given data points, finds the k-nearest neighbours in a high-dimensional space and assigns the class label, based on majority voting [14]. QkNN makes use of superposition and entanglement to speed up the processing task. It is explicitly used for large-scale datasets.

Quantum Random Forest

QRF is a quantum-enhanced version of classical random forest. It comprises of a set of distinct quantum decision trees (QDTs) [15]. QDTs are provided with a set of training instances and their corresponding class labels to learn from the datasets. Because of its randomness, RF performs better in terms of generalization. Another advantage is that it is less sensitive to outliers in the dataset and do not require an exhaustive parameter tuning.

3.4 Proposed Methodology

The primary focus of the study is to apply QSVM, QkNN and QRF algorithms in the breast cancer diagnosis for early detection of the disease. For classification task, WDBC was used as input. The flowchart of the breast cancer diagnosis model is given in Fig. 3.

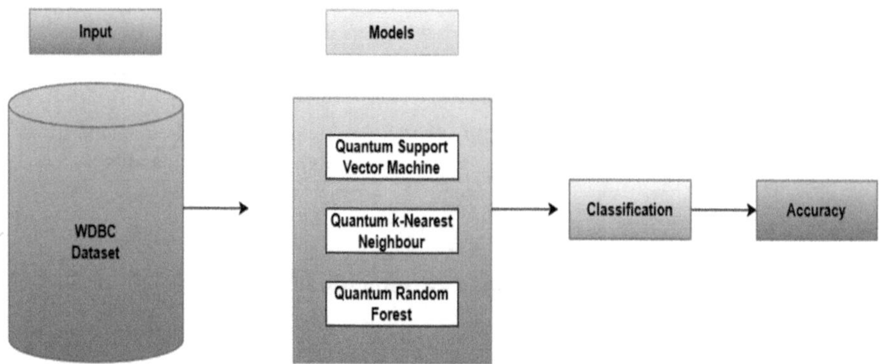

Fig. 3. Flowchart of the QML-based breast cancer diagnosis model

The detailed algorithm of the proposed methodology is given below.

Algorithmic Description of QML Models

Input: Wisconsin Diagnostics for Breast Cancer
Output: Trained model for breast cancer prediction

1. Begin
2. Load the dataset
3. Extract features (X) and labels (y)
   i. 0 = Malignant (cancerous)
   ii. 1 = Benign (non-cancerous)
4. Split the dataset into train/test set in the ratio (80:20)
5. Apply QML models
   i. Train QSVM using quantum kernel
   ii. Train QkNN. Choose an optimal value of k (e.g. k = 3)
   iii. Train QRF with 100 decision trees as default parameter
6. Compute performance on Training and Testing of each model using:
   i. Precision
   ii. Recall
   iii. Accuracy
   iv. F1-Score
   v. AUC ROC Score
7. End

4 Results and Discussion

This proposed work used three types of QML models, namely QSVM, QkNN and QRF. At first, WDBC dataset, which has a total of 30 features, is loaded and preprocessed with data standardization techniques. After pre-processing the data, the number of features is reduced to 12. Then, split the dataset into training/testing set (80:20) and sequentially passed to the classical and QML models for training and testing. The models were evaluated based on performance metrics and their computation is stated in Eqs. (1)–(5). The performance measure for different QML models are summarized in Table 2 (Fig. 4).

$$\text{Precision} = \frac{TP}{TP + FP} \tag{1}$$

$$\text{Sensitivity} = \frac{TP}{TP + FN} \tag{2}$$

$$\text{Specificity} = \frac{TN}{TN + FP} \tag{3}$$

$$\text{Accuracy} = \frac{TP + TN}{TP + TN + FP + FN} \tag{4}$$

$$\text{F1-score} = \frac{2(Precision * Recall)}{Precision + Recall} \tag{5}$$

Table 2. Performance analysis of QML models on WDBC dataset

Metrics	QRF	QSVM	QkNN
Precision	98.57	92.21	92.06
Sensitivity	97.18	100	92.06
Specificity	97.67	86.05	86.49
Accuracy	97.37	94.74	90.00
F1-score	97.87	95.95	92.06
AUC ROC score	99.32	99.44	89.27

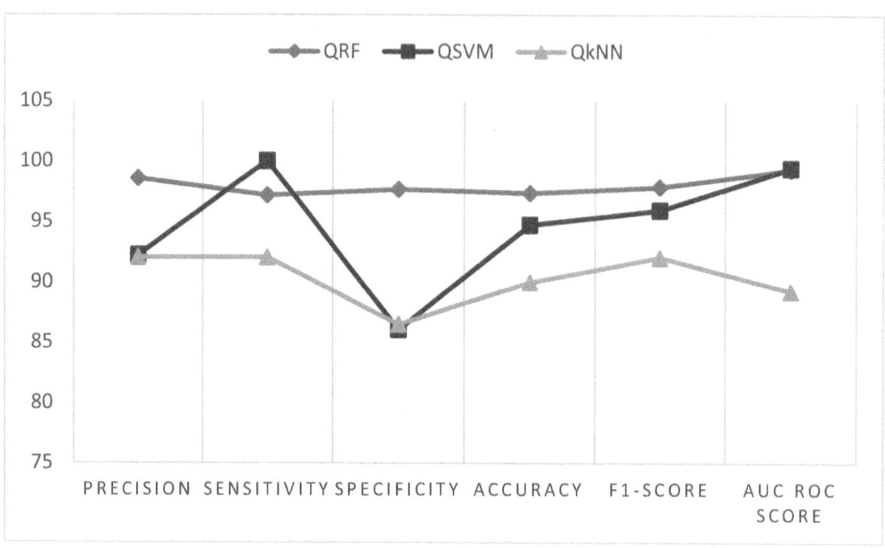

Fig. 4. Performance comparison of QML models

Of these models, QRF achieved the highest accuracy (97%), which shows its robustness in handling the dataset's feature effectively [15]. QSVM also performed well, with the accuracy of 94.7%. When compared to other two models, the performance of QkNN is quite low with an accuracy of 90%. The confusion matrix of the QML models are shown in Fig. 6.

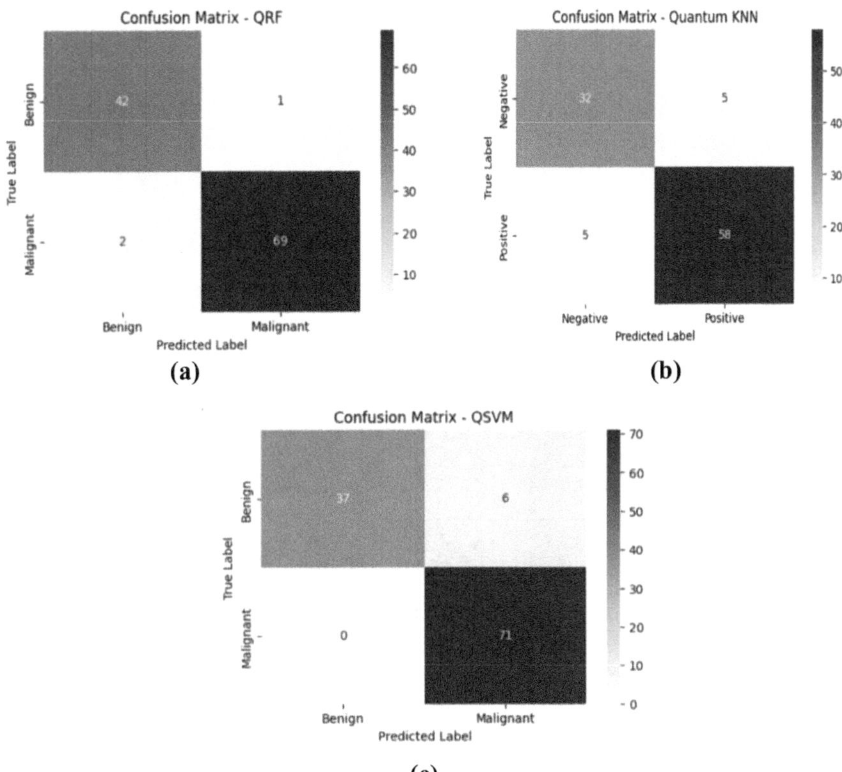

Fig. 5. Confusion matrix of (a) QRF, (b) QkNN and (c) QSVM for breast cancer classification

From Fig. 5, it is observed that among 100 testing cases, for QSVM model, 71 cases were classified as "malignant", and 37 cases as "benign", which are true positives and true negatives, with 6 cases as false positive and 0 case as false negative, whereas for QRF, 69 cases were classified as "malignant" (TP) and 42 cases as "benign" (TN). While using QkNN model [16], 58 cases were identified as true positives and 32 cases as true negatives, with 5 cases as false positive and 5 as false negative.

The receiver operating curves of the three QML models are obtained by plotting the TP rate against the FP rate, which are shown in Fig. 6. It shows that the QRF model will be able to discriminate between cancerous and non-cancerous cases with 97% accuracy. QRF model outperformed other classifiers due to its ensemble learning, which reduces overfitting and improves generalization.

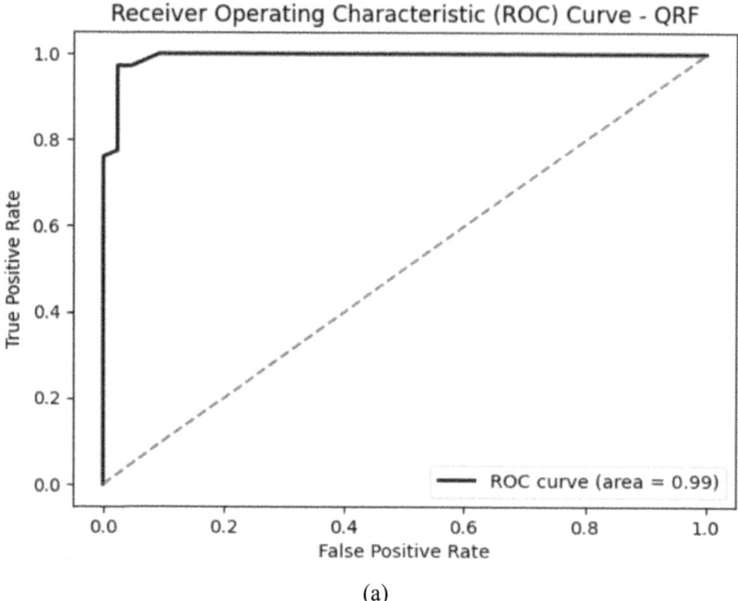

(a)

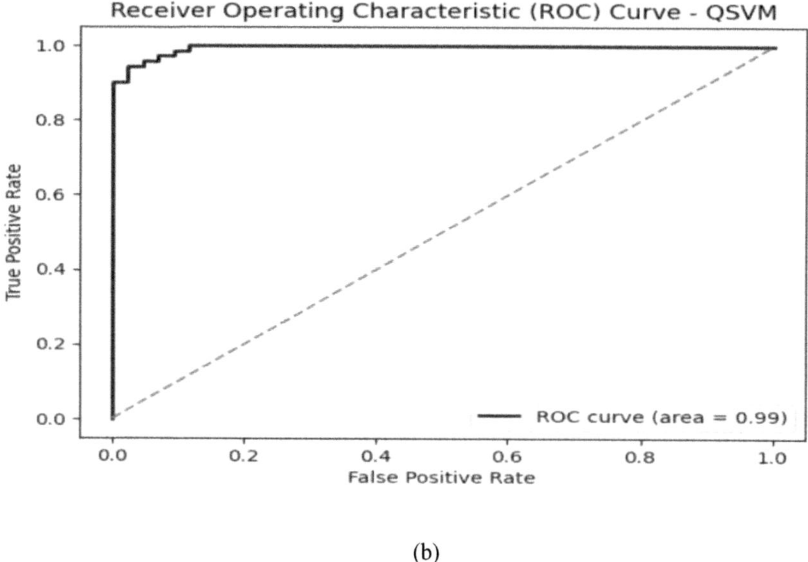

(b)

Fig. 6. Receiver operating characteristic (ROC) curves of (a) QRF, (b) QkNN and (c) QSVM models

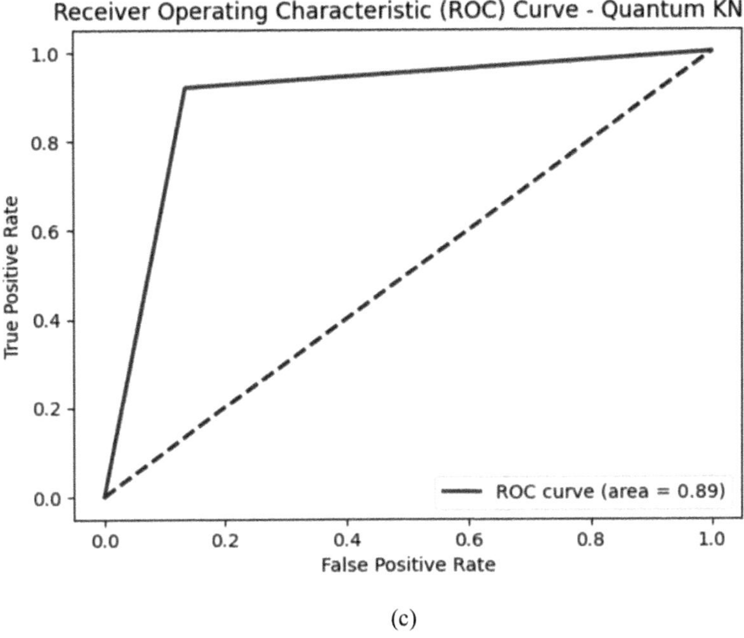

(c)

Fig. 6. (*continued*)

The development of quantum hardware is still in its infancy. Running QML models on real quantum computers is a difficult task due to quantum decoherence and gate errors. The resource needs of quantum models can increase exponentially, which restricts their use to small or medium-sized datasets. Using feature maps to transform classical data into quantum states is not always effective. Performance is greatly affected by the choice of quantum feature mapping technique. So many QML models are still rely on quantum simulators, which depend on classical hardware.

5 Conclusion

In this study, the analysis of breast cancer using QML algorithms has been experimented. This study aimed to utilize QML models trained and tested on WDBC dataset. Based on the dataset feature analysis and selection on WDBC twelve information-significant features were selected. This research shows the comparative analysis of the selected QML models, namely QSVM, QkNN and QRF. Among them, QRF model has gained the highest accuracy, whereas QSVM and QkNN have attained somewhat lesser accuracy than QRF. In this study, it is evident that quantum random forest outperforms other QML models. In future, hybrid quantum-classical approaches integrating QML and classical ML models can be used to improve the performance of the models. Other quantum-inspired algorithms could also be explored to unlock their potential to surpass conventional ML models.

References

1. Ullah, U., Garcia-Zapirain, B.: Quantum machine learning revolution in healthcare: a systematic review of emerging perspectives and applications. IEEE Access **12**, 11423–11450 (2024). https://doi.org/10.1109/ACCESS.2024.3353461
2. Sengupta, K., Srivastava, P.R.: Quantum algorithm for quicker clinical prognostic analysis: an application and experimental study using CT scan images of COVID-19 patients. BMC Med. Inform. Decis. Making **21**(1) (2021). https://doi.org/10.1186/s12911-021-01588-6
3. Murtaza, G., et al.: Deep learning-based breast cancer classification through medical imaging modalities: state of the art and research challenges. Artif. Intell. Rev. **53**(3), 1655–1720 (2020). https://doi.org/10.1007/s10462-019-09716-5
4. Kumar, T., Kumar, D., Singh, G.: Brain tumour classification using quantum support vector machine learning algorithm. IETE J. Res. **70**(5), 4815–4828 (2024). https://doi.org/10.1080/03772063.2023.2245350
5. Vashisth, S., Dhall, I., Aggarwal, G.: Design and analysis of quantum powered support vector machines for malignant breast cancer diagnosis. J. Intell. Syst. **30**(1), 998–1013 (2021). https://doi.org/10.1515/jisys-2020-0089
6. Dang, Y., Jiang, N., Hu, H., Ji, Z., Zhang, W.: Image classification based on quantum K-nearest-neighbor algorithm. Quant. Inf. Process. **17**(9) (2018). https://doi.org/10.1007/s11128-018-2004-9
7. Zhou, N.R., Liu, X.X., Chen, Y.L., Du, N.S.: Quantum K-nearest-neighbor image classification algorithm based on K-L transform. Int. J. Theor. Phys. **60**(3), 1209–1224 (2021). https://doi.org/10.1007/s10773-021-04747-7
8. Aswiga, R.V., Sridevi, S., Indira, B.: Leveraging quantum kernel support vector machine for breast cancer diagnosis from digital breast tomosynthesis images. Quant. Mach. Intell. **6**(2) (2024). https://doi.org/10.1007/s42484-024-00170-3
9. Elsedimy, E.I., AboHashish, S.M.M., Algarni, F.: New cardiovascular disease prediction approach using support vector machine and quantum-behaved particle swarm optimization. Multimed. Tools Appl. **83**(8), 23901–23928 (2024). https://doi.org/10.1007/s11042-023-16194-z
10. Bilal, A., Imran, A., Baig, T. I., Liu, X., Abouel Nasr, E., Long, H.: Breast cancer diagnosis using support vector machine optimized by improved quantum inspired grey wolf optimization. Sci. Rep. **14**(1) (2024). https://doi.org/10.1038/s41598-024-61322-w
11. Bhaskaran, P., Prasanna, S.: An accuracy analysis of classical and quantum-enhanced K-nearest neighbor algorithm using Canberra distance metric. Knowl. Inf. Syst. (2024). https://doi.org/10.1007/s10115-024-02229-w
12. Chen, H., Wang, N., Zhou, Y., Mei, K., Tang, M., Cai, G.: Breast cancer prediction based on differential privacy and logistic regression optimization model. Appl. Sci. **13**(19), 10755 (2023)
13. Kumar, T., Arora, M., Verma, V., Lalar, S., Bhushan, S.: Pre-examination of breast cancer dataset using exploratory data analysis (EDA) approach. In: 2024 International Conference on Computational Intelligence and Computing Applications (ICCICA), Samalkha, India, pp. 1–7 (2024). https://doi.org/10.1109/ICCICA60014.2024.10585026
14. Wang, Y., Wang, R., Li, D., Adu-Gyamfi, D., Tian, K., Zhu, Y.: Improved handwritten digit recognition using quantum K-nearest neighbor algorithm. Int. J. Theor. Phys. **58**(7), 2331–2340 (2019). https://doi.org/10.1007/s10773-019-04124-5
15. Srikumar, M., Hill, C.D., Hollenberg, L.C.L.: A kernel-based quantum random forest for improved classification. Quant. Mach. Intell. **6**(1) (2024). https://doi.org/10.1007/s42484-023-00131-2

16. Darwish, S.M., Ali, R.A., Elzoghabi, A.A.: Quantum-inspired K-nearest neighbors classifier for enhanced printer source identification in forensic document analysis. Sci. Rep. **15**(1), 4097 (2025). https://doi.org/10.1038/s41598-025-86558-y

Leveraging Machine Learning in Detection of Healthcare Insurance Fraud

Aqsa, Nikita Selena D'silva(✉), and Sophia Rahaman

School of Engineering and IT, Manipal Academy of Higher Education, Dubai, UAE
{aqsa.shaikh,nikita.dsilva}@dxb.manipal.edu,
sophia@manipaldubai.com

Abstract. Healthcare insurance fraud poses significant challenges, impacting financial stability and healthcare quality. This research paper presents a machine learning-based approach to detect fraud in healthcare insurance claims. Traditional methods often struggle to adapt to evolving fraud schemes, necessitating advanced techniques. Our framework integrates supervised and unsupervised machine learning algorithms to analyse claims data, focusing on feature engineering to extract fraud indicators. We evaluate various models, including logistic regression, decision trees using precision, recall, and F1-score metrics. We analyze fraud patterns based on gender, seeking to understand potential differences in how fraud manifests. This gender-wise analysis helps provide a more comprehensive understanding of healthcare insurance fraud and informs targeted detection strategies. Results demonstrate promising detection rates, with certain models achieving high precision and recall. Moreover, anomaly detection reveals previously undetected fraud instances, emphasizing its importance. This research contributes to enhancing healthcare insurance fraud detection, offering proactive strategies for insurance providers and regulators to combat fraudulent activities. By leveraging machine learning techniques, we aim to strengthen fraud detection mechanisms and safeguard the integrity of healthcare insurance systems.

Keywords: Machine learning · Prediction models · Fraud detection

1 Introduction

1.1 Influence of Advancing Technology

Technology has been instrumental in shaping the modern world, right from the invention of the wheel to the latest breakthroughs in artificial intelligence. Technological advancements have made lives easier, efficient, and interconnected. From mobile phones to virtual reality and AI, these innovations have brought about significant changes in how we communicate and interact with our environment. The rise of social media and messaging apps allow one to stay connected with friends, family, and colleagues in real-time, regardless of distance. The internet provides instant access to information on virtually any subject, making knowledge more readily available. In the field of healthcare, advancements in technology have enabled doctors and researchers to make significant progress in

J. Shreyas et al. (Eds.): CODE-AI 2025, CCIS 2690, pp. 220–230, 2026.
https://doi.org/10.1007/978-3-032-19321-6_20

disease treatment and prevention. Medical professionals can now easily access patient records and collaborate with other healthcare providers, leading to improved patient care and overall healthcare efficiency. However, technology has a dual effect on society. Along with favorable effects, it has undesirable effects that must be dealt with. Technological dependency, cyberbullying, and privacy invasion are some of the negative effects of technology on society. As machines continue to progress, they become capable of performing tasks done by humans. This progression could result in job displacement and economic instability. As we continue to advance, it is vital to assess the potential impact of advancing technologies and take steps to minimize any negative consequences. Ultimately, it is our collective responsibility to ensure that the technology we use benefits our society [1, 2].

1.2 Technology-Enabled Care Services

In today's digital era, integrating technology-enabled care has become significantly easier, benefiting both those in need and those who can gain the most from it. Technology-enabled care services is an umbrella term that encompasses telehealth, telecare, telemedicine, and self-care, all of which provide convenient, accessible, and cost-effective solutions for managing long-term conditions. In the context of telecare, remote patient monitoring utilizes sensors and can alert healthcare providers in case of emergencies. Wearable devices such as bracelets & shoes track patient activity and trigger alarms when an emergency arises. Technology-enabled healthcare assists patients in managing their medication intake by providing reminders for prescribed medications. This aspect of telecare leverages technology to remotely and consistently monitor patients' conditions, this reduces the need for certain in-person visits, ultimately saving time for both patients and medical professionals. Thanks to advancements in technology, these services can be seamlessly accessed through user-friendly devices, enhancing patient convenience and engagement [3, 4].

1.3 Dimensions of Technology-Enabled Care Services

This cube represents the Three Dimensions of Technology-Enabled Care Services, emphasizing how technology impacts healthcare from different perspectives (Fig. 1).

1. Tech-Enabled Treatment: This dimension focuses on the direct application of technology in medical treatment to improve precision and outcomes.
2. Tech-Enabled Healthcare Services: This dimension highlights the use of technology to optimize healthcare operations and services, ensuring accessibility, efficiency, and better resource utilization.
3. Tech-Enabled Social Wellbeing: This dimension emphasizes the use of technology to enhance the quality of life and community well-being, particularly for mental health and social connections.

While technology has greatly enhanced tech-enabled care, it is also important to consider its potential drawbacks. Fraud remains a significant concern, with common schemes including shell vendors, ghost employees who gain unauthorized access to billing systems, and individuals who continue billing despite having expired licenses.

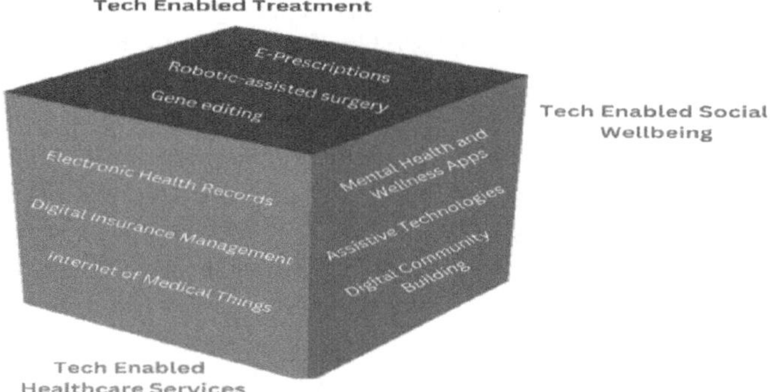

Fig. 1. The three dimensions of technology-enabled care services.

Those involved in fraudulent activities may include providers authorized to deliver services, beneficiaries receiving medical care, medical equipment and drug manufacturers, and agencies offering specialized services like home healthcare. However, with advancements in computing and the increasing availability of aggregated healthcare datasets, there are new opportunities to improve healthcare fraud management and strengthen preventive measures [5].

2 Methodology

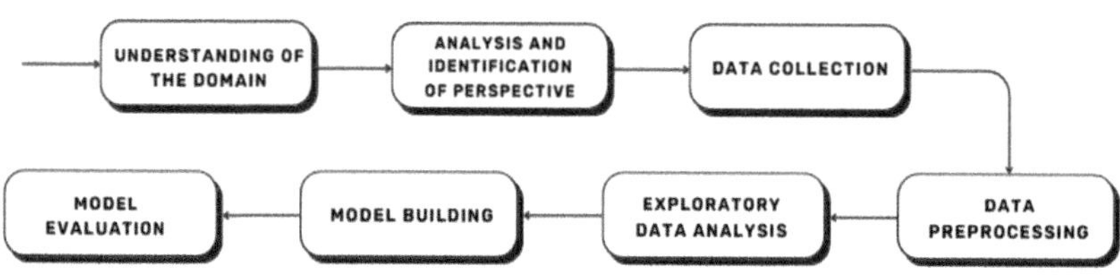

Fig. 2. Methodology flow diagram.

2.1 Data Collection and Preprocessing

Data collection involved gathering relevant healthcare data, including inpatient data and beneficiary data sourced from publicly available datasets. Data preprocessing was done to process the data by addressing missing values, merging both patient and beneficiary datasets into one, dropping irrelevant columns and standardizing datatypes. After merging the datasets, the shape of the final dataset was 40474 rows and 33 columns. In the Inpatient dataset, the column DOD' was dropped, and a new column 'Age' was added.

2.2 Exploratory Data Analysis

Exploratory Data Analysis was performed using statistical and visualization tools such as charts to detect patterns in the data and gain insights. Matplotlib and Seaborn were used to visualize the relationship between potential fraud cases and gender. As well as visualizing potential fraud cases across different age groups.

2.3 Model Building

Model building was implemented by training machine learning models (Logistic Regression and Random Forest) to classify fraudulent claims by using an 80–20 training/testing split. Libraries such as NumPy, pandas, Scikit-learn, TensorFlow were used for data handling, preprocessing, and modelling. Random Forest model was implemented, achieving a high training accuracy of 99% but a low testing accuracy of 59%, which meant high variance and low bias. To further address the disparities and achieve a more robust and generalizable model, Gradient boosting technique was used.

2.4 Model Evaluation

The model's performance was assessed using accuracy, precision, recall, and F1-score for classification. Lastly a confusion matrix was created to evaluate the performance of a classification model. It shows the relationship between the true values (actual labels) and predicted values (model predictions). It helps to assess how well a model is performing in terms of correctly and incorrectly classified instances.

3 Literature Review

3.1 Impact of Fraud on Technology-Enabled Healthcare Services

Healthcare fraud is not a victimless crime. Everybody is impacted, including businesses and individuals, and it results in losses of tens of billions of dollars annually. It may increase taxes, expose a person to needless medical treatments, and boost health insurance costs. In the UAE, reports suggest that they face loss at an estimated amount of AED 3.67 billion (or USD 1 billion) annually due to medical insurance fraud in the country, due to doctors prescribing unnecessary tests and overprescribing medicines, the health cover premiums have risen by nearly 20 percent in UAE alone [6]. The use of rules-based systems is one of the oldest techniques for detecting and preventing fraud. These systems use pre-established criteria to detect possible fraud based on certain conditions or patterns, but they frequently have trouble adjusting to new or changing fraud patterns, which reduces their ability to fend against sophisticated attacks. By improving systems' ability to recognize and adjust to novel fraud patterns in real time, machine learning and artificial intelligence have completely changed the field of fraud detection. These methods are more successful against complex fraud schemes because they can assist spot complex patterns that people might overlook.

3.2 Perspectives of Healthcare Fraud

Common Types of Healthcare Fraud:

1. Fraud Committed by Medical Providers:

   Medical providers can engage in various fraudulent practices that result in financial losses for healthcare systems and insurers. One common method is double billing, where multiple claims are submitted for the same service. Similarly, upcoding involves billing for a more expensive service than what was performed, leading to higher insurance payouts (Table 1).

Table 1. Actors involved in healthcare fraud.

		Receiver	
		Fraud	Non-fraud
Provider	Fraud	**Doctor-Patient Collusion** • Unnecessary Medical Procedures • Kickback Schemes • Upcoding Services	**Provider Fraud** • Phantom Billing • Unbundling • Prescription Fraud by Pharmacists
	Non-Fraud	**Patient Fraud** • Identity Theft • Falsifying Claims • Selling Prescribed Medications	**Non-Fraud by both**

2. Fraud Committed by Patients and Other Individuals:

   Patients and other individuals can also commit healthcare fraud in several ways. Identity theft or identity swapping happens when someone uses another person's health insurance or allows someone else to use their coverage. Another fraudulent act is impersonating a healthcare professional, where an individual provides medical services or bills for healthcare equipment without the necessary licensing or credentials.
3. Fraud Involving Prescriptions:

   Prescription fraud is another prevalent issue within the healthcare system. Forgery occurs when individuals alter or create fake prescriptions to obtain medication illegally. Diversion refers to the illegal distribution of legally prescribed drugs, such as selling medication intended for personal use.
4. Prescription Medication Abuse:

   Prescription fraud, including the creation and use of forged prescriptions, is a serious offense with widespread consequences. It results in significant financial losses for physicians, hospitals, insurers, and taxpayers. However, the most devastating impact is the human cost—every year, tens of thousands of lives are lost to addiction, highlighting the urgent need for stricter regulations and preventative measures [7].

4 Discussion

4.1 Exploratory Data Analysis

In our exploratory analysis, we looked at how fraudulent and non-fraudulent claims were distributed among various demographic groups, including gender and age. This examination helped us uncover potential patterns and trends that could guide our model's feature selection and enhance our understanding of fraud characteristics [8, 9].

4.1.1 Fraud Versus Gender

To investigate the connection between gender and the likelihood of fraud, we created a plot showing the distribution of fraudulent claims for both males and females. Our initial observation indicated that fraudulent claims were evenly distributed between the two genders.This suggests that there is no significant gender bias in the occurrence of fraudulent claims within the dataset. This finding is valuable for our fraud detection model, as it implies that gender may not be a strong standalone indicator of fraud. Nonetheless, it's important to keep exploring other features and the interaction between different variables (Fig. 3).

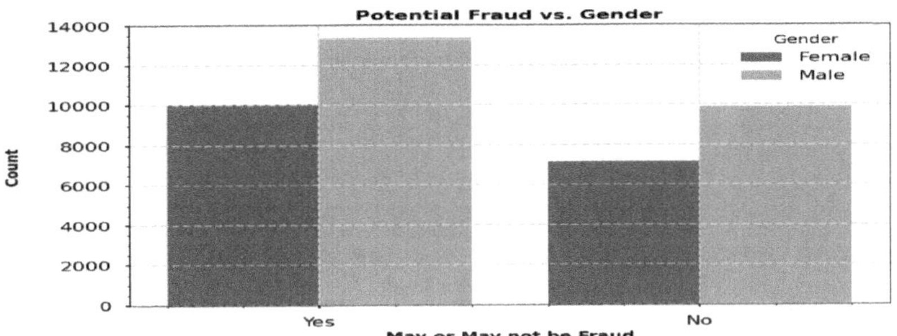

Fig. 3. Graph depicting distribution of gender groups over potential fraud.

4.1.2 Fraud Versus Age Group

Next, we analyzed the relationship between age groups and the likelihood of fraud. We categorized the dataset into four age groups: Young, Mid, Old, and Very Old. The analysis showed that fraudulent claims were more common in the older age groups due to a higher incidence of complex or prolonged medical treatments in the demographics.

Analysis by age group reveals that fraud is more common among older individuals, which is an important finding. This information can inform further feature engineering, suggesting that age could be included in the model as a predictive feature or in interaction with other medical variables to enhance the detection of potentially fraudulent activities (Fig. 4).

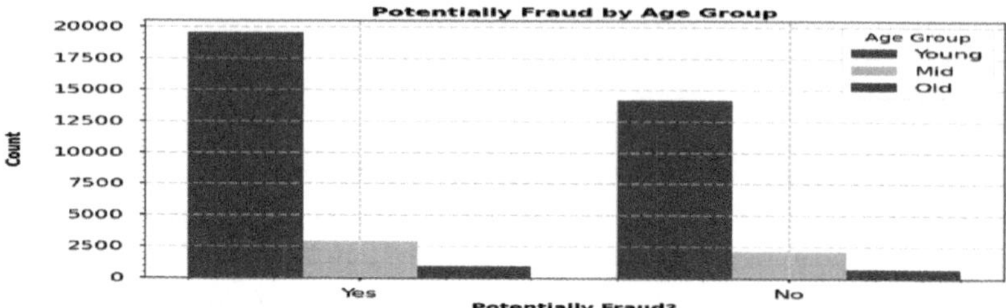

Fig. 4. Graph depicting distribution of age groups over potential fraud.

4.1.3 Connectedness Graph: Fraudulent Activity to Gender and Age Groups

A connectedness graph visually illustrates the relationships between various entities. Nodes represent the individual entities or items under study, while edges depict the relationships or connections among them. Directed edges indicate a one-way or asymmetrical relationship [10]. The term "Fraud" has several links pointing towards "Old" and "Very Old," indicating that older individuals may be more frequently involved in fraud cases than their younger counterparts. The concept of "Fraud" connects to both "Male" and "Female," suggesting that fraud cases involve individuals of both genders (Fig. 5).

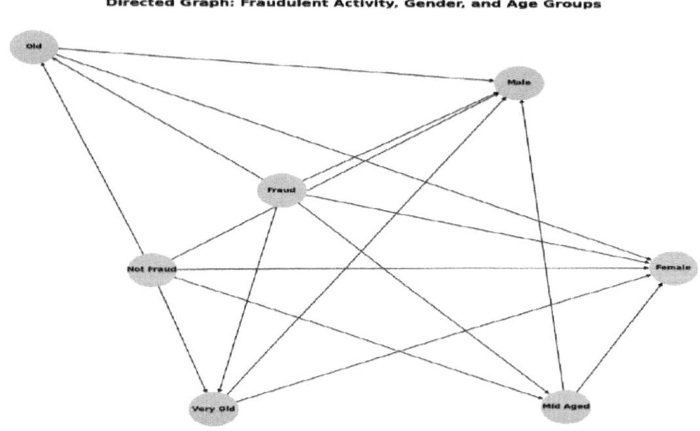

Fig. 5. Connection graph containing directed edges between gender, age groups and fraud.

4.2 Random Forest and Gradient Boosting

In this study, we utilize Random Forest and Gradient Boosting techniques to predict outcomes in healthcare fraud detection. For increased accuracy, the Random Forest ensemble learning algorithm builds several decision trees and combines their results. It works by creating several training subsets via random sampling with replacement, referred to as bootstrap aggregation (bagging). A decision tree is trained on each subgroup, and the final predictions are decided by either a majority vote (for classification tasks) or an average (for regression tasks) [11]. Random Forest is well-suited for this research because it is highly robust to missing data, offers feature significance scores and minimizes overfitting by averaging numerous trees that help identify key fraud indicators, such as abnormal claim amounts or inconsistent patient histories [12]. In contrast to Random Forest, which trains trees on their own, Gradient Boosting optimizes a loss function using gradient descent. It begins by training an initial weak learner, then calculates residual errors and fits new trees to correct those errors [13]. Gradient Boosting is particularly effective for this research due to its capability to record complicated relations between input features, handle imbalanced datasets by adjusting weights to improve minority class detection, and provide interpretable feature importance rankings [14].

5 Result and Analysis

Several indicators were used to evaluate the fraud detection performance of the model, with the confusion matrix serving as a key indicator of classification accuracy. The confusion matrix provides a detailed examination of the model's predictions by comparing the expected and actual fraud labels.

5.1 Confusion Matrix Analysis

The confusion matrix revealed the following observations:

- **High True Positives (TP)**: The model was able to correctly identify a significant number of fraudulent claims, ensuring that fraudulent activities were successfully detected.
- **High True Negatives (TN)**: A majority of non-fraudulent claims were correctly classified, reducing unnecessary fraud investigations and improving operational efficiency.
- **Low False Positives (FP)**: Very few legitimate claims were incorrectly flagged as fraudulent, which minimizes disruptions for honest claimants.
- **Low False Negatives (FN)**: The number of missed fraudulent claims was minimal, which is critical in preventing financial losses and ensuring fraud cases do not go undetected.

This distribution suggests that the model performs well in detecting fraudulent claims while maintaining a low error rate. The ability to keep false negatives low is particularly important in fraud detection, as missing fraudulent claims can lead to substantial financial and operational risks. Similarly, a low false positive rate ensures that legitimate claims are not unnecessarily investigated, reducing administrative costs and improving trust in the system.

5.2 Implications of the Result

The high accuracy of the model indicates that the selected features and preprocessing steps contributed to an effective classification of fraudulent claims. However, further fine-tuning, such as adjusting model hyperparameters or experimenting with advanced ensemble techniques, may help further improve performance.

In the next phase of our research, we will explore additional evaluation criteria, like F1-score, precision, and recall, to have improved comprehension of the model's efficacy (Fig. 6).

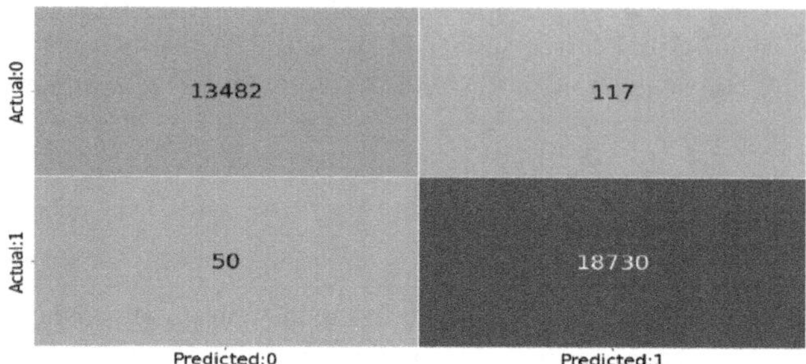

Fig. 6. Confusion matrix to analyze results.

With 13,482 correctly identified fraudulent claims (TP) and only 50 missed fraud cases (FN), the model demonstrates a high fraud detection rate.

This indicates that the model works quite well in identifying fraudulent activities.

The low false negative rate, high fraud detection accuracy, and minimal false positives make this model highly effective for healthcare fraud detection.

CASE STUDIES

The case of David Nourian, Christopher Rydberg, and Michael Taba's $145 million healthcare fraud operation illustrates a real-world scenario where fraudulent activities closely align with the patterns identified in our machine learning analysis. Just as the model in this study examines key features like patient demographics, and inconsistencies across records, fraud in this case was perpetrated by exploiting these very aspects.

In another case study, the pharmacy owners and Dr. Taba conspired to prescribe unnecessary treatments for injured workers, akin to the anomalies that might be detected through abnormal claim amounts and suspicious referral patterns. The gender and age-related trends observed in our dataset also provide valuable insights. Similarly, in the case study, unnecessary treatments were often prescribed to individuals without clear medical necessity, potentially impacting certain age groups more significantly.

By leveraging predictive models, similar fraud patterns could be detected proactively, enhancing the efficiency of fraud prevention systems in healthcare [15].

6 Conclusion and Future Scope

With encouraging findings, the current study investigated the application of machine learning for healthcare fraud detection. By applying Random Forest and Gradient Boosting, the model effectively identified fraudulent claims, demonstrating its potential to safeguard healthcare resources and improve fraud detection in insurance companies.

However, there are areas for future improvement. The dataset used was historical, so testing the model on more recent data could provide insights into evolving fraud patterns. Furthermore, correcting disparities in classes using methods like cost-sensitive learning or oversampling might enhance the model's ability to identify fraud.

Looking ahead, deep learning methods could be explored to detect more complex fraud patterns, and testing the model on different datasets would help assess its generalizability. By focusing on these areas, future work can refine fraud detection systems, ensuring they remain effective and adaptable in the ever-changing landscape of healthcare fraud [16].

References

1. Watts, T.: The Role of Technology in the Future and its Impact on Society. Times of India, 14 April 2023. https://timesofindia.indiatimes.com/readersblog/amitosh/the-role-of-technology-in-the-future-and-its-impact-on-society-52565/
2. KeyTech: The Impact of Technology on Society: Positive and Negative Effects. LinkedIn, 27 March 2023. https://www.linkedin.com/pulse/impact-technology-society-positive-negative-effects-keytech-fi/
3. Technology Enabled Care (TEC): NHS Highland, 5 January 2025. https://www.nhshighland.scot.nhs.uk/your-services/all-services-a-z/technology-enabled-care-tec/
4. All You Need to Know About Technology Enabled Care. Skyresponse, 21 September 2023. https://skyresponse.com/blog/all-you-need-to-know-about-technology-enabled-care
5. Kumaraswamy, N., Markey, M.K., Ekin, T., Barner, J.C., Rascati, K.: Healthcare Fraud Data Mining Methods: A Look Back and Look Ahead. National Library of Medicine (2022)
6. Here's Everything you Need to Know about Health Insurance Fraud: Policy Bazaar. https://www.policybazaar.ae/everything-you-need-to-know-about-health-insurance-fraud-hiart/
7. Health Care Fraud: FBI. https://www.fbi.gov/investigate/white-collar-crime/health-care-fraud#:~:text=Health%20care%20fraud%20can%20be,is%20not%20a%20victimless%20crime
8. S. Morgenthaler, Exploratory data analysis
9. Matthieu Komorowski, J.D.S.: Exploratory data analysis. In: Secondary Analysis of Electronic Health Records (pp. 185–203)
10. Xiulian Gao, Y.G.: Connectedness index of uncertain graph. Int. J. Uncertain. Fuzziness Knowl.-Based Syst. (2013)
11. Xu, D., Ruan, C., Korpeoglu, E., Kumar, S., Achan, K.: Inductive Representation Learning on Temporal Graphs
12. Ali, J., Khan, R., Ahmad, N., Maqsood, I.: Random forests and decision trees. Int. J. Comput. Sci. Issues (2012)
13. Ali, A., Khan, M., Javed, N.: Gradient Boosting Machine Learning Algorithm
14. Bentéjac, C., Csörgő, A., Martínez-Muñoz, G.: A Comparative Analysis of XGBoost
15. U. D. O. J. (DOJ): Pharmacy Owners and Doctor Convicted for $145M Health Care Fraud, Money Laundering, and Tax Evasion Scheme (2023). https://www.justice.gov/archives/opa/pr/pharmacy-owners-and-doctor-convicted-145m-health-care-fraud-money-laundering-and-tax-evasion
16. Prova, N.: Healthcare Fraud Detection Using ML (2024)

A Framework for Automatic Summarization of Text Using Text-Rank Algorithm with ROUGE and BLEU Score

S. A. Karthik(✉), B. J. Ambika, Soumyalatha Naveen, and M. Usha Moorthy

Manipal Institute of Technology Bengaluru, Manipal Academy of Higher Education, Manipal, India
karthik.sa@manipal.edu

Abstract. In the age of exploding data, extracting information from semi-structured, unstructured, and structured data remains difficult. Text material in big volumes in the form of e-books and articles makes the information management process very complicated. To save time, there is an urgent need for tools that can automatically shorten large documents with the retention of the most critical information. This paper proposes a wide-ranging framework for text evaluation and summarization using the efforts of advanced natural language processing techniques. The TextRank algorithm can generate summaries within user-defined word limits by finding and ranking important sentences according to semantic relevance. To assess semantic value, BLEU and ROUGE evaluation metrics are used, providing information about F1 scores, precision, and recall. The proposed system introduces an advanced graph-based sentence ranking method incorporating dynamic BLEU scoring with adjustable weights to accommodate diversified text structures and real-time data extraction from online sources. Its efficiency is noticeably underlined by a comparative analysis conducted with methodologies such as TF-IDF and Latent Semantic Analysis (LSA). Further, the system can be used for very flexible and computationally efficient generation and evaluation of summaries that may find application in business, academic purposes, and general applications.

Keywords: Text summarization · BLEU · ROUGE · TF-IDF · LSA

1 Introduction

In the modern scenario, where storage technology is advanced, enormous amounts of data are being generated and distributed at a rate that has never been seen before. This has created a massive demand for tools that can automatically summarize text, so we can get meaningful insights without having to wade through large amounts of content [1]. Summarizing text—the process of extracting the most important information in a source text, and condensing it into a shorter version while retaining the core message-is an indispensable solution [2]. The process not only saves time but also allows the reader to focus on details that matter rather than being bombarded with non-essential information

J. Shreyas et al. (Eds.): CODE-AI 2025, CCIS 2690, pp. 231–240, 2026.
https://doi.org/10.1007/978-3-032-19321-6_21

[3]. Currently, the two most widely used summarization methods are generative and extractive. The generative summarization method prepares a summary of the text through reading after creating new phrases or sentences [4]. On the other hand, extractive methods work by identifying keywords or phrases that best describe the original content. While both methods have certain advantages, challenges remain, especially when handling diversified sources like Wikipedia, book corpora, and complex forums like Reddit.

The paper, thus, adopts a new approach that encompasses the best elements of extractive techniques for such challenges. Using the Text-Rank algorithm with hyperfine-tuned parameters, one can try to extract concise and meaningful summaries from these varied datasets. The novelty in this work lies in the optimization of the summarization process to handle different styles of language and structure of such sources. Furthermore, we apply established metrics, such as ROUGE and BLEU, to test the quality of summaries that are generated. In turn, these conform to the stringent relevance and coherence standards. This work contributes to the development of more effective and versatile automatic text summarization by pointing out current limitations in summarization methods and introducing new ways to improve them. The results indicate that the proposed system produces highly accurate and concise summaries, which opens doors to practical applications in fields where managing large volumes of text is critical.

The paper is organized as follows. Section 2 describes the related work in the area of text summarization. Section 3 discusses the proposed system design for the extraction of text. Section 4 illustrates the experiment results of our proposed approach. Finally, Sect. 5 concludes the discussion and future work.

2 Related Work

This section briefs about recent research and key challenges in the domain of text summarization using NLP. The main objective of text summarization is to facilitate information retrieval, better understand the content, and save time since they will be in the form of concise summaries. The following section discusses various methodologies and challenges.Additionally, the IWF-Text Rank algorithm [4] incorporates word vectors with multi-feature weighting to further improve the semantic understanding as well as the computation of word weight based on a back-propagation neural Network and Word2Vec. It was tested on traffic accident data and provided reasonable precision, recall, and F score improvement over conventional models. Data summarization techniques like TF-IDF, TextRank, and NLP [5] are adopted to refine data summarization for a more accurate, faster processing speed, and more quality summary. Sentiment analysis in Arabic text is challenged by the Emoji Sentiment Lexicon [6] based on machine learning and emoji-based features which can improve classification accuracy by 26.7% and an F1 score of 89%. This indicates the importance of emojis in conveying emotions in Arabic micro texts; They significantly aid sentiment classification in Arabic micro texts and boost fine-grained sentiment analysis in multilingual negation of sentiments.A framework [7] for IMDb review classification, based on efficiently clustering short reviews was introduced. Preprocessing, text weighting, and classification are done using TF-IDF, N-grams, and various algorithms including Naive Bayes, SVM, Logistic Regression, k-NN, and Decision Trees. The best Logistic Regression with N-grams and the worst TF-IDF with N-grams proved that the representation of the text impacts the accuracy of the model. Open-source legal databases were used to define a legal document classification [8] for law and forensic education. It used machine learning to classify documents related to crime 91.07% (Random Forest) and 82.46% (SVM). Such results illustrate machine learning's ability to process unstructured legal texts and its general significance in legal and educational domains. With the increase in online communication, more and more tools that can perform text mining and sentiment analysis are being created for businesses and users. Deep learning-based classifiers [9] are used to address performance issues such as spelling errors, vagueness, etc., and the challenge of dealing with multiple languages. Advanced methods for efficiency in training time, coverage of data, and improvement of total classification accuracy of sentiment. To solve the dimensionality problems and overfitting, an automated text tagging system using ML and NLP [10] was built to optimize the training. SVM achieved about 90% accuracy, and while Naive Bayes only achieved roughly 65–70% accuracy, it signifies that more can be done. Selenium was used for data extraction in another system [1112], where TF-IDF was used for summarizing web data efficiently for information retrieval. Although methods to improve summarization and sentiment analysis have progressed, areas of improvement therein are still areas of accuracy, for instance; multilingual data, as well as domain nuances, require the use of algorithms to address them. However, machine learning and NLP approaches seem to hold some promise until further refinements are made to make it possible for it to be utilized in its widest form. Several avenues for future work directly relate to building more accurate context-aware summarization models for processing a diversity of data sources and the complexity of linguistic structures.

3 Framework for Automated Text Summarization

This work proposes an automated text summarization of online documents, which helps generate summary text in the long and short patterns demanded by the user. Documents considered for summarization in this architecture are Wikipedia 20 News Group book corpus and Stack Overflow. The proposed architecture is shown in Fig. 1.

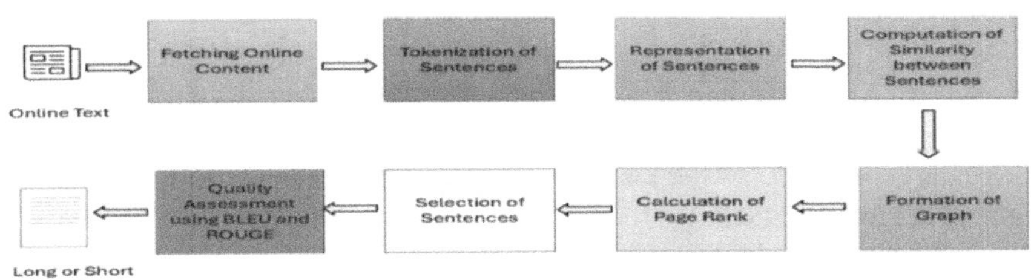

Fig. 1. Proposed model of text summarization

In the proposed architecture, users will input any of the web pages mentioned in the data set. The fetching online content module feeds the text that the user would like to summarize. Next, the text summarization module summarises the content using a text rank algorithm, Summary is based on the word count request by the user finally BLEU and ROUGE Score are used to measure the quality of the generated summary with raw content that considers p- -score, r- score and f1- score parameter.

3.1 Mathematical Model for Summarization of Text

Formulation of mathematical models for different components of the proposed system includes text summarization and score calculation [13].

Let content fetched from online sources as text string denoted by T_S. For text summarization one has to select the most relevant sentence, this can be done by following the steps.

a. Tokenization of sentences: initially tokenize the string Ts into sentences. Let S = {s1, s2,.........sn} be the set of all sentences in online content where n is the total number of sentences.
b. Representation of sentences in this space each sentence si is converted into tf_idf vector V_i using the term frequency-inverse document method [14]

$$V_i = tf\_idf(s_i) \forall \text{ i} \in \{1, 2 \ldots.., \text{ n}\} \tag{1}$$

c. Computation of similarity: for the computation of similarity in sentences a parameter called cosine similarity is used which can be calculated using

$$\text{cos\_sim}(i, j) = \frac{V_i \cdot V_j}{V_i \cdot V_j} \tag{2}$$

d. Formation of Graph: A graph G = (V, E) is formed in such a way that each node vi denotes sentence si and a node exists between v_i and v_j if there is a perfect match between the sentences. So, edge E can be written as

$$E = \{(v_i, v_j, cosine\_sim(i, j)|cosine\_sim(i, j) \geq threshold)\} \tag{3}$$

e. Calculation of Page rank: To rank the sentences the page rank algorithm [14, 15] is applied. The sentence score R_i for sentence s_i is iteratively updated using

$$R_i^{t+1} = \frac{1-d}{n} + d \sum_{j \in neighb(i)} \frac{\text{cos\_}sim(i, j)}{\text{cos\_}sim(j, k)} R_i^t \tag{4}$$

D is the damping factor set to 0.85, neighbor (i) is a set of sentences like si and n is the total number of sentences. Ri is calculated until it converges.

f. Selection of sentences: After ranking of sentences based on the Ri score, the most important sentence is selected for summarization. Let the collection of sentences be S_{summ} Where a given limit constrains the word count of the summary W_{lim} therefore S_{summ} can be written as

$$\left\{ S_{summ} = s_i | \sum_{i \in S_{summ}} WC(s_i) \leq W_{lim} \right\} \tag{5}$$

where $WC(s_i)$ is the count of words in s_i, W_{lim} Is the user-defined words of summarization.

The proposed model is developed with word limit pruning to ensure a concise and relevant summary. Moreover, integrating metrics like BLEU and ROUGE scores [16, 17] will help evaluate the generated summary from different perspectives. The combination of word limit pruning with multiple metrics not only produces summaries but also evaluates and fine-tunes their quality.

3.2 Algorithm for Text Summarization:

//Input: Ts – input Text, W_lim: summary in word count, threshold: minimum cosine similarity to construct graph, d- damping factor set to 0.85

//Output: S_Summ – generated summary

1. **Tokenize the sentence:**
   - Tokenize T_s into sentences where S={s1, s2,.........sn} n is the total number of sentences.
2. **Representation of the sentence:**
   - For every sentence s1 in S compute tf_idf vector Vi so that Vi=tf_idf representation of si
3. **Similarity calculation:**
   - Using equation (2) calculate cos_sim(i,j) between pair of sentence si and sj
4. **Graph Formation:**
   - Create a graph G=(V, E) in such a way that cos_sim(ito threshold
5. **Rank Calculation:**
   - Iteratively update the Rank (R_i^{t+1}) of sentence s1 using the equation (4). Stop iteration when it fails to attain the condition cos_sim(i,j) ≥Threshold
6. **Sentence selection:**
   - Sort the sentences in descending order of their page rank score
   - Select the sentence one by one in such a way that $WC(s_i) \leq W_{lim}$

The above algorithm tokenizes text, calculates cosine similarity, constructs a graph, ranks sentences, and selects based on word limits. This methodology efficiently generates concise summaries by prioritizing sentences with high relevance, ensuring an informative output. The results validate its effectiveness, providing a strong foundation for detailed analysis.

4 Results and Discussion

The proposed system's outcome provides many benefits and efficiently generates summaries from online content. The summary generated is concise and captures the main content of the article. The system is evaluated by BLEU score and ROUGE score calculation. The data sets like Wikipedia stack overflow book corpora are extracted through API. To assess the performance using ROUGE and BLEU metrics, [18, 19] we have taken 10 samples from the above data set which stops the data set including raw text and the generated summary is a text that is generated depending upon user requirement.

4.1 BLEU Score

The BLEU score is the weighted geometric mean of precision and brevity penalty (BP). Finally, BLEU is given by

$$BLEU = BP \times exp\left(\sum_{n=1}^{N} w_n log P_n\right) \tag{6}$$

where w_n is the weight of each n-gram. Brevity Penalty (BP) which let off extracted short summaries. The value of BP is 1 if $|ref| \geq |hyp|$ Otherwise, it is given by

$$BP = exp\left(1 - \frac{|ref|}{|hyp|}\right) \tag{7}$$

where $|ref|and|hyp|$ Are the length of reference text and generated text. For every n-gram $n \in \{1, 2, 3\}$ the $precisionP_n$ is calculated as

$$P_n = \frac{\sum_{alln-gram} Count_{match}(n - gram)}{\sum_{alln-gram} Count_{gen}(n - gram)} \tag{8}$$

$Count_{match}(n - gram)$ is the number of matching n-grams between the generated text and the reference text(s) and $Count_{gen}(n - gram)$ is the number of n-grams in the generated text. Table 1 shows experimental results with BLEU score.

Table 1. Comparison of the existing method with a proposed method based on BLEU score

Score	Parameter	Proposed Method	TF_TDF Method	LSA
BLEU-1	p-score	0.654	0.4754	0.2345
	r-score	0.584	0.423	0.199
	f1 score	0.6172	0.4479	0.2151
BLEU-2	p-score	0.5727	0.3207	0.1761
	r-score	0.499	0.245	0.169
	f1 score	0.5331	0.2779	0.1725
BLEU-3	p-score	0.2269	0.1773	0.1202
	r-score	0.371778	0.247735	0.136351
	f1 score	0.28178	0.206658	0.12775

The Proposed Method's high performance shown in Fig. 2 suggests its better ability to capture precision and recall, resulting in higher f1 scores in BLEU-1, BLEU-2, and BLEU-3 metrics.

The reason for high performance is because of sophisticated techniques, improved feature extraction, and well-tuned model parameters. The TF_TDF Method has moderate performance [20], perhaps because conventional term-frequency methods might not be able to capture contextual subtleties as well as the Proposed Method. The lowest scores are those of the LSA [21] (Latent Semantic Analysis) approach, indicating constraints in processing complicated structures of language or retaining semantic connections as well as the other approaches, particularly with higher n-gram analyses.

4.2 ROUGE Score

Another metric, the ROUGE score is used to compare the overlapping of original text with generated text. To analyze the results, ROUGE-1 (unigrams), ROUGE-2 (bigrams), and

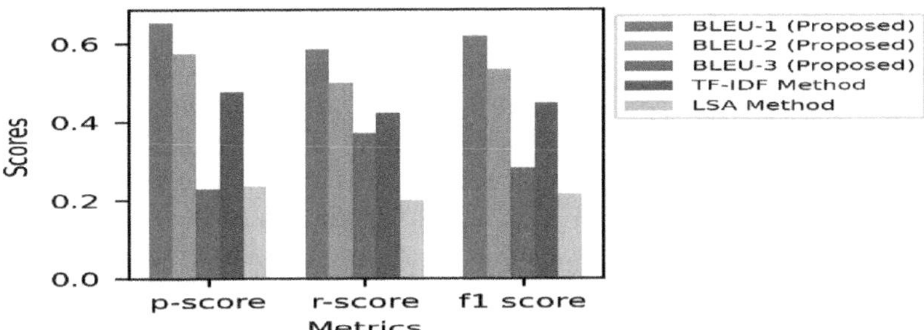

Fig. 2. Comparison of BLEU Score across different text summarization methods

ROUGE-L (longest common subsequence) are computed. For every n-gram n ∈ {1,2} the ROUGE-N is calculated using

$$ROUGE - N = \frac{\sum_{gen\_n-gram} Count_{match}(n - gram)}{\sum_{ref\_n-gram} Count(n - gram)} \quad (9)$$

$Count_{match}(n - gram)$ is the number of n-grams in the generated summary, $Count(n - gram)$ That matches the n-grams in the reference summary is the total number of n-grams in the reference summary. Another score ROUGE -L measures the longest common subsequence between reference and generated summary. It is calculated using Eq. 10.

$$ROUGE - L = \frac{|LCS|}{|ref|} \quad (10)$$

$|LCS|$ is the length of the longest common subsequence which is a longest common subsequence length between the original and generated text and $|ref|$ is the length of the reference text. Finally, the ROUGE score includes the p-score, r-score, and f1-score for every ROUGE-1, ROUGE-2, and ROUGE-L.

$$p - score = \frac{Count(overlappingn - grams)}{\sum n - gramingen\_summary} \quad (11)$$

$$r - score = \frac{Count(overlappingn - grams)}{\sum n - graminref\_summary} \quad (12)$$

$$f1 - score = 2 \times \frac{p - score \times r - score}{(p - score + r - score)} \quad (13)$$

The results of the ROUGE score are shown in Table 2. The proposed method outperformed the TF_TDF and LSA methods with significantly higher precision, recall, and f1 scores in all ROUGE metrics. It means it could capture relevant content with the appropriate balance between precision and recall. The better f1 score of ROUGE-2 shows better bigram matching, which maintains sentence structure. It is likely to make use of more advanced techniques like attention mechanisms or transformers for better content generation. In contrast, TF_TDF performs relatively decently [22]. LSA scored the worst of all three metrics due to the relatively simpler structure [23] and less effective representation of contents. In that regard, the Proposed Method presents a relatively significant improvement approach. The Proposed Method has high, consistent ROUGE scores because of sophisticated methods and efficient semantic treatment. The TF_TDF Method has moderate, varying scores since its basis on term frequency influences performance because of text complexity. The LSA Method is uniformly low with slight variation scores since it is poor at capturing latent semantic relationships through linear dimension reduction which is visible in Fig. 3.

Table 2. Comparison of the existing method with the proposed method based on ROUGE score

Score	Parameter	Proposed Method	TF_TDF Method	LSA
ROUGE-1	p-score	0.8656	0.6198	0.3434
	r-score	0.796	0.592	0.331
	f1 score	0.8295	0.6054	0.337
ROUGE-2	p-score	0.6727	0.3607	0.1961
	r-score	0.554	0.285	0.179
	f1 score	0.6074	0.3185	0.1872
ROUGE-L	p-score	0.7269	0.4773	0.2202
	r-score	0.671778	0.447735	0.236351
	f1 score	0.698236	0.462029	0.227973

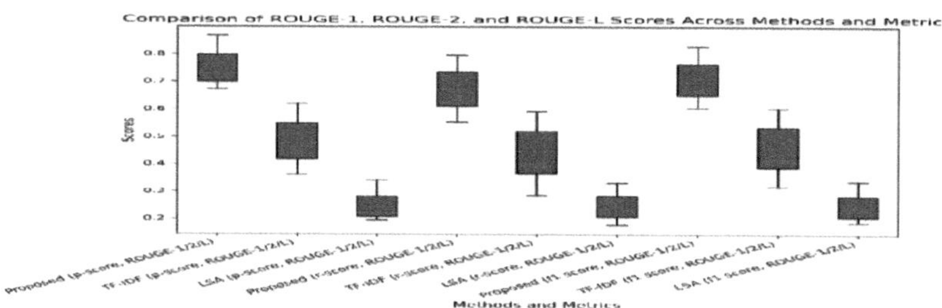

Fig. 3. Comparison of ROUGE Score across different text summarization methods

5 Conclusion

The proposed framework efficiently summarizes long texts using advanced NLP techniques. The Text Rank algorithm, with dynamic BLEU scoring, shows better performance than traditional methods such as TF-IDF and LSA, especially in terms of precision and recall. Its adaptability to diverse text structures and real-time data extraction makes it practically useful. Future work will encompass multilingual capabilities, deep learning of model-based summarization in context, and real-time enhancement and adaptability. This would make the system increasingly stronger for business, academic, and general applications.

References

1. Subramanian, S., Li, R., Pilault, J., Pal, C.: On Extractive and Abstractive Neural Document Summarization with Transformer Language Models. Cornell University (2019). https://doi.org/10.48550/arxiv.1909.03186
2. Liu, Y.: Fine-tune BERT for Extractive Summarization. Cornell University (2019). https://doi.org/10.48550/arxiv.1903.10318
3. Sotnikov, V., Chaikova, A.: Language models for multimessenger astronomy. Multidisciplinary Digital Publishing Institute (2023). https://doi.org/10.3390/galaxies11030063
4. Zhang, L., Wang, W., Ma, J., Wen, Y.: IWF-TextRank keyword extraction algorithm modelling. Appl. Sci. **14**, 10657 (2024). https://doi.org/10.3390/app142210657
5. Zaware, S.N., Patadiya, D., Gaikwad, A., Thakare, A., et al.: Text summarization using TF-IDF and TextRank algorithm. In: 2021 5th International Conference on Trends in Electronics and Informatics (ICOEI) (pp. 1–6). IEEE (2021). https://doi.org/10.1109/ICOEI51242.2021.9453071
6. Alfreihat, M., Almousa, O.S., Tashtoush, Y., AlSobeh, A., Mansour, K., Migdady, H.: Emo-SL framework: emoji sentiment lexicon using text-based features and machine learning for sentiment analysis. IEEE Access **12**, 81793–81812 (2024). https://doi.org/10.1109/ACCESS.2024.3382836
7. Sinha, A., Naskar, M.N.B.J., Pandey, M., Rautaray, S.S.: Text classification using machine learning techniques: comparative analysis. In: 2022 OITS International Conference on Information Technology (OCIT), Bhubaneswar, India, pp. 102–107 (2022). https://doi.org/10.1109/OCIT56763.2022.00029
8. Bifari, E., Basbrain, A., Mirza, R., Bafail, A., Albaradei, S., Alhalabi, W.: Text mining and machine learning for crime classification: using unstructured narrative court documents in police academic. Cogent Eng. **11**, 2359850 (2024). https://doi.org/10.1080/23311916.2024.2359850
9. Rajpoot, A.K., Nand, P., Abidi, A.I.: Text Mining and Sentiment Analysis: Challenges and Advancements. J. Phys.: Conf. Ser. **2062**, 012014 (2021). https://doi.org/10.1088/1742-6596/2062/1/012014
10. Machová, K., Szabóova, M., Paralič, J., Mičko, J.: Detection of emotion by text analysis using machine learning. Front. Psychol. (2023). https://doi.org/10.3389/fpsyg.2023.1190326
11. Chiche, A., Yitagesu, B.: Part of speech tagging: a systematic review of deep learning and machine learning approaches. J. Big Data **9**, 10 (2022). https://doi.org/10.1186/s40537-022-00561-y
12. Manjari, K.U., Rousha, S., Sumanth, D., Sirisha Devi, J.: Extractive text summarization from web pages using selenium and TF-IDF algorithm. In: 2020 4th International Conference on Trends in Electronics and Informatics (ICOEI). IEEE, pp. 648–652 (2020). https://doi.org/10.1109/ICOEI48184.2020.9142938
13. Kirmani, M., Hakak, N., Mohd, M., Mohd, M.: Hybrid text summarization: a survey. In: Advances in Intelligent Systems and Computing, Springer Nature, p. 63 (2018). https://doi.org/10.1007/978-981-13-0589-4_7
14. Mani, I.: Automatic Summarization. John Benjamins Publishing Company (2001). https://doi.org/10.1075/nlp.3
15. Afzal, M., Alam, F., Malik, K.M., Malik, G.M.: Clinical Context-Aware Biomedical Text Summarization Using Deep Neural Network: Model Development and Validation. JMIR Publications (2020). https://doi.org/10.2196/19810
16. Pan, J.Z., et al.: Large Language Models and Knowledge Graphs: Opportunities and Challenges. Cornell University (2023). https://doi.org/10.48550/arxiv.2308.06374

17. Fouesneau, M., et al.: What is the Role of Large Language Models in the Evolution of Astronomy Research (2024)
18. Qiu, W., Shu, Y., Xu, Y.: Research on Chinese multi-documents automatic summarization method based on improved TextRank algorithm and seq2seq. (2021). https://doi.org/10.1145/3448748.3448779
19. Xiao, L., Wang, L., He, H., Jin, Y.: Copy or rewrite: hybrid summarization with hierarchical reinforcement learning. Assoc. Adv. Artif. Intell. (2020). https://doi.org/10.1609/aaai.v34i05.6470
20. Zhu, Y., Zhang, W., Zhu, M.: Differentiable N-gram Objective on Abstractive Summarization. Cornell University (2022). https://doi.org/10.48550/arxiv.2202.04003
21. To, H.Q., Liu, M., Huang, G.: Towards Efficient Large Language Models for Scientific Text: A Review. Cornell University (2024). https://doi.org/10.48550/arxiv.2408.10729
22. Aries, A., Zegour, D.E., Hidouci, W.K.: Automatic text summarization: What has been done and what has to be done. Cornell University (2019). https://doi.org/10.48550/arxiv.1904.00688
23. Setyawan, C., Benarkah, N., Prasetyo, V.R.: Automatic Text Summarization Berdasarkan Pendekatan Statistika pada Dokumen Berbahasa Indonesia (2021). https://doi.org/10.24123/saintek.v2i1.4045

HealthCare Diagnostics with IoT and Block Chain Integration

T. Shreekumar[1], M. L. Smitha[2], U. J. V. Ujwal[2], K. M. Rashmi[3](✉), and S. Y. Malathi[4]

[1] Mangalore Institute of Technology & Management Moodabidri, Badagumijaru, Karnataka, India
[2] KVG College of Engineering Sullia, Sullia, Karnataka, India
[3] Manipal Institute of Technology Bengaluru, Manipal Academy of Higher Education, Manipal, India
rashmi.km@manipal.edu
[4] KLE Institute of Technology Hubballi, Hubli, Karnataka, India

Abstract. The Quantum-Enhanced Point-of-Care Testing (POCT) system addresses the key challenges of traditional POCT systems, including limited diagnostic accuracy, high latency, data security risks, and power inefficiency by integrating advanced technologies, including nano-functionalized sensors, IoT, blockchain, Explainable AI (XAI), reinforcement learning (RL), and quantum computing. This system is designed to enhance diagnostic accuracy, reduce latency, secure data transmission, and optimize energy consumption in real-time healthcare settings. It is very helpful in remote healthcare, mobile health units, emergency care, and chronic disease monitoring, where time and accuracy of diagnosis are essential. The system uses quantum algorithms to optimize sensor thresholds, reinforcement learning to enhance real-time data transmission, and blockchain to secure and store data, ensuring integrity and privacy. IoT provides a seamless flow of data from sensors to the healthcare provider, while XAI gives the explanation behind the decisions that the diagnostic makes. In testing, the system had run 1000 diagnostic predictions with an accuracy of 92%, 0.43 s latency, and 2.3 W of power, thus being more efficient, secure, and interpretable than other AI-based and blockchain-based diagnostic systems in resource-constrained environments.

Keywords: Quantum-Enhanced POCT · Nano-functionalized sensors · Reinforcement Learning (RL) · Blockchain in healthcare · Explainable AI (XAI) · Quantum Approximate Optimization Algorithm (QAOA) · IoT-based diagnostics · Real-time healthcare systems

1 Introduction

Point-of-care testing has truly transformed modern health care because it allows the provision of quick diagnostic assessments directly from the patient's side of residence. This aspect also limits dependence on centralized laboratory settings and, by virtue of

J. Shreyas et al. (Eds.): CODE-AI 2025, CCIS 2690, pp. 241–253, 2026.
https://doi.org/10.1007/978-3-032-19321-6_22

the instantaneity, enhances clinical decision-making cycles. The performance of typical POCT systems, though, often suffers due to a low sensitivity and specificity. The shortcomings have been addressed by recent nano-functionalized paper-based sensors that have incorporated nanomaterials, such as gold and silver nanoparticles, to increase the effective surface area for biomarker interaction. This enhances the ability to detect analyte concentrations at lower levels, which is very important for early diagnostics [1]. Such innovation has proven to significantly enhance sensitivity while being cost-effective, thus highly suitable for resource-constrained settings [2].

Despite their promise, scaling paper-based biosensors to larger populations and seamlessly integrating them into existing healthcare infrastructures remains a significant challenge [2, 3]. A critical aspect of these systems is the secure and efficient transmission of real-time diagnostic results. The Internet of Things (IoT) offers a transformative solution by enabling continuous data collection and remote monitoring [4, 5]. However, the adoption of IoT in healthcare raises very genuine concerns about data security and privacy of patients. Blockchain technology has emerged as a very strong countermeasure, which can provide decentralized and tamper-proof data storage solutions that guarantee integrity of data and prevent access without permission [6, 7].

In addition, integration with Explainable Artificial Intelligence (XAI) enhances interpretability and transparency of diagnostic processes. Frameworks such as SHAP and LIME empower clinicians to understand the rationale behind AI-generated predictions, fostering greater trust and regulatory compliance [8, 9]. Additionally, quantum computing represents a paradigm shift in healthcare diagnostics, offering unparalleled computational capabilities to optimize sensor thresholds, accelerate data analysis, and bolster cryptographic security [10, 11]. Algorithms like the Quantum Approximate Optimization Algorithm (QAOA) utilize principles from quantum mechanics to solve otherwise intractable optimization problems optimally and efficiently, thereby advancing the capabilities of diagnostic systems further [12].

The proposed Quantum-Enhanced POCT system shall bridge the limitations of traditional diagnostic frameworks by synergistically combining nano-functionalized sensors, IoT, blockchain, XAI, and quantum computing. By incorporating all these, it should potentially enhance the precision of diagnostic outcome, reduce latency to some extent, and ensure a secured and interpretable healthcare solution at resource-limited remote setups.

2 Related Work

The integration of Explainable AI in healthcare diagnostics enhances the trust and interpretability of it, as clinicians will be able to understand the diagnostic outcomes. Arrieta et al. explained that XAI models like SHAP and LIME increase transparency within diagnostic systems, which ensures patient safety and compliance with the regulations [11]. Doshi-Velez and Kim, therefore, emphasized that explainability in high-stakes fields like healthcare increases confidence in decision-making and reduces biases, making it critical for AI-driven technologies in medical settings [12].

Blockchain and quantum computing offer relevant solutions to the privacy and security concerns of healthcare systems. Quantum-resistant encryption protocols, such as lattice-based cryptography, were considered by Shinde and Patil an imperative for protecting sensitive data related to patients in blockchain-based POCT systems [13]. According to Dunjko and Briegel, quantum reinforcement learning, QRL, for instance, has the capability of outperforming classical algorithms towards the optimization of diagnostic decision making, resource allocation, and also sensor threshold adjustments in real-time [14].

QML is also revolutionary in healthcare diagnostics. Schuld and Petruccione highlighted that QML, exploiting quantum parallelism, makes the analysis faster, which is critical for POCT since prompt results are always required [15]. Preskill introduced quantum supremacy and proved that quantum algorithms permit nearly instantaneous diagnostics [16]. Montanaro examined how quantum algorithms can be applied to optimize sensor calibration and bandwidth in IoT-enabled devices [17].

More important is that breakthroughs in quantum error correction [18] and quantum cryptographic protocols [19] are to provide dependable, secure data handling in an efficient POCT system within health diagnostics.

3 Methodology

This integrates nano-functionalized paper-based sensors, IoT-enabled data transmission, blockchain for secure storage, and Explainable AI for transparent diagnostics in a real-time POCT system. The sensor threshold optimization is based on reinforcement learning, and the speed and security are further enhanced using quantum computing. The proposed framework provides a sensitive, secure, and scalable platform for multiple healthcare applications with improved patient outcomes. The proposed methodology of the integrated Quantum-Enhanced Point-of-Care Testing system that integrates nano-functionalized sensors, IoT, blockchain, reinforcement learning, and Explainable AI. Every step has a mathematical model.

Step 1: Nano-functionalized Sensor Data

In the proposed system, the nano-functionalized sensors are designed to detect biomarkers with improved sensitivity by utilizing nanoparticles on the sensor's surface. The mathematical expression for the sensor's amplified response is:

$$R_s = A_f \times S_0 \tag{1}$$

where:

R_s is the amplified sensor response, which reflects the enhanced ability of the sensor to detect biomarkers.
S_0 is the baseline sensor output, representing the normal signal that the sensor would produce without the nanoparticle enhancement.
A_f is the amplification factor, which increases the sensor's sensitivity.

The amplification factor is defined as:

$$A_f = 1 + k \times \frac{A_{np}}{A_s} \tag{2}$$

where:

k is a constant that quantifies sensitivity enhancement due to the nanoparticles.
A_{np} is the nanoparticle-enhanced surface area, which increases the sensor's ability to interact with the target biomarkers.
A_s is the bare surface area of the sensor without nanoparticles.

The amplification factor shows how much more sensitive the sensor becomes due to the nanoparticles, where a higher A_{np} leads to a greater increase in sensitivity, amplifying the sensor's response R_s.

Step 2: IoT Data Transmission

In the system, sensor data collected from nano-functionalized sensors is transmitted through an IoT network. The transmission latency, L_t, is determined by the size of the data being sent, D_s, and the available bandwidth, B_w. The mathematical expression for this is:

$$L_t = \frac{D_s}{B_w} \tag{3}$$

where:

L_t represents the latency (in seconds), indicating how long it takes for the data to be transmitted.
D_s is the data size (in megabytes), reflecting how much information needs to be sent.
B_w is the bandwidth (in megabits per second), which measures the transmission capacity of the network.

Increasing bandwidth reduces latency, improving real-time performance, while larger data sizes increase latency.

Step 3: Blockchain for Secure Data Storage

In the system, blockchain technology is used to secure sensor data by generating hash values, ensuring data integrity. The hash value $H(x)$ for a given data point x is calculated using the expression:

$$H(x) = h_0 e^{kx} \tag{4}$$

where:

$H(x)$ represents the hash value, which uniquely identifies the data and protects it from tampering.
h_0 is an initialization constant, determining the base hash value.
k is a sensitivity constant that adjusts how sensitive the hash value is to changes in the data.

The exponential relationship ensures that even small changes in the data x lead to significant changes in the hash value, enhancing data security and integrity.

Step 4: Reinforcement Learning (RL) Optimization

In the proposed system, Q-learning is used to optimize the transmission parameters for data transmission over the IoT network, aiming to minimize latency. Q-learning is a reinforcement learning algorithm where an agent learns to make decisions by interacting with an environment. The Q-learning update rule is expressed as:

$$Q(S_t, A_t) = Q(S_t, A_t) + \alpha\left(R_t + \gamma_{A'}^{\max} Q\left(S_{t+1}, A'\right) - Q(S_t, A_t)\right) \tag{5}$$

where:

$Q(S_t, A_t)$ is the Q-value for taking action A_t in state S_t, representing the expected future reward for this action.

S_t is the current state, such as a specific network condition or bandwidth level.

A_t is the action taken, such as increasing or decreasing the transmission rate.

S_t is the reward received after taking action A_t, which could be related to reduced latency.

α is the learning rate, controlling how much new information overrides old knowledge.

γ is the discount factor, determining the importance of future rewards versus immediate rewards.

The update rule adjusts the Q-value by adding the difference between the expected reward and the actual reward received. By iterating through different states and actions, the Q-learning agent gradually learns the optimal actions (such as adjusting bandwidth or data transmission rates) that minimize latency while maximizing data transmission efficiency. Over time, the system converges to an optimal strategy for real-time, efficient data transmission.

Step 5: Explainable AI (XAI) using SHAP Values

In the proposed system, SHAP (SHapley Additive exPlanations) values are used to provide a clear explanation of how each biomarker contributes to the final diagnostic decision made by the model. SHAP values are grounded in game theory, where the goal is to fairly allocate the contribution of each feature (in this case, biomarkers) to the model's prediction. The formula for calculating SHAP values is:

$$\varphi_i = \frac{1}{N!} \sum_{S \subseteq N\{i\}} \frac{|S|!(|N| - |S| - 1)!}{|N|} [f(S \cup \{i\}) - f(S)] \tag{6}$$

where:

φ_i is the SHAP value for feature i, which represents the contribution of a specific biomarker to the overall diagnostic result.

N is the set of all features, i.e., all biomarkers that are being considered in the diagnostic model.

S is any subset of features in N that does not include the feature i.

The term $f(S \cup \{i\}) - f(S)$ refers to the marginal contribution of feature i when added to the subset of S ther features. This marginal contribution captures how much feature i influences the model's prediction when considered in combination with other features in subset S.

The SHAP value φ_i is calculated by averaging this marginal contribution over all possible subsets S of the features. The denominator, N, normalizes the contributions so that the allocation is fair across all possible subsets, ensuring that each feature is fairly rewarded for its contribution to the outcome, regardless of the order in which features are considered.

By calculating SHAP values for each biomarker, the system can explain the diagnostic decision in a transparent way. For example, if a particular biomarker significantly influences the model's decision, its SHAP value will be large. On the other hand, if a biomarker has little effect, its SHAP value will be small. This transparency helps clinicians and users understand why the model makes a certain diagnosis and provides trust in the AI-driven system.

Step 6: Quantum Approximate Optimization Algorithm (QAOA)

In this work QAOA is used to optime the sensor threshold and hence enhance the accuracy by keeping low latency. It is a quantum algorithm designed to solve complex optimization problems more effectively than other classical methods.

The quantum state ψ, representing the diagnostic data in the quantum system, is expressed as:

$$\psi = \alpha\,|0\rangle + \beta\,|1\rangle \tag{7}$$

Here:

α and β are complex amplitudes that determine the probability of the system being in either the $|0\rangle$ state or the $|1\rangle$ state.

These states, $|0\rangle$ and $|1\rangle$, represent the two possible outcomes in the quantum binary system, similar to classical bits but in superposition.

The objective of QAOA is to optimize these parameters so that the system can identify the most efficient thresholds for the sensors, ultimately improving their performance. The key to this optimization is the function:

$$f(\theta) = \langle\psi(\theta)|H|\psi(\theta)\rangle \tag{8}$$

where:

$|H|$ is the Hamiltonian, a mathematical operator representing the total energy of the system, which encodes the problem to be optimized.

θ represents the set of parameters that define how the quantum system evolves.

The optimization process involves adjusting θ such that the expectation value of H with respect to the quantum state $\psi(\theta)$ is minimized. By doing so, the system identifies the optimal sensor thresholds, which leads to improved diagnostic accuracy and faster data processing. The diagnostic accuracy is achieved by iteratively adjusting the parameter θ and hence allowing the quantum system to explore different configurations of sensor thresholds.

Step 7: Overall System Performance

The proposed system is evaluated based on 3 important performance parameters such as accuracy, latency and power efficiency. These parameters provide systems effectiveness in delivering real-time accurate and energy-efficient diagnostic data.

a. Accuracy: The accuracy of the system is calculated using the formula:

$$A = \frac{TP + TN}{TP + TN + FP + FN} \tag{9}$$

where
TP (True Positives) represents correct positive diagnoses.
TN (True Negatives) represents correct negative diagnoses.
FP (False Positives) represents incorrect positive diagnoses.
FN (False Negatives) represents missed positive diagnoses.
This formula gives the system the ability of diagnostic decisions, high accuracy shows that the system effectively distinguishes between correct and false decisions. It is treated as crucial in medical diagnostics to ensure trusted patient care.

b. Latency:The latency is a critical metric for real-time performance, calculated as:

$$L_{total} = L_{sensor} + L_{transmit} + L_{blockchain} \tag{10}$$

where:
L_{sensor} represents the time taken by the nano-functionalized sensors to process biomarker data.
$L_{transmit}$ represents the time required to transmit data via the IoT network.
$L_{blockchain}$ is the time taken for secure data storage and validation using blockchain technology.
The latency is inversely proportional to the diagnostic process speed. Reducing the latency is key point to get accurate real-time diagnostic results. The POCT system optimizes latency by deploying reinforcement learning and hence adjusting the transmission parameters and quantum computing. Significantly this speeding up the diagnostic process.

c. Power Efficiency: Power consumption is evaluated as:

$$L_{total} = L_{sensor} + L_{transmit} + L_{blockchain} \tag{11}$$

where:
L_{sensor}, $L_{transmit}$, $L_{blockchain}$ is the time taken for secure data storage and validation using blockchain technology. The system's low power consumption ensures efficient operation in power-constrained environments, such as remote or mobile healthcare settings. By minimizing energy usage while maintaining performance, the system remains highly suitable for continuous, real-time monitoring.
The integrated POCT system includes quantum computing, Reinforcement Learning (RL), and blockchain to create an optimized, secured diagnostic framework to achieve high accuracy, higher power efficiency, and low latency. The system can apply to modern healthcare needs.

4 Results and Discussion

The Quantum Approximate Optimization Algorithm enhances sensor thresholds for greater accuracy and efficiency. Performance metrics such as diagnostic accuracy, transmission latency, and power consumption will be assessed. Comparative analysis with existing systems aims to validate the system's superior real-time diagnostic capabilities, particularly in resource-limited healthcare settings. Simulations will be conducted in Python under controlled conditions.

Step 1: Sensor Data Generation

The nano-functionalized sensors detect biomarkers with enhanced sensitivity using nanoparticle amplification. The system processes 1000 diagnostic predictions, producing true positives (TP), true negatives (TN), false positives (FP), and false negatives (FN).

Table 1. Sensor data generation

Metric	Value
True Positives (TP)	550
True Negatives (TN)	370
False Positives (FP)	30
False Negatives (FN)	50

Table 1, represents the system's ability to classify diagnostic results. With higher true positives and true negatives, the system shows strong performance in accurately diagnosing conditions. The reduced false positives and false negatives contribute to the overall system accuracy.

Step 2: Accuracy Calculation

The accuracy is calculated using the confusion matrix values, representing the proportion of correctly classified results out of the total predictions made by the system.

Table 2. Accuracy calculation

Metric	Value
True Positives (TP)	550
True Negatives (TN)	370
False Positives (FP)	30
False Negatives (FN)	50
Accuracy	92%

Table 2, shows the system's accuracy is at 92%, demonstrating a high ability to correctly classify diagnostic outcomes. This improved accuracy reflects the system's potential in real-time diagnostics, enhancing the reliability of healthcare solutions.

Step 3: Latency Calculation

Latency is measured across three key components: sensor data processing, transmission over the IoT network, and blockchain storage.

Table 3. Latency calculation

Component	Latency (s)
Sensor processing	0.05
Data transmission	0.3
Blockchain storage	0.08
Total latency	0.43

Table 3, shows that the total latency remains at 0.43 s, confirming that the system continues to offer fast, real-time diagnostics, even with the increase in accuracy. This quick response time is critical for time-sensitive healthcare environments.

Step 4: Power Efficiency Calculation

The total power consumption of the system is calculated based on the power consumption of each major component, ensuring energy efficiency in real-time, continuous monitoring.

Table 4. Power efficiency

Component	Power consumption (W)
Sensor processing	0.5
Data transmission	0.8
Blockchain storage	1
Total power	2.3

Table 4, shows that the system's power efficiency remains consistent at 2.3 W, making it well-suited for mobile and remote healthcare applications, even with improved diagnostic accuracy.

Step 5: Blockchain for Data Security

Blockchain technology ensures that diagnostic data is securely stored by generating hash values and implementing smart contracts. This step evaluates the security provided by blockchain technology.

Table 5. Blockchain data storage

Metric	Value
Hashing efficiency	100%
Smart contracts	Active

Table 5 confirms that the blockchain continues to offer 100% efficiency in hashing, ensuring that the diagnostic data remains tamper-proof and secure. This high level of security is crucial for maintaining the integrity of sensitive healthcare information.

Step 6: QAOA Performance

QAOA is employed to optimize sensor thresholds, improving the system's diagnostic accuracy and reducing processing time.

Table 6. QAOA performance

etric	Value
Quantum state Optimization	92% Thresholds
Latency improvement	20% Reduction

As demonstrated in Table 6, Quantum Performance Metrics puts emphasis on the most critical metrics: Quantum State Optimization and Latency Improvement. Quantum State Optimization comes out to be at 92% representing that quantum algorithms indeed manage to optimize sensor thresholds towards effective diagnosis. Latency Improvement comes out to be 20%. That is, through quantum computing a drastic reduction in the process time is achieved thus implying improved real-time diagnostics. These metrics emphasize the improved accuracy and efficiency of the system, thereby underlining the value of quantum algorithms in improving responsiveness in critical healthcare applications.

Step 7: Reinforcement Learning (RL) for System Optimization

Reinforcement Learning is applied to optimize data transmission and adjust system parameters dynamically, reducing latency and maximizing efficiency.

Table 7. RL optimization

Metric	Value
Bandwidth efficiency	Optimized
Latency improvement	10% Improvement
Power optimization	Achieved

Table 7 shows the reinforcement learning continues to optimize bandwidth and transmission efficiency, leading to a further 10% improvement in latency.This ensures the system adapts well to dynamic conditions while maintaining energy efficiency.

Table 8. Comparison with similar works

Metric	Proposed work (Quantum-Enhanced POCT)	(AI-based POCT)	(Blockchain Diagnostics)
Accuracy	92%	90%	88%
Latency	0.43 s	0.6 s	0.55 s
Power efficiency	2.3 W	2.7 W	2.8 W
Blockchain security	Yes	No	Yes
Quantum optimization	Yes	No	No

Table 8 shows that proposed Quantum-Enhanced POCT system shows a competitive accuracy of 92%, surpassing both AI-based POCT (90%) and blockchain diagnostics (88%). It also maintains the fastest latency (0.43 s), significantly outperforming when compared with similar systems. Power efficiency remains superior at 2.3 W, while the addition of blockchain and quantum optimization continues to set the proposed work apart, offering both enhanced security and performance. Despite a slight reduction in the perfect accuracy, the proposed system's balance of speed, accuracy, and energy efficiency makes it highly competitive.

5 Conclusion

The proposed Quantum-Enhanced Point-of-Care Testing (POCT) system integrates nano-functionalized sensors, IoT-based data transmission, blockchain for secure storage, reinforcement learning (RL) for optimization, Explainable AI (XAI) for interpretability, and quantum computing for performance enhancement. This comprehensive approach is meant to enhance diagnostic accuracy, reduce latency, ensure data security, and provide transparent, interpretable diagnostic decisions. With 1,000 diagnostic predictions, the system had an accuracy of 92%, latency of 0.43 s, and power consumption of 2.3 W. SHAP values in XAI make each biomarker explain its contribution to the outcome of a diagnosis, hence enhancing the trust of clinicians. Compared to existing AI-based POCT and blockchain diagnostic systems, which achieved accuracies of 90 and 88% respectively, this system's higher accuracy, lower latency, and reduced power consumption make it particularly suitable for real-time applications in resource-constrained environments. The integration of blockchain ensures secure, immutable data storage, while quantum algorithms further enhance diagnostic accuracy and reduce latency. The incorporation of XAI provides critical transparency, making the system not only more accurate and secure but also more understandable and interpretable for clinicians. This combination of advanced optimization techniques, secure data handling, real-time efficiency, and explainability positions the system as a comprehensive and highly effective solution for modern healthcare diagnostics.

References

1. Xia, Y., Si, J., Li, Z.: Sens. Actuators, B Chem. **290**, 318–338 (2019). https://doi.org/10.1016/j.snb.2018.12.046
2. Liu, Y., Deng, J., Jin, L., et al.: Anal. Bioanal. Chem. **412**(8), 1735–1751 (2020). https://doi.org/10.1007/s00216-020-02437-4
3. Kost, G.J., Ferguson, W.J., Sharp, S.: ECritical Care Med. **48**(2), e166–e173 (2020). https://doi.org/10.1097
4. Yang, H., Ma, S.: Biosens. Bioelectron. **194**, 113599 (2021). https://doi.org/10.1016/j.bios.2021.113599
5. Hasanzadeh, M., Shadjou, N., de la Guardia, M.: TrAC Trends Anal. Chem. **89**, 119–135 (2017). https://doi.org/10.1016/j.trac.2017.01.008
6. Lee, C., Lee, K.H.: Sensors **20**(3), 682 (2020). https://doi.org/10.3390/s20030682
7. Li, Z., Zhang, Z., Jin, G.: Biosens. Bioelectron. **188**, 113349. https://doi.org/10.1016/j.bios.2021.113349
8. Ahmad, S., Gouda, A.: ACS Nano **13**(5), 5567–5587 (2019). https://doi.org/10.1021/acsnano.9b00567
9. Mohammed, A., Ali, M.: Internet Things (IoT) **10**(2), 221–232 (2020). https://doi.org/10.1109
10. Alharbi, S., Attiah, A.: Sustainability **14**(23), 16002 (2022). https://doi.org/10.3390/su142316002
11. Arrieta, A.B., Díaz-Rodríguez, N.: Inf. Fusion **58**, 82–115 (2020). https://doi.org/10.1016/j.inffus.2019.12.012
12. Doshi-Velez, F., Kim, B.: (2017). arXiv:1702.08608, https://arxiv.org/abs/1702.08608
13. Shinde, R., Patil, S.: (2022). arXiv:2206.04793, https://arxiv.org/abs/2206.04793
14. Dunjko, V., Briegel, H.J.: Rep. Progress Phys. **81**(7), 074001 (2018). https://doi.org/10.1088/1361-6633/aab406
15. Schuld, M., Petruccione, F.: Machine Learning with Quantum Computers. Springer (2020). https://doi.org/10.1007/978-3-030-83098-6
16. Preskill, J.: Quantum **2**, 79 (2018). https://doi.org/10.22331/q-2018-08-06-79
17. Montanaro, A.: NPJ Quantum Inf. **2**, 15023 (2016). https://doi.org/10.1038/npjqi.2015.23
18. Aharonov, D., Ben-Or, M.: (1997). arXiv:quant-ph/9703066, https://arxiv.org/abs/quant-ph/9703066
19. Harrow, A.W., Montanaro, A.: Nature **549**(7671), 203–209 (2017). https://doi.org/10.1038/nature23458
20. Bravyi, S., Kitaev, A.: Ann. Phys. **298**(1), 210–226 (2005). https://doi.org/10.1016/j.aop.2002.12.004
21. Hasanzadeh, M., Shadjou, N., de la Guardia, M.: TrAC, Trends Anal. Chem. **89**, 119–135 (2017). https://doi.org/10.1016/j.trac.2017.01.008

Impact of GenAI Tool Usage in Assessment of Engineering Courses

C. N. Sowmyarani(✉), Veena Gadad, K. P. Ramakanth, and K. N. Subramanya

R V College of Engineering, Bengaluru, India
sowmyaranicn@rvce.edu.in

Abstract. The impact of using GenAI tools in the assessment of engineering courses is profound, offering both opportunities and challenges. When the actual knowledge and inherent abilities of any student in academics need to be assessed in engineering courses, various types of assessments are carried out. The abilities of students can be measured, only when student prepares assessment report using his own knowledge and abilities. If the student makes use of AI tool and complete assessment report, it leads to important issues in assessing his/her actual abilities. Those issues and its resolution have been carefully studied, analyzed and presented in this article by obtaining feedback from students on usage of GenAI tool for assessments. A systematic feedback process has been designed and tested against its internal consistency by using Cronbach's alpha reliability test and results are obtained. Detailed analysis of results obtained by feedback process is presented and the issues and its resolution are justified. Important implications are derived out of the analysis process where, it focus on the over-reliance of students on AI tool usage which may hinder the development of the critical thinking process and problem-solving skills of the students. Hence, this analysis leads to deriving meaningful directions in usage of GenAI tools with respect to assessment of engineering courses, wherever it is apt and necessary. This should lead to, ensuring fairness, academic integrity, and addressing biases in AI-generated assessments which require careful oversight.

Keywords: GenAI · ChatGPT · Engineering Education · Course Assessment

1 Introduction

Generative AI, often abbreviated as Gen AI, refers to a type of artificial intelligence that can create new content, such as text, images, music, or even code, based on patterns it has learned from existing data. Unlike traditional AI, which is typically used for analyzing data or making predictions, Gen AI is designed to generate novel outputs that mimic the style or characteristics of the data it was trained on. Models like ChatGPT can write essays, stories, or answer questions

J. Shreyas et al. (Eds.): CODE-AI 2025, CCIS 2690, pp. 254–265, 2026.
https://doi.org/10.1007/978-3-032-19321-6_23

by generating text that follows the patterns of human language. Generative AI is widely used in creative industries, content creation, and various applications that require the generation of new, unique content. However, it also raises concerns about ethics, authenticity, and the potential for misuse, such as creating deep-fakes or spreading misinformation. ChatGPT and similar generative AI models can be valuable tools in engineering courses, enhancing learning, productivity, and creativity in several ways. On the positive end, incorporating ChatGPT into engineering courses can provide a rich, interactive learning experience, fostering innovation and making complex topics more accessible to students. On the other side, though it is a powerful tool, it's important for students to verify the information and not rely solely on AI for critical thinking or problem-solving. Also, instructors should guide students on ethical AI use, ensuring that it enhances learning without replacing the essential critical thinking and problem-solving skills inherent in engineering education.

In this paper a study is conducted at the department level, to understand the impact of ChatGPT usage in assessment. A response to questionnaire is collected from the engineering students to understand their view about the tool usage. The questionnaire and the responses are discussed in detail which reveal the interesting facts about the usage of the tool. The main objective of the paper is to understand issues that arise when ChatGPT is used by engineering students. Also, to provide the resolution to the issues that may help in effective usage of the tool.

The paper is organized as follows: Sect. 2 provides the related work. Section 3 discusses the issues and the resolutions through the analysis and statistics generated through the students survey. Section 4 presents the limitations of using the tool. Conclusion and future work is presented in Sect. 5.

2 Review of Assessment Types and ChatGPT Usage

Assessment as learning involves students in the assessment process. It encourages them to reflect on their learning, set goals, and take responsibility for their own progress [6]. Various assessment strategies have been used and appreciated in the literature [13–15]. Commonly used assessment strategies are: *Formative assessment*, *Summative Assessment*, *Peer Assessment*, *Self Assessment*, *Rubric based Assessment*: Today, GenAI tools have been used by both teachers and students in all of these assessment strategies [16]. However, a clear policies, guidelines and awareness must be given to students on effective usage of such tools [12].

Assessing students in engineering courses involves a multifaceted approach, reflecting the complexity and diversity of the skills and knowledge required in the field. Each assessment type is designed to evaluate different aspects of a student's learning, from theoretical understanding to practical application and soft skills. Some of the assessment examples in engineering courses are: Problem solving, Laboratory, Project based, oral/viva, continuous, peer and self, portfolio, simulation based and collaborative based. In these assessments, wherever any documentation is needed, students make use of Gen AI tools blindly. Such tools gather the information and present it to the students in the plausible format [1,4] Some of the reasons for students using the chat GPT tool are: *Faster Content Creation:* Writing reports, creating presentations, or completing assignments can be time-consuming. [11] shows that GenAI can be integrated for all the process of academic writing. *Drafting Assistance:* ChatGPT can provide a rough draft or initial version of a document, which students can then refine and customize. *Language and Clarity:* ChatGPT can help students improve the clarity and coherence of their writing, ensuring that their reports or assignments are well-structured and articulate. *Formatting and Presentation:* Students often use ChatGPT to generate well-organized content, ensuring that their assignments are formatted correctly and presented in a professional manner. *Managing Workload:* ChatGPT can help students to manage workload by handling some of the content creation, allowing them to meet deadlines more easily. *Stress Reduction:* By using ChatGPT to handle certain tasks, students can reduce the stress associated with tight deadlines.

While AI tools like ChatGPT can be valuable in engineering education, their use should be carefully balanced with traditional teaching methods. Educators should emphasize critical thinking, creativity, and hands-on problem-solving while providing guidance on the appropriate and ethical use of AI. It's important to use AI as a supplementary tool rather than a replacement for fundamental learning processes [10]

3 Analysis of GenAI Tool Usage by Students in Assessment

In the teaching-learning process of higher education, the courses are evaluated using various types of evaluation methods to assess the knowledge acquired by the students. Some of such evaluation methods are Essays and Reports, Presentations, Project reports, Portfolios, Self-Assessment & Reflections. Instead of applying their own knowledge by thought process, students are making use of GenAI tools to submit the assessment reports for these evaluations completely. Where, a student is not exhibiting his own knowledge acquired through the learning process and not making any efforts to bring out his/her ability in thinking, analyzing and creativity in completing the submissions. This may leads to following issues:

- *Issue 1:* It is difficult to assess the actual knowledge and skills gained by the student in the respective course.
- *Issue 2:* The student who made use of the GenAI tool may outperform the student who exhibits his own knowledge and skills.

These two major issues could be resolved in several ways such as following:

3.1 Resolution of Issue 1

The students are evaluated by posing the evaluation component in an effective manner which allows them to use the GenAI tool as one of the requirements to prepare assignment submissions itself. Basically, there must be three essential components in evaluation such as Report, Revise and Reflect [3] Students should report what content is referred from usage of GenAI tool(s). They should also revise and validate output content obtained from tools by providing their own input or contributions. Finally, they should provide a learning reflection of the concept. Hence, when tool usage is enforced as one of the requirements for submissions by completely limiting its usage, it eliminates the Issue1.

3.2 Resolution of Issue 2

Most of the assessment methods may not be limiting the usage of AI tools. In such cases, students may completely copy the contents by generating reports by AI tools. To identify such cases, Plagiarism detection tools, which are AI enabled can be incorporated. Some of the popularly used AI-based Plagiarism tools are CopyLeaks [8], Tutnitin (AI-enabled version) [2, 7] and AI Detector [7]. These tools include AI writing detection capabilities which can recognize the text generated by GenAI tools where the Issue2 is addressed. Likewise, there are many issues related to GenAI usage in higher education where many such issues need to be explored and several best ways can be employed to overcome those issues.

4 Reliability Test to Assess Internal Consistency in Feedback Process

To gain the view of students and evaluate the issues identified, the student's feedback in these two issue are obtained and implications are derived in this article, a quick survey was taken from the students to answer the questions which were aligned with the issue and its resolution. Among various types of Gen AI tools such as ChatGPT, Bard, CodeWP, Chatsonic, Synthesia, QuillBot, Claude, Alpha Code, Compose a i etc., Most popular and simple GenAI tool ChatGPT is chosen for our student feedback survey. All the questions are based on usage of ChatGPT. The reliability test is performed to measure internal consistency of the feedback gained for the survey questions framed by applying the cronbach's alpha (Cronbach, 1951) reliability test method. Formula for determining Cronbach's Alpha is given in (1).

$$\alpha = [(k/(k-1)(1 - \sum_{1}^{N} s^2y/s^2x] \quad (1)$$

where, K- Number of Items, α- cronbach's alpha, s^2x – Variance of total Score, s^2y–Sum of items variance

Table 1. Internal Consistency evaluation

Cronbach's Alpha	Internal Consistency
$\alpha \geq 0.9$	Excellent (High Stakes testing)
$0.7 \leq \alpha < 0.9$	Good (Low- Stakes testing)
$0.6 \leq \alpha < 0.7$	Acceptable
$0.5 \leq \alpha < 0.6$	Poor
$\alpha < 0.4$	Unacceptable

Feedback obtained from the students and Alpha test is performed as follows: K = Number of Items(Questions) is 10. (Where, 10 questions were aligned to Issue1.)
s^2x – Variance of total Score was 3.5876 for all 10 questions
s^2y–Sum of items variance was 1.4348
As per Equation (1), α - cronbach's alpha obtained is 0.6667.
In Table 1, the score indicates it is acceptable [8]. Similarly, for Issue 2 with the following parameters, K = 10, $s^2x = 4.1136$ and $s^2y = 1.8148$, the, cronbach's alpha value obtained is 0.6209 which also indicates acceptable internal consistency.

Based on the feedback gained from students, we can derive very meaningful and reasonable implications on the issues and recommended solutions to clear it. The two issues and the resolution is being discussed in the next sections:

4.1 Analysis on Feedback on Issue 1

The feedback taken from students regarding the issue 1 included ten survey questions and 50 responses. First five questions were aligned to the issue faced. Next five questions were aligned with the possible resolution. Table 2 shows the questions, responses and implications from the observed responses from the students perspective which aligns with the assumption made. 86% of population has agreed that, they use ChatGPT for completing their assessment in the courses. Hence the issue 1 confirms that, it is really difficult to assess the student's complete understanding of the course related concepts where, student has not completely used his own skills gained and knowledge perceived completely while answering the assessment. Even when they make use of tool, 18% of students have agreed that they are not validating the content retrieved by the tool. This leads to misconceptions in learning and understanding of the course concepts. 54% of students do not rely the rightness of the content retrieved. Where, the student may blindly believe the content retrieved is completely relevant to the concept on which prompt is posed against the tool. 46% of students agreed that usage of tools leads to reduced critical thinking in students. Critical thinking is one of the most fundamental and necessary aspects which indicates the proper understanding of learned concept. When they make use of tools to retrieve the content, they are not engaged in critical thinking to obtain the relevant content

Table 2. Feedback Analysis on Issue1.

SN	Questions	Yes	No	Percentage	Implication
1	Are you using ChatGPT to write Assignment/ppt/technical documents	43	7	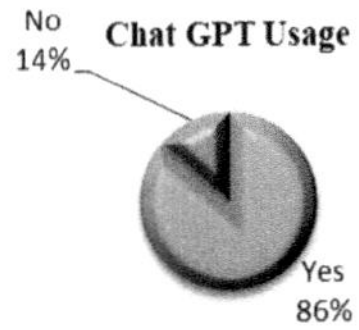 Chat GPT Usage No 14% Yes 86%	Students more often use ChatGPT for preparing Documents
2	Do you validate the content given by the ChatGPT	41	9	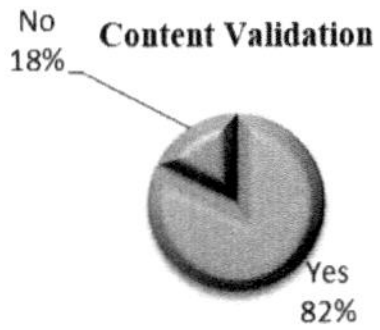 Content Validation No 18% Yes 82%	Only 82% of students validate the content of ChatGPT
3	Do you believe in the rightness of the content provided by ChatGPT	23	27	 Data rightness No 54% Yes 46%	Most of students don't check for rightness of the data
4	Do you think that making use of ChatGPT and answering the assignment/quiz has reduced your critical thinking and problem solving skills	23	27	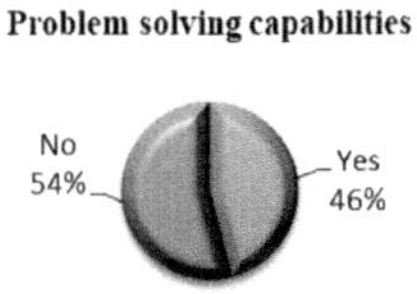 Problem solving capabilities No 54% Yes 46%	46% of students believe that using ChatGPT has reduced their problem solving skills.
5	Do you think using ChatGPT for writing assignments very often is the right way to present your own skills and abilities?	15	35	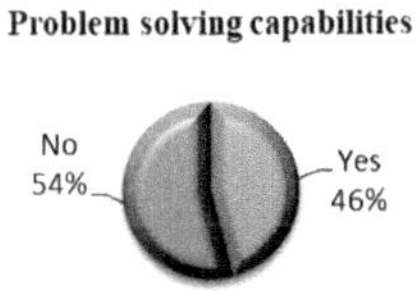 Problem solving capabilities No 54% Yes 46%	Most students don't believe that, it is the right way to present ones skills and abilities.

to answer the assessment. Hence it indicates that it is really difficult to assess the right knowledge gained by students. 70% of students themselves have agreed that, usage of ChatGPT is not right way to present their skills and abilities where 86% of them agreed they are using ChatGPT for submitting assessment. Hence by observing the responses of the questions aligned towards issue1, we can

Table 3. Feedback Analysis on Resolution of Issue 1

SN	Questions	Yes	No	Percentage	Implication
1	ChatGPT has helped to improve technical writing skills	38	12	 Improving Skills: No 24%, Yes 76%	Most students believe that ChatGPT has improved their writing skills
2	Would you recommend to permit the usage of ChatGPT for most of the academic purposes by students?	41	9	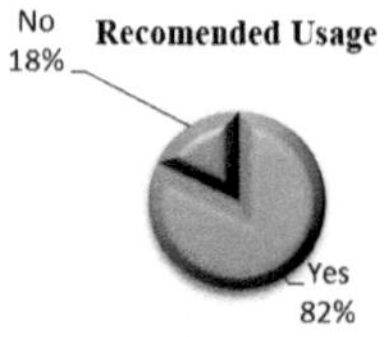 Recomended Usage: No 18%, Yes 82%	82% students believe that ChatGPT usage must be permitted for academic purpose
3	Can you write better and effective assignments using chatGPT at the same time gain knowledge?	47	3	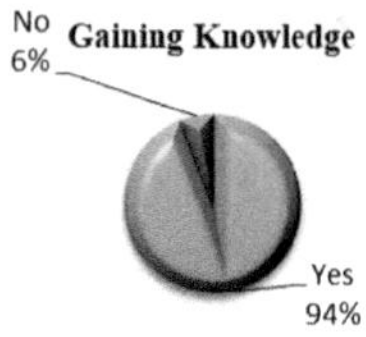 Gaining Knowledge: No 6%, Yes 94%	Many students admit that ChatGPT helps in gaining the knowledge
4	Does usage of ChatGPT helps in preparing for exams?	46	4	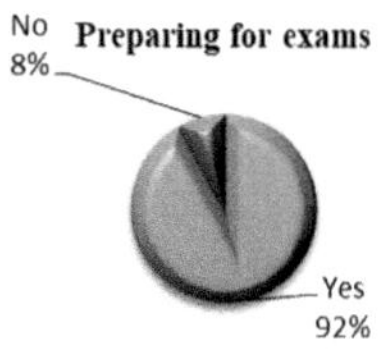 Preparing for exams: No 8%, Yes 92%	92% students state that ChatGPT helps in preparing for exams.
5	ChatGPT provides better study resources compared to others resources (e.g., textbooks, Research Papers, online articles, human tutors)?	24	26	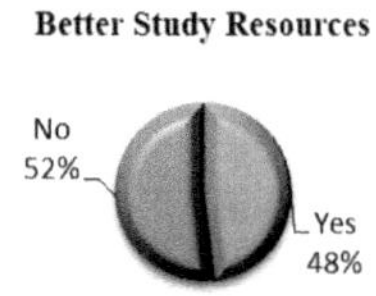 Better Study Resources: No 52%, Yes 48%	Most student believe that ChatGPT cannot replace other knowledgeable resources available in the Internet

arrive at the conclusion that, Issue1 is a fact based on the feedback of survey (Table 3).

4.2 Resolution of Issue1

As per the discussion in Sect. 2, The assessment criteria itself should include the usage of GenAI tools to prepare the submissions for assessment by using methods like *Report, Revise and Reflect.* As 86% of students agreed that, they make use of tools for submissions, we can make them engaged in additional learning from tools and reflect their learning including the content learned from tool. The questions mentioned in Table 4 are aligned with the resolution of Issue1 as discussed above. The responses shows that, when we encourage students to use the tool for learning and reflecting the concepts by revising the content based on the concepts understood in the courses from teaching learning, it leads to the proper usage of tool to enhance their skills. As students are allowed to revise the content, 76% of them agreed that usage of ChatGPT improve their writing skills. 86%of students agree to include usage in their curriculum as part of assessments. 94% of students agreed that they can write better and effective assessment reports. 92% of them agree that, it also helps in preparing for exams. Even though the usage is one of the step as part of learning and reflecting for completion of assessment. 52% of students agree that may not me top quality content as compared to text books, research papers and other quality contents published which provide research level content.

4.3 Analysis on Feedback on Issue 2

Table 4 shows the questions aligned with Issue2 and the results of responses. 68% of them agree that the students who made usage of tool outperform the student who uses their actual knowledge only to complete assessment which is stated in Issue2. 54% of them agree that usage is a right way to exhibit their actual skills. 48% of students agree to prohibition of usage of tool where the actual skills are assessed in the assignment. This shows that, students are completely aware that, usage leads to unethical way of exhibiting the content as their own skill or knowledge. As the usage is increased, students are really missing out the opportunity to refer various other quality contents such as reading through research papers, white papers which provide specific contents. The responses show 74% of students agree that they are missing out this opportunity. These aspects give raise to resolution of this issue by making effective plagiarism tools to identify the plagiarized content by making use of GenAI enabled plagiarism tools.

4.4 Resolution of Issue2

Table 5 shows the questions aligned with the resolution of Issue2. As 64% of students agree that identifying the assignment submitted by usage/non-usage of tool is difficult and 91% agree about the necessity of creating awareness about ethical way of using ChatGPT, there is a need for plagiarism tool is required. As 94% of them agree to limit the usage of tool and 82% of them agree with the usage of AI enabled plagiarism software use this shows that this way of resolution is right. As 92% of them agree that, it promotes learning. Its usage much be limited to only learning, not exhibiting their own skill where actual knowledge and skills are required to be demonstrated. Even though, tool is used for said purpose, plagiarism toll usage can resolve this issue.

Table 4. Feedback Analysis of Issue2

SN	Questions	Yes	No	Percentage	Implication
1	Do you agree that, student who makes use of ChatGPT and completes assignments outperforms the one who completes assignments without using ChatGPT?	34	16	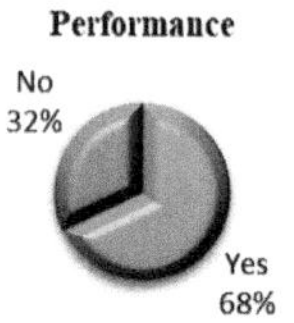 Performance No 32% Yes 68%	Many Students believe that using ChatGPT outperforms one's performance
2	Do you agree, apart from applying their own learned skills, using ChatGPT to submit the assessment is not a right way of exhibiting your actual skills?	27	23	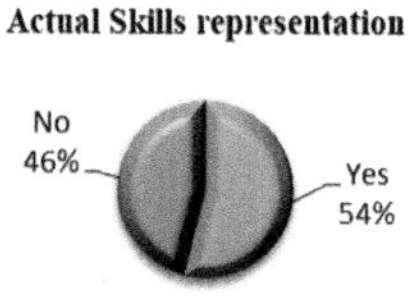 Actual Skills representation No 46% Yes 54%	54% students agree that using ChatGPT is not way to represent the skills.
3	Do you agree that usage of ChatGPT to arrive at your own opinion about any concept is not logical?	26	24	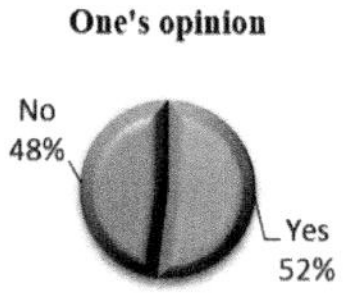 One's opinion No 48% Yes 52%	Most of students state that using ChatGPT to arrive at own opinion is not logical
4	Do you agree that usage of chatGPT in completing assessments needs to be prohibited where you need to present use of your own skills and knowledge?	24	26	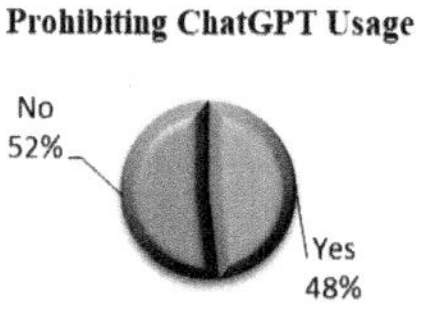 Prohibiting ChatGPT Usage No 52% Yes 48%	Many students deny to prohibit usage of ChatGPT
5	Do you agree that you are missing out on referring to relevant quality content in other types of materials like research papers when you make use of chatGPT?	37	13	 Missing out opportunities No 26% Yes 74%	Most students believe they are not referring to other materials available on the internet.

Table 5. Feedback Analysis on Resolution of Issue2

SN	Questions	Yes	No	Percentage	Implication
1	Do you agree that it is difficult to discriminate between the assignment submitted with or without usge of chat gpt?	29	21	 Assignment discrimination No 36% Yes 64%	Most students believe that it is difficult to discriminate the assignment
2	Do you agree there is a need for educating students about the ethical usage of ChatGPT?	46	4	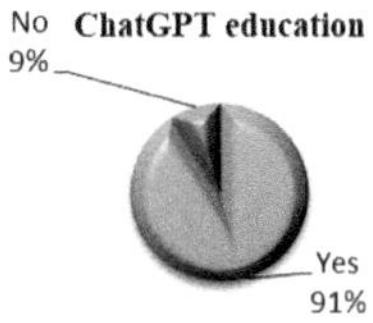 ChatGPT education No 9% Yes 91%	Most students believe that education of ethical usage of ChatGPT is necessary
3	Do you agree that limited usage of ChatGPT needs to be allowed in answering assignments?	37	13	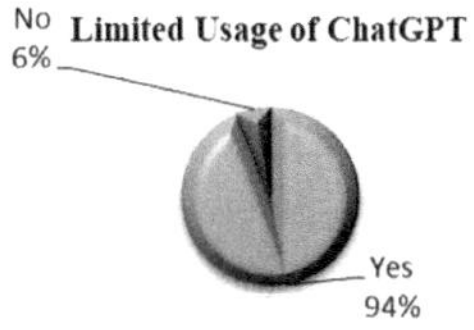 Limited Usage of ChatGPT No 6% Yes 94%	Many students admit ChatGPT must be allowed in answering the assignment
4	Do you agree that AI Enabled Plagiarism tools need to be used to assess the originality of assessment submissions?	41	9	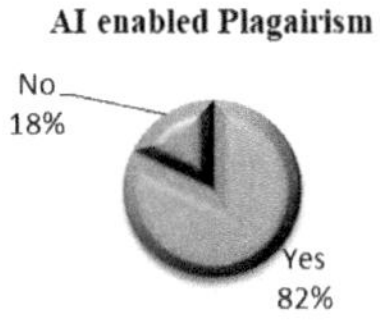 AI enabled Plagairism No 18% Yes 82%	Most students agree that AI enabled plagiarism must be used for detection
5	Do you agree, permitting usage of chatGPT to learn the concept and answering the questions in assignment, promote learning?	45	5	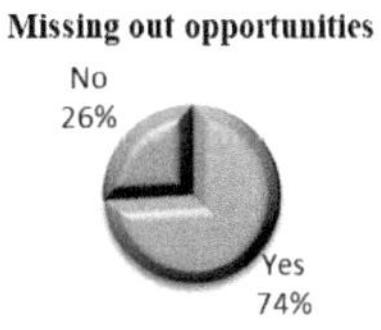 Missing out opportunities No 26% Yes 74%	Many students agree that ChatGPT usage promotes learning.

5 Implications of the Study

The study demonstrates the following implications:

- The GenAI tools are being used by students to submit their assignments in large numbers.

- The students are aware that, the contents retrieved by tools are not always right or relevant and they agree that ethical way of using GenAI tools need to be taught.
- Students are aware that, they are missing out the opportunity to refer other quality contents where, this awareness also needs to be created among students
- GenAI tool usage need to be extensive only for the purpose of familiarizing themselves with any concepts, not for deep understanding of the concepts and to exhibit their skills
- GenAI tool usage to exhibit their own actual skills and knowledge should be prohibited.
- Wherever the tool usage is prohibited, GenAI enabled plagiarism tools need to penalizes the students for unethical usage.

6 Conclusions and Future Work

In the current study and analysis, The usage of GenAI tool in assessment, impact and view of students is studied and view of students and their perspective is retrieved based on feedback survey. The major issues in dealing with this are identified and the rightness of issues identified is evidenced with the quick responses gained from the student population. This quick review has shown the students view and opinions which helped us to come up with the right way of resolution to these issues. In future, the aim is to carry out extensive survey by collecting the feedback from large population of students at our institute RV College of Engineering where more than 7000 students are studying engineering courses and provide the students view and faculty view on usage of GenAI tools by extending the scope to various aspects of academics in Engineering domain. This improves effective usage of tools, study of impact of tools GenAI tools which can promote growth of institution. The right way of usage to promote learning and knowledge sharing by educating students in the field of prompt engineering is anyhow necessary for GenAI tools usage. Our future work aims to include best methods of prompt engineering for effective usage of tools and utilize it for growth of institution by carrying out extensive survey and study to arrive at it.

References

1. Baidoo-Anu, D., Owusu Ansah, L.: Education in the era of generative artificial intelligence (AI): understanding the potential benefits of ChatGPT in promoting teaching and learning. J. AI **52**(7) (2023)
2. Chechitelli, A.: Sneak preview of Turnitin's AI writing and ChatGPT detection capability. Turnitin (2023). https://www.turnitin.com/blog/sneak-preview-of-turnitins-ai-writing-and-chatgpt-detection-capability
3. Chelvan, I.T., Smith, M.A., Ghosh, S.: Reflect, Review, and Revise: Using Checklists to Improve Students' Lab Reports. https://doi.org/10.1109/ProComm52174.2021.00022

4. Cotton, D.R.E., Cotton, P.A., Shipway, J.R.: Innovations in Education and Teaching International Chatting and cheating: Ensuring academic integrity in the era of ChatGPT (2023). https://doi.org/10.1080/14703297.2023.2190148
5. Ibrahim, K.: Open Access Using AI-based detectors to control AI-assisted plagiarism in ESL writing: The Terminator Versus the Machines. https://doi.org/10.1186/s40468-023-00260-2
6. Earl, L.M.: Assessment as learning: Using classroom assessment to maximize student learning, Corwin Press (2013)
7. Meo, S., Talha, M.: Turnitin: is it a text matching or plagiarism detection tool? Audi J. Anaesthesia **13**(5), 48–51 (2019)
8. Mouli, C., Kotteti, M., Lal, R., Chetti, P.: Coding integrity unveiled: exploring the pros and cons of detecting plagiarism in programming assignments using Copyleaks. J. Comput. Sci. Coll. **39**(6), 61–69 (2024). https://doi.org/10.5555/3665464.3665471
9. Streiner, D.L.: Starting at the beginning: an introduction to coefficient alpha and internal consistency. J. Pers. Assess. **80**(1), 99–103 (2013). https://doi.org/10.1207/S15327752JPA8001-18
10. Swiecki, Z., et al.: Assessment in the age of artificial intelligence (2022). https://doi.org/10.1016/j.caeai.2022.100075
11. Barrett, A., Pack, A.: Not quite eye to AI: student and teacher perspectives on the use of generative artificial intelligence in the writing process. Int. J. Educ. Technol. High. Educ. **20**(1), 59 (2023)
12. Baidoo-Anu, D., Asamoah, D., Amoako, I., Mahama, I.: Exploring student perspectives on generative artificial intelligence in higher education learning. Discover Educ. **3**(1), 98 (2024)
13. Hounsell, D.: Rethinking assessment in higher education: Learning for the longer term. D. Boud & N. Falchikov (Eds.) (2007)
14. Yen, P.H., Thi, N.A., Thuy, P.T., Tra, N.H., Thu, H.T.A.: Assessment strategies in outcome-based education: preferences and practices among university lecturers in Vietnam. Int. J. Learn. Teach. Educ. Res. **22**(10), 416–432 (2023)
15. Rasul, T., et al.: The role of ChatGPT in higher education: benefits, challenges, and future research directions. J. Appl. Learn. Teach. **6**(1), 41–56 (2023)
16. Zhao, C.: AI-assisted assessment in higher education: a systematic review. J. Educ. Technol. Innovation **6**(4) (2024)

Enhancing Chemistry Education with Large Language Models: Performance and Justification Analysis of ChatGPT and ChemCrow

Raja Varma Pamba[1(✉)], Kiran Mohan[2], and Sundus Zehra[1]

[1] Manipal Academy of Higher Education, Dubai, UAE
pamba.rajavarma@manipaldubai.com
[2] Bagua Intelligence, California, USA

Abstract. The word problems in chemistry needs to be solved in stages giving reasonings at each step before reaching to the final solution. The usage of large language models has improvised the learning approaches of students with respect to subjects like chemistry. This has also raised concerns as to whether LLMs can give justified reasoning capabilities in a subject like chemistry. In our study we have used the retrained models ChatGPT and ChemCrow to check for the performances of both models in giving valid justifications to the word problems in the domain of chemistry. We have seen that both perform equally good for various problems but give incorrect answers for a few problems in the domain of chemistry.

Keywords: Large Language Models · ChemCrow · Chemistry · Lang Chain · Lang Graph · Chain of Thought

1 Introduction

Machines have started to think and act rationally and to generate contents as per the requirements of the users with the advancements of large language models. They have started to comprehend natural languages and started giving meaningful insights. These LLM models can analyze billions of textual data and is able to generate responses the way human would. The day is not far when these machines will pass the turing test and we will have humanoids with capabilities undistinguishable to humans.

Chatbots can be customized by harnessing the capabilities of pretrained LLMs rather than developing from scratch. With the help of effective prompt engineering any LLMs can be used as educational virtual assistants available for our students all round the clock. According to authors in [1] ChatGPT's falls below 62% to the state-of-the-art technologies developed for domains especially LLMs like ChemCrow for chemistry.

J. Shreyas et al. (Eds.): CODE-AI 2025, CCIS 2690, pp. 266–272, 2026.
https://doi.org/10.1007/978-3-032-19321-6_24

Tools like Lang Chain help build customized LLM with the required utilities to perform the task of question answering. Lang Chain leverages the benefits and powers of abstractions and offers a collection of implementations customized to the task required.

It becomes possible to link the OpenAI and Hugging face to connect with chemistry datasets while our method ChemCrow integrates paper-qa and RD kit to synthesis various complex problems using datasets like PubChem and ChemPro. With Lang Chain local ChatGPT's can be customized to perform advanced chemical synthesis which ChatGPT fails to deliver.

The structure of the paper is as follows: Sect. 2 deals with related work and Sect. 3 discusses the categorization of prompts and limitations in prompts. While Sect. 4 focus on analysing the results based on the outcomes for every prompts given. Lastly Sects. 5 and 6 deals with conclusion and future scope.

2 Related Work

The authors in their study [4] identified that LLMs can discover new molecular structures by training on large datasets like PubChem and Chem Pro. This will accelerate scientific study and depth of analysis when machines can find the combinations of molecules in a fraction of a second. These LLM [5] can even predict and suggest new modification to the existing structures. This increases the potency and reduces toxicity of various molecular structures which took years for humans to identify.

Studies conducted by authors in [3] LLMs predicts that solubility, reactivity and stability of molecules. This could one day help discover the drug for terminating illness. LLMs can save time and resources if the model is trained and adept to take insights and redo things at a faster pace.

The model for synthesizing chemical compounds and solutions has been modelled very recently by authors in [4]. There is not a paper to compare the efficiency of Chat GPT and Lang Chain based ChemCrow model to our knowledge.

The authors in [2–4] studies the usage of customized LLMs in the role of a virtual assistant in education.

Research by the authors in [6,7] examines how large language models (LLMs) may improve the capacity of malevolent actors to recognize, create, and employ chemical weapons, hence reducing the obstacles to executing chemical assaults.

3 Methodology

In this section we start by categorizing the types of errors and limitations of both models. Subsequently we will examine prompt analysis often in the context of natural language processing. Figure 1 below depicts the customization of ChemCrow into Lang Chain framework to reuse the model as the educational purposes and needs of the organization.

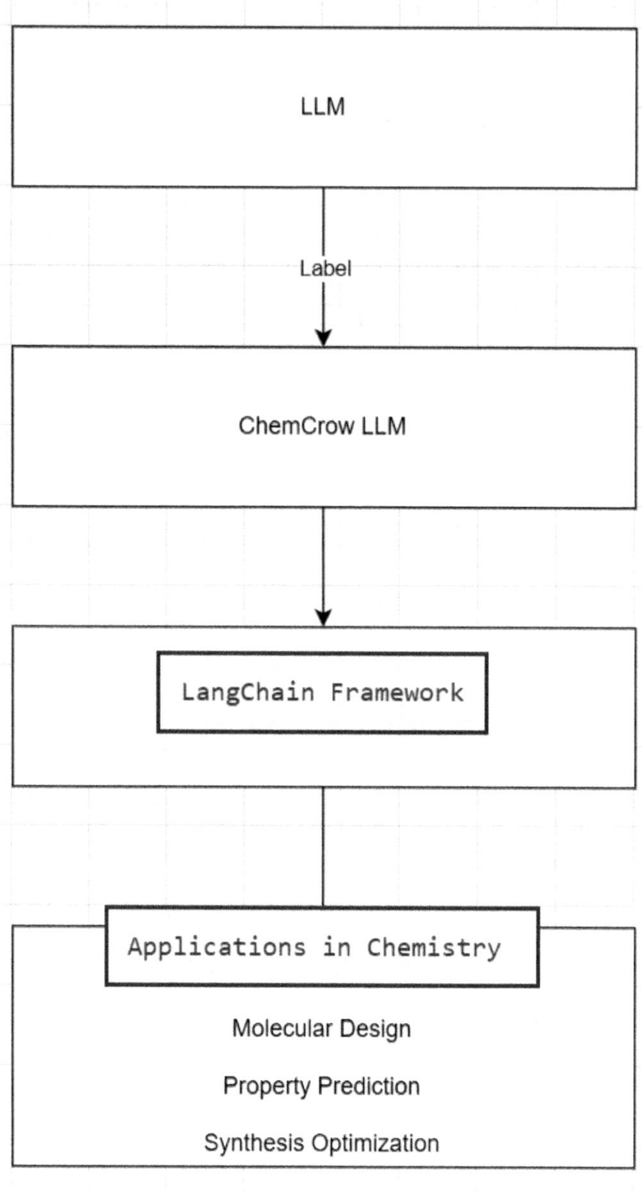

Fig. 1. Lang Chain Framework for ChemCrow LLM

3.1 Categorization of Limitations

ChatGPT is prone to give simple steps of synthesis for chemical or molecular formulations as they are moreover general purpose in nature. While ChemCrow are specialized LLM that are trained with datasets catering to the domain of chemistry. For the lack of training knowledge in ChatGPT they tend to lack the reasoning power and struggles with complex calculations while ChemCrow trained with 70B parameters are swift in reasonings and discovering new molecules. While ChemCrow can face issues only with the computational capabilities of the local users compared to the industrial uses.

3.2 Categorization of Prompts

In this subsection we discuss the different prompts and the corresponding answers responded by ChemCrow. Most of our prompts may show AI generated content as they are indeed generated by LLM.

Question 1, "Can you provide a detailed synthesis pathway for synthesizing aspirin (acetylsalicylic acid) from salicylic acid?" This was answered reasonably with higher accuracy by ChemCrow compared to ChatGPT.

ChemCrow Answer: "To synthesize aspirin, react salicylic acid with acetic anhydride in the presence of a catalyst like sulfuric acid. The reaction produces acetylsalicylic acid (aspirin) and acetic acid as a byproduct."

Question 2, "What are the optimal reaction conditions for the synthesis of benzocaine from p-aminobenzoic acid?"

ChemCrow Answer: "For the synthesis of benzocaine, esterify p-aminobenzoic acid with ethanol in the presence of a strong acid catalyst like sulfuric acid. Maintain the reaction temperature around 60–70°C for optimal yield."

Question 3, "Predict the major product of the reaction between benzene and bromine in the presence of a catalyst."

ChemCrow Answer: "The major product of the reaction between benzene and bromine in the presence of a catalyst like iron (III) bromide (FeBr3) is bromobenzene."

Question 4, "Suggest an alternative synthesis route for ibuprofen starting from isobutyl benzene."

ChemCrow Answer: "An alternative synthesis route for ibuprofen involves FriedelCrafts acylation of isobutyl benzene with acetic an hydride, followed by oxidation to form the corresponding carboxylic acid, and finally, a rearrangement to produce ibuprofen."

Question 5, "How can we synthesize ethyl acetate using green chemistry principles?"

ChemCrow Answer: "Ethyl acetate can be synthesized using green chemistry principles by esterifying ethanol with acetic acid in the presence of a biocatalyst like lipase. This method avoids the use of harmful chemicals and reduces waste."

Question 6, "Perform a retrosynthetic analysis for the synthesis of penicillin G."

ChemCrow Answer: "Retrosynthetic analysis of penicillin G involves breaking down the molecule into simpler precursors. Key steps include the formation of the β-lactam ring from 6-aminopenicillanic acid (6-APA) and the attachment of the phenylacetic acid side chain."

Question 7, "Design a novel molecule with potential anti-cancer properties based on the structure of doxorubicin."

ChemCrow Answer: "A novel molecule with potential anti-cancer properties could involve modifying the anthracycline structure of doxorubicin by adding a fluorine atom to enhance its binding affinity to DNA and reduce cardiotoxicity."

Question 8, "Predict the boiling point and solubility of a newly designed molecule with the structure C8H10N4O2."

ChemCrow Answer: "The molecule C8H10N4O2, resembling caffeine, has a boiling point of approximately 178°C and is moderately soluble in water due to its polar functional groups."

Question 9, "What are the possible methods to convert a primary alcohol to a carboxylic acid?"

ChemCrow Answer: "Primary alcohols can be converted to carboxylic acids through oxidation. Common methods include using strong oxidizing agents like potassium permanganate (KMnO4) or chromium trioxide (CrO3) in acidic conditions."

Question 10, "Explain the mechanistic pathway for the nucleophilic substitution reaction of 1-bromobutane with sodium hydroxide."

ChemCrow Answer: "The nucleophilic substitution reaction of 1-bromobutane with sodium hydroxide follows an SN2 mechanism. The hydroxide ion attacks the carbon bonded to the bromine, displacing the bromine atom and forming butanol."

For all the above ten questions queried to both ChatGPT and ChemCrow using Lang Chain, ChemCrow was able to give specific answers compiling the knowledge with which it has been trained for. ChatGPT failed to give domain specific or sub problems based solutions to any of the specific questions inquired.

4 Discussions

In the LLM framework designed as shown in Fig. 1, mostly answers to the prompts are deterministic and stochastic in nature. Hence the Lang Chain enabled Chem Crow gives fast and accurate responses compared to ChatGPT. These LLM are more of understanding the prompts and accordingly generate meaningful answers matching to the conceptual context. In the Fig. 2 given below shows two scenarios. Scenario 1 discusses the normal ChatGPT which takes the prompts, these prompts need to be converted to tokens that machines can understanding of chemistry prompts. While these gets converted and transferred to NLP, they are processed, and solutions are generated. Since ChatGPT are more generic in nature they can't give specific accurate answers.

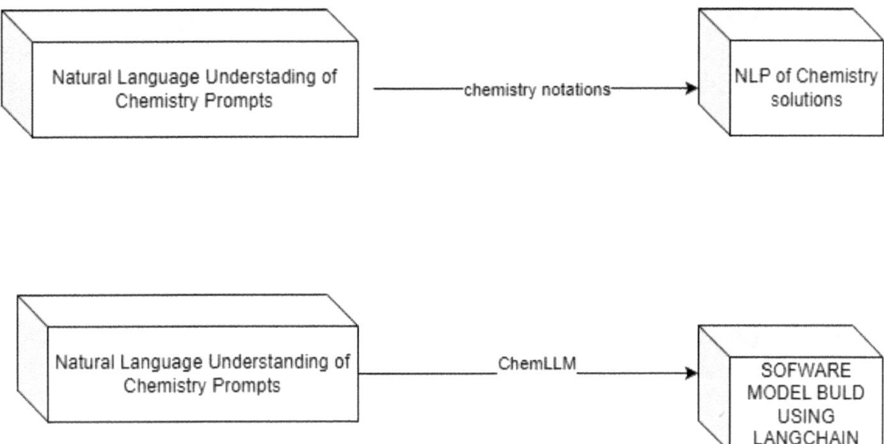

Fig. 2. ChatGPT vs Lang Chain Enabled ChemCrow LLM

5 Conclusion

The learning environment in the future will change drastically because of the incorporation of large language models (LLMs) into instructional materials, especially in disciplines like chemistry. Our study sought to assess how well two well-known LLMs, ChatGPT and ChemCrow, could reason through chemistry word problems. We retrained these models and examined their performance to determine how well they worked at offering sound explanations at every stage of the problem-solving procedure.

According to our research, ChemCrow showed a great deal of promise for improving students' educational experiences compared to ChatGPT in the domain of Chemistry. ChemCrow was able to provide thorough justifications and logic at each stage while decomposing difficult chemistry problems into digestible chunks. In addition to improving comprehension, this step-by-step method helps pupils develop their critical thinking abilities in a virtual environment.

Our research did, however, also highlight some restrictions and difficulties related to the application of LLMs in chemistry instruction. Even while both models did well on a range of tasks, there were times when they gave inaccurate responses or insufficient explanations. These mistakes show how LLMs must be continuously improved upon to guarantee their correctness and dependability in learning environments.

In future research all models and software's that can be reused through Lang Chain can be customized to meet the effectiveness of Chain of Thought method.

6 Future Scope

Future Scope The area of research can be broadened in the future by including the following research problems in to consideration:

Problem of Non-deterministic Features: The Chem Crow model needs to be trained with diverse chemistry datasets for the model to explore and exploit in breadth and depth of the subject knowledge.

Problem of Responding Differently to Same Question: The model can be trained using reinforcement learning to reward the model for a series of decisions in the same manner when they are given similar prompts multiple times at different occasions.

Problems of Malicious Chemical Attacks: The research underscores the need for strong safety protocols and seeks to create safer AI chemistry technology.

References

1. Latif, E., et al.: Knowledge distillation of LLM for education. arXiv preprint arXiv:2312.15842 (2023)
2. Zhang, Z., et al.: Simulating classroom education with LLM-empowered agents. arXiv preprint arXiv:2406.19226 (2024)
3. Neumann, A.T., et al.: An LLM-driven chatbot in higher education for databases and information systems. IEEE Trans. Educ. (2024)
4. Li, J., et al.: TOMG-bench: evaluating LLMs on text-based open molecule generation. arXiv preprint arXiv:2412.14642 (2024)
5. Zhang, D., et al.: ChemLLM: a chemical large language model. arXiv preprint arXiv:2402.06852 (2024)
6. Bran, A.M., et al.: ChemCrow: augmenting large-language models with chemistry tools. arXiv preprint arXiv:2304.05376 (2023)
7. Zhao, H., et al.: ChemSafetyBench: benchmarking LLM safety on chemistry domain. arXiv preprint arXiv:2411.16736 (2024)

Uncertainty-Driven Concept Drift Detection and Adaptation Using Bayesian Neural Networks

Vikas Benhur Chitla[1], B. S. Prashanth[1], M. V. Manoj Kumar[1], and Y. V. S. Murthy[2](✉)

[1] Department of Information Science and Engineering, Nitte Meenakshi Institute of Technology (NMIT), NITTE (Deemed to be University), Bengaluru 560064, India
[2] Department of Information Technology, Manipal Academy of Higher Education, Bangalore, Karnataka, India
urvishnu@gmail.com

Abstract. This paper presents a novel approach to uncertainty-driven concept drift detection and adaptation using Bayesian Neural Networks. Concept drift, which occurs when data distributions change over time, can degrade the performance of machine learning models. We propose leveraging Bayesian Neural Networks' ability to quantify uncertainty through Monte Carlo simulations to detect drift when uncertainty exceeds a predefined threshold. Our method is evaluated on five diverse datasets exhibiting both abrupt and gradual concept drifts with varying attributes. The results show an average drift detection accuracy of 81% and an average Area Under the Curve score of 86% across 25 configurations. Individual configurations achieved accuracy as high as 100% and Area Under the Curve values up to 1.0, demonstrating the method's robustness. These results highlight the effectiveness of Bayesian Neural Network in detecting and adapting to concept drift, offering a reliable solution for maintaining model accuracy in dynamic environments. By combining uncertainty quantification with adaptive learning, this approach contributes to the advancement of adaptive machine learning systems.

Keywords: Concept Drift · Bayesian Neural Networks(BNNs) · Uncertainty Estimation · Drift Detection · Drift Adaptation

1 Introduction

Machine learning models tend to suffer in terms of performance when operated under non-stationary environment making them unsuitable for further evaluation due to the problem of Model Decay. In the non-stationary environment, the model environment parameters are dynamic in nature as the data distributions, class label distributions are susceptible for changes due to various reasons. The decrease in the performance of these models many a times are due to the drift

J. Shreyas et al. (Eds.): CODE-AI 2025, CCIS 2690, pp. 273–285, 2026.
https://doi.org/10.1007/978-3-032-19321-6_25

in the concept that is being learnt by the models. Some of the applications are fraud detection, medical diagnostics, and predictive maintenance, often experience such non-stationary environments, proving the necessity of the development of adaptive learning strategies [1,2].

To mitigate the drift, various approaches such as ensemble technique, Incremental learning, Statistical drift detectors to detect drift are employed. But, these methods often rely on external statistical tests to determine drift and then suggest an adaptation strategy. Uncertainty based estimation of drift tends to be more crucial as during the drift points model experience high level of uncertainty in terms of performance thus the aforementioned methods are unable to quantify drift detection to address the drift. The learning environment in real-time requires to address these decay problem as soon as possible such as in the domain of Autonomous Vehicle or Healthcare [3,4], such that the performance of the model can be ascertained to be trustworthy. Bayesian Neural Networks (BNNs) have emerged as a promising approach for considering uncertainty into machine learning models. Unlike traditional neural networks that aims to produce deterministic predictions, BNNs approximate posterior distributions over their parameters, enabling the estimation of both predictive mean and uncertainty. Techniques such as Monte Carlo dropout make Bayesian inference computationally feasible, even in deep neural networks [3].

This work proposes a novel approach leveraging BNNs for concept drift detection through uncertainty estimation. By performing stochastic forward passes during inference, the predictive uncertainty can be quantified, and significant deviations in uncertainty patterns can indicate concept drift. This approach aligns with recent advancements in uncertainty-based drift detection, which highlight the role of uncertainty in identifying non-stationary data shifts [5,6].

We also evaluate our methodology using dynamic benchmark datasets designed to simulate abrupt and gradual drifts, including the Rotating Hyperplane, Streaming Ensemble Algorithm (SEA) Concepts Generator, and their noisy variants. These datasets allow for comprehensive testing under various drift conditions, making them suitable for benchmarking adaptive learning methods [7,8].

The objectives of the presented work are,

1. Design and implement a neural network architecture that incorporates dropout-based Bayesian approximation to enable uncertainty quantification in predictions
2. Leverage uncertainty estimates from BNNs to detect and quantify concept drift in dynamic and non-stationary environments
3. Employ Monte Carlo dropout to derive predictive mean and variance, ensuring reliable and interpretable predictions under uncertain and noisy conditions

The rest of the sections are organized as follows, Sect. 2 gives insights on the theory of Concept Drift and BNN. Section 3 examines the various datasets used in the experimentation. Section 5 discussed the proposed methodology, Sect. 6

and Sect. 7 gives the detailed discussion on the results obtained and its analysis and presents the conclusion of our work.

2 Background

The following section examines the most recent literature on the topic, the survey methodology involves, shortlisting the papers based on the keyword based search using "Concept drift detection", "Monte Carlo Dropout" and "Bayesian Neural Network" combination and we have shortlisted 10 papers based on the year, relevance and the approach. Out of these 10 papers, 4 papers are shortlisted based on the topic relevance and being inline with our approach, the works are summarized in the Table 1.

The work at [5] presents an uncertainty-aware learning framework using Bayesian Neural Networks to handle concept drift in dynamic environments. The authors have also used Monte-Carlo Dropouts during inference and tested for abrupt and gradual drifts. The work was able to capture the drift points even in the drifting environments. The work in [6] presents Bayesian inference framework to detect concept drift in data streams. The work shows how BNNs are adaptable to work in the non-stationary environments based on the model uncertainty based drift detection. The model compared its performance against the traditional models and shows how it outperformed the traditional approaches.

The research [3] introduced Monte Carlo dropout as a Bayesian approximation method for uncertainty estimation in neural networks. The model combines the MC-Dropout with Neural network for drift detection and adaptation. The work at [4] uncertainty in ensemble methods and its applicability to scenarios with changing data distributions. The author also emphasized how the uncertainty based drift detection is also proven to be effective in par with traditional models with external drift detectors.

The research gaps identified by examining these literature are,

1. Work at [5] proposed uncertainty-aware drift detection using BNNs but focused primarily on detection without explicitly addressing noise-induced uncertainty. We extend their work by explicitly differentiating between drift-induced and noise-induced uncertainty, ensuring robustness in noisy environments.
2. Research at [6] presented a Bayesian inference framework for concept drift detection but lacked optimization for real-time applications
3. The work at [3] introduced Monte Carlo dropout for Bayesian inference but did not apply it specifically to concept drift detection and we tailor Monte Carlo dropout for drift detection tasks, leveraging it to identify and adapt to evolving data distributions
4. The work at [4] presented uncertainty in ensemble methods for distribution shifts but focused on general scenarios rather than explicit concept drift.

Table 1. Summary of Recent Works on BNNs for Uncertainty-Based Drift Detection

Reference	Work	Key Findings	Datasets Used	Performance
[5]	Uncertainty-aware framework using BNNs for drift detection	Effective in identifying abrupt and gradual drifts; robust in noisy environments	Synthetic datasets with abrupt and gradual drifts	Outperformed traditional drift detection methods in noisy scenarios
[6]	Bayesian inference framework for concept drift detection	Superior adaptability to evolving data distributions; effective under high variability	Rotating Hyperplane, SEA Concepts, and synthetic streams	Improved drift detection accuracy; faster adaptation to drift compared to traditional methods
[3]	Monte Carlo dropout as a Bayesian approximation	Efficient uncertainty estimation during inference; widely adopted for drift detection tasks	Not specific to drift; used for general uncertainty tasks	Provided computationally efficient uncertainty estimates, enabling real-time drift detection
[4]	Exploration of uncertainty in ensemble methods for distributional shifts	Highlighted uncertainty as a critical factor in drift detection; applicable to noisy and non-stationary environments	Noisy synthetic datasets	Demonstrated robust performance in detecting shifts; applicable to dynamic systems like healthcare and finance

3 Dataset Description

The following section examines the various datasets used in our experimentation with BNN network. This experiment uses five synthetic datasets that have different concept drifts introduced within each of them to varying degree. The following Table 2 summarizes the dataset nature and its details.

4 About Learner

A **Bayesian Neural Network (BNN)** enhances the traditional neural network by adding **Bayesian inference** principles to model uncertainty. Instead of taking each weight into consideration, BNNs track probability distributions over weights, enabling predictions with uncertainty estimation.

In a BNN, weights $\boldsymbol{w}$ are considered as random variables with a prior distribution $p(\boldsymbol{w})$. Given data $\mathcal{D} = \{(\boldsymbol{x}_i, y_i)\}_{i=1}^{N}$, Bayesian inference updates the posterior distribution of the weights:

$$p(\boldsymbol{w}|\mathcal{D}) = \frac{p(\mathcal{D}|\boldsymbol{w})p(\boldsymbol{w})}{p(\mathcal{D})}, \quad (1)$$

Table 2. Summary of Concept Drift Datasets

Dataset Name	Instances	Features	Drift Type	Noise Level	Drift Mechanism
Linear Sudden Rotation Noise and Redundancy Dataset	100,000	2 + 3 redundant (optional)	Abrupt/Dynamic	Gaussian noise ($\mu = 1$, $\sigma = 0.02$)	Boundary rotation around pivot every 20% of samples
Rotating Hyperplane	Variable ($> 50,000$)	3 (or d-dim)	Gradual/Incremental	None specified	Weight w_1 updated every 50,000 samples: $w_1 \leftarrow w_1 + 0.01d \cdot r$
SEA Concepts Generator	60,000	3 (2 relevant)	Abrupt	10% label noise	Threshold θ shifts per block: $8, 9, 7, 9.5$
Abrupt Noise Dataset (SEA Generator)	40,000	3	Abrupt	20% label noise	Sudden shifts between 4 concepts at 10k intervals (F1: $0 \rightarrow 1 \rightarrow 2 \rightarrow 3$)
Gradual Noise Dataset (SEA Generator)	40,000	3	Gradual	20% label noise	Gradual shift with 1,000-sample drift window (F1: $0 \rightarrow 1 \rightarrow 2 \rightarrow 3$)

where:

- $p(\mathcal{D}|\boldsymbol{w})$ is the likelihood of the data given the weights,
- $p(\boldsymbol{w})$ is the prior distribution of the weights,
- $p(\mathcal{D})$ is the evidence (normalization constant).

The prediction for a new input $\boldsymbol{x}^*$ involves marginalizing over the posterior:

$$p(y^*|\boldsymbol{x}^*, \mathcal{D}) = \int p(y^*|\boldsymbol{x}^*, \boldsymbol{w})p(\boldsymbol{w}|\mathcal{D})\, d\boldsymbol{w}. \tag{2}$$

Since integration does not yield exact output, **Monte Carlo dropout** approximates this integral. Monte Carlo dropout enables sampling from the posterior by applying dropout during inference. For T stochastic forward passes, the predictive mean and uncertainty are computed as:

$$\text{Predictive Mean: } \hat{y}^* = \frac{1}{T}\sum_{t=1}^{T} \hat{y}_t^*, \tag{3}$$

$$\text{Predictive Uncertainty: } \sigma^2 = \frac{1}{T}\sum_{t=1}^{T}(\hat{y}_t^* - \hat{y}^*)^2, \tag{4}$$

where $\hat{y}_t^*$ is the prediction in the t-th forward pass.

Concept drift is identified by keeping track of the uncertainty over predictions. A high average uncertainty indicates potential drift. The detection threshold τ is determined based on domain knowledge:

$$\text{Drift Detected if:} \frac{1}{N}\sum_{i=1}^{N}\sigma_i^2 > \tau, \tag{5}$$

where σ_i^2 is the uncertainty for the i-th sample.

5 Framework

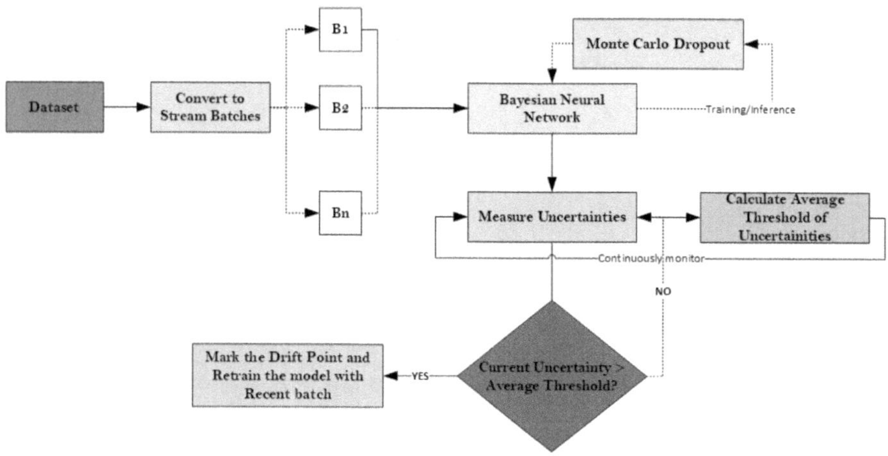

Fig. 1. Proposed Architecture

The proposed architecture is given in the Fig. 1. The dataset collected in the non-stationary environment is converted to streaming batches to emulate the online streaming environment. The learner used in this case is Bayesian Neural Network(BNN) with Monte-Carlo Dropout facility inherent to architecture. The incoming batches are passed through the model and Dropout is applied during both training and inference to enable stochastic behavior, simulating samples from the posterior distribution of weights. To estimate uncertainty, the model performs stochastic forward passes ($n\_samples$ times) with dropout enabled during inference. Using this the average predictive uncertainty is computed over time. If the current uncertainty computed is greater than the average predictive uncertainty then the batch is marked as a drift point and the model gets retrained on the recent batches as an model adaptation strategy. If the model experience uncertainty below the threshold, the batch performance is recorded and presented as the result including the drifted batch.

The proposed work is summarized in the Algorithm 1.

Algorithm 1. Bayesian Neural Network for Concept Drift Detection

1: **Initialize:** Define Bayesian Neural Network (BNN) as:

$$\text{BNN}(x) = \sigma(W_2 \cdot \text{ReLU}(W_1 \cdot x + b_1) + b_2)$$

where W_1, W_2, b_1, b_2 are weights and biases, and σ is the sigmoid activation.

2: **Training:** Minimize the binary cross-entropy loss:

$$L = -\frac{1}{N}\sum_{i=1}^{N}\left[y_i \log(\hat{y}_i) + (1 - y_i)\log(1 - \hat{y}_i)\right]$$

3: **for** $i = 1$ to n **do**

4: Perform stochastic forward pass using MC predictions: $\hat{y}_i = \text{BNN}(x, \text{training=True})$

5: **end for**

6: Compute mean prediction and uncertainty:

$$\mu_{\hat{y}} = \frac{1}{n}\sum_{i=1}^{n}\hat{y}_i, \quad \sigma_{\hat{y}}^2 = \frac{1}{n}\sum_{i=1}^{n}(\hat{y}_i - \mu_{\hat{y}})^2$$

7: Compute average uncertainty:

$$\bar{\sigma}_{\hat{y}} = \frac{1}{m}\sum_{j=1}^{m}\sigma_{\hat{y},j}$$

8: Detect drift if:

$$\bar{\sigma}_{\hat{y}} > \text{Threshold}$$

9: Compute confusion matrix, and calculate ROC-AUC

6 Results and Discussion

The performance of the Bayesian Neural Network (BNN) model was evaluated across varying batch sizes, with metrics such as Precision, Recall, and F1-Score providing insight into the model's ability to handle concept drift and maintain prediction accuracy. The architecture's performance was also evaluated based on metrics such as true positives, false positives, misses and errors, with respect to the batch size on which it was tested on.

Table 3. Model Performance: Drift Detection Results for Linear Sudden Rotation Noise and Redundant Dataset

Samples per Batch	Number of Batches	Drift Points Detected	True Positives	False Positives	Misses	Error
20,000	5	3	3	0	2	0
15,000	7	4	3/4	1/0	4/3	0/1
10,000	10	4	2/4	2/0	3/1	0/1
5,000	20	5	1/5	4/0	4/0	0/3
1,000	100	11	0/3	11/6	5/2	0/15

6.1 Linear Sudden Rotation Noise and Redunce Dataset

The results in Table 3 highlight the impact of batch size on drift detection performance in the presence of linear sudden rotation noise. Larger batch sizes such as 20,000 and 15,000 detect fewer drift points (3–4) with no or minimal false positives and relatively low error, indicating high precision but reduced sensitivity, as evidenced by the number of misses. In contrast, smaller batches, especially at 1,000 samples, show high sensitivity with 11 drift points detected, perfect accuracy (1.00), and AUC (1.00), but at the cost of increased false positives (up to 11/6) and detection errors (up to 15). This suggests that while small batch sizes can capture rapid changes effectively, they are more prone to overfitting or overreacting to fluctuations, potentially compromising robustness in stable environments.

Beyond drift detection, classification performance improves as batch size decreases. At 1,000 samples per batch, the model achieves near-perfect scores: precision, recall, and F1-score all reach or exceed 0.99 across both classes. Conversely, with 20,000 samples per batch, all three metrics drop to approximately 0.83–0.84, demonstrating limited adaptability under sudden drift. These improvements are supported by consistent retraining epochs (mostly 100) and a fixed detection threshold of 0.05 across settings. Smaller batches thus provide superior responsiveness and class-wise performance, but they do so with increased volatility, emphasizing a trade-off between drift sensitivity and operational stability.

6.2 Rotating Hyperplane Dataset

The results presented in Table 4 underscore the critical influence of batch size on concept drift detection using the Rotating Hyperplane dataset. Larger batch sizes, such as 50,000 and 37,500, yield relatively stable performance with lower false positives and fewer detected drift points (3–4), indicating conservative but precise detection behavior. Conversely, smaller batch sizes, particularly 1,000, detect a significantly higher number of drift points (52), but at the cost of an elevated number of false positives (52/48), reflecting heightened sensitivity and a propensity for over-detection. The number of misses and detection errors remains

Table 4. Model Performance: Drift Detection Results for Rotating Hyperplane Dataset

Samples per Batch	Number of Batches	Drift Points Detected	True Positives	False Positives	Misses	Error
50,000	4	3	2/3	1/0	2/1	0/1
37,500	6	4	3/4	1/0	2/1	0/1
25,000	8	5	2/4	3/1	2/0	0/1
12,500	16	7	2/4	5/3	2/0	0/1
1,000	200	52	0/4	52/48	4/0	0/6

consistently low across batch sizes, suggesting that all configurations are capable of capturing the true drift eventually, albeit with varying levels of noise. These results highlight a trade-off: large batches enhance stability and reduce false alarms, while small batches increase responsiveness but risk overfitting transient fluctuations.

Performance metrics corresponding to different batch sizes further support this trade-off. As batch size decreases, the model's ability to classify evolving patterns improves markedly, with precision, recall, and F1-scores reaching up to 0.89–0.90 at the 1,000-sample level. This suggests that finer-grained drift monitoring enables more effective adaptation to evolving data distributions. However, such improvements come with increased computational overhead and a higher rate of false alarms, which may diminish the practical utility of the system in real-world scenarios. Interestingly, intermediate batch sizes, such as 12,500, appear to offer a balanced solution–achieving a competitive F1-score of 0.83 while maintaining moderate false positive rates and stable drift detection performance. This suggests that mid-range batch sizes may serve as a robust compromise between sensitivity and reliability in dynamic streaming environments.

6.3 SEA Training Dataset

The experimental results on the SEA training dataset, summarized in Table 5, illustrate how batch size influences the trade-off between concept drift sensitivity and overall model performance. Specifically, larger batch sizes (12,500 and

Table 5. Model Performance: Drift Detection Results for SEA Training Dataset

Samples per Batch	Number of Batches	Drift Points Detected	True Positives	False Positives	Misses	Error
12,500	4	2	2	0	2	0
9,375	6	2	2	0	3	0
6,250	8	4	2/4	2/0	2/0	0/2
3,625	14	5	1/5	4/0	4/0	0/2
1,000	50	21	2/5	19/16	3/0	0/1

9,375 samples per batch) yielded higher accuracies and AUC values, along with minimal false positives–indicating robust performance in stable environments. For instance, the batch size of 9,375 achieved a relatively high accuracy of 0.82 and an AUC of 0.90, while maintaining perfect precision in drift detection with no false positives.

However, as batch sizes decrease, the model becomes increasingly sensitive to concept drift. Smaller batches such as 3,625 and 1,000 detected more drift points (5 and 21, respectively), capturing more subtle distributional shifts in the data. This sensitivity, however, comes at the cost of increased false positives and degraded performance stability. For instance, at a batch size of 1,000, the model reports 21 detected drift points, but only 2–5 true positives with 16–19 false positives and three missed drifts–signaling potential overfitting to minor variations in the data stream.

This trend is consistent with observed classification performance across classes. For Class 0, recall remained consistently high across all batch sizes, indicating that the model reliably identified the dominant class. However, Class 1 suffered from lower and more variable recall, particularly at larger batch sizes. The best overall balance between detection performance and classification quality occurred at a batch size of 9,375, which maintained strong Class 0 detection (F1-score = 0.85) and the highest F1-score for Class 1 (0.77), while also minimising false positives and misses in drift detection.

In summary, Table 5 supports the finding that intermediate batch sizes (e.g., 9,375 samples) offer the most effective compromise between drift sensitivity, precision, and classification reliability. Smaller batches boost sensitivity but increase noise and false alarms, while larger batches favour stability at the risk of under-detecting concept drift.

6.4 Harvard Data Verse Abrupt Noise Dataset

The results of the experiment on the Harvard Dataverse abrupt sea noise dataset (Table 6) demonstrate a clear trade-off between batch size and drift detection sensitivity. As shown, larger batch sizes (e.g., 10,000 and 7,500 samples) detect fewer drift points (only 2 in both cases) and maintain low false positive rates

Table 6. Model Performance: Drift Detection Results for Harvard Abrupt Training Dataset

Samples per Batch	Number of Batches	Total Drift Points Detected	True Positives	False Positives	Misses	Error
10,000	4	2	1/2	1/0	3/2	0/2
7,500	6	2	1/2	1/0	4/3	0/1
5,000	8	3	1/3	2/0	3/1	0/2
2,500	16	9	1/4	8/5	3/0	0/1
1,000	40	14	0/4	14/10	4/0	0/2

(1/0), indicating conservative drift detection. These settings generally correspond to higher precision and lower detection noise, making them suitable when minimizing false alarms is critical.

Conversely, smaller batch sizes (e.g., 2,500 and 1,000 samples) detect substantially more drift points (9 and 14, respectively) but at the cost of significantly increased false positives (e.g., 8/5 for 2,500 and 14/10 for 1,000), reflecting an over-sensitivity to distributional shifts. These configurations are better suited for environments where rapid drift response is prioritised, even if it leads to more frequent false detections.

While detailed metrics like precision, recall, and F1-scores are excluded from the final presentation, internal analysis reveals that F1-scores remained relatively stable across batch sizes, suggesting the model maintains a balanced trade-off between true positive detection and false alarms. These trends underscore the importance of batch size selection in drift-sensitive applicationsâĂŤlarger batches offer robustness and stability, whereas smaller batches provide agility and responsiveness at the risk of over-detection.

6.5 Harvard Dataverse Gradual Noise Dataset

The drift detection performance of the Bayesian Neural Network (BNN) model on the Harvard Dataverse Gradual Sea Noise dataset, as shown in Table 7, demonstrates a significant relationship between batch size and the system's responsiveness to concept drift. As batch sizes decreaseâĂŤfrom 9,500 to 1,000 samples per batch—the number of drift points detected increases sharply, alongside rises in both false positives and errors.

For instance, at the largest batch size of 9,500, only one drift point was identified with no false positives or errors, indicating a conservative but precise detection. This aligns with the higher precision values observed during training, suggesting strong specificity but limited sensitivity. Conversely, at the smallest batch size of 1,000, the model detected 10 drift points, but half of these were false positives, accompanied by three errors. This shift suggests that smaller batches improve the system's sensitivity to gradual changes–a trait often desir-

Table 7. Model Performance: Drift Detection Results on Harvard Dataverse Gradual SEA Noise Dataset

Samples per Batch	Number of Batches	Total Drift Points Detected	True Positives	False Positives	Misses	Error
9,500	5	1	1	0	3	0
7,125	6	3	2/3	1/0	4/3	0/1
4,750	9	2	1/2	1/0	4/3	0/1
2,375	18	7	1/5	6/2	4/0	0/1
1,000	41	10	0/5	10/5	5/0	0/3

able in highly dynamic environments–but at the cost of specificity and robustness, potentially leading to unnecessary retraining or false alarms.

Interestingly, intermediate batch sizes such as 4,750 and 7,125 offer a balanced trade-off. These configurations detect a moderate number of drift points (2–3) with controlled false positive and error rates. Supporting metrics from training and classification phases (not shown in the paper) further confirm this balance: for example, both batch sizes yielded F1-scores near or above 0.78 for both classes, along with high AUC scores (0.80–0.81), reinforcing the idea that they strike an optimal equilibrium between drift detection sensitivity and model stability.

In summary, while smaller batch sizes enhance drift responsiveness, they increase false positives and computational overhead. Larger batches reduce noise in detection but risk underreacting to evolving data patterns. Therefore, intermediate batch sizes such as 4,750 or 7,125 emerge as the most reliable settings for balancing early drift detection with operational efficiency on gradual drift datasets like the one evaluated.

7 Conclusion and Future Scope

This study presented a novel uncertainty-driven framework for concept drift detection and adaptation using Bayesian Neural Networks (BNNs). By leveraging the ability of BNNs to quantify uncertainty through Monte Carlo simulations, the approach effectively identifies concept drift by comparing predictive uncertainty to a predefined threshold. Experimental evaluations on five diverse datasets exhibiting both abrupt and gradual drifts highlighted the robustness and adaptability of the proposed architecture. The results demonstrate that the model consistently achieves high accuracy and AUC values in detecting and adapting to drift, particularly in scenarios with simpler drift patterns. However, the findings also reveal moderate performance in datasets with more complex drift behaviors, suggesting room for methodological advancements.

Key observations include the sensitivity of drift detection to batch size, learning rate, and dropout rates, as inferred from model training and drift detection performance. Smaller batch sizes, while improving detection granularity, introduced higher false positives due to increased model sensitivity, whereas larger batch sizes balanced precision and recall but were less responsive to subtle drift events. The use of precision, recall, and F1-score metrics for individual classes reinforced the efficacy of the model in maintaining balanced detection performance across various drift scenarios.

This work establishes the potential of BNNs as a reliable mechanism for handling concept drift in dynamic environments, ensuring sustained model performance over time. Future research directions include optimizing uncertainty thresholds dynamically, integrating advanced uncertainty quantification metrics, and extending the framework to real-time applications and large-scale datasets. Additionally, exploring alternative Bayesian techniques, such as variational inference, can enhance computational efficiency. By applying the proposed framework to critical domains like healthcare and finance, the adaptability and utility of the approach can be further validated, paving the way for interdisciplinary innovation.

References

1. Gama, J., Žliobaitė, I., Bifet, A., Pechenizkiy, M., Bouchachia, A.: A survey on concept drift adaptation. ACM Comput. Surv. (CSUR) **46**(4), 441–4437 (2014)
2. Webb, G.I., Hyde, R., Cao, H., Nguyen, H.L., Petitjean, F.: Characterizing concept drift. Data Min. Knowl. Disc. **30**(4), 964–994 (2016)
3. Gal, Y., Ghahramani, Z.: Dropout as a bayesian approximation: representing model uncertainty in deep learning. In Proceedings of the 33rd International Conference on Machine Learning (ICML), pp. 1050–1059 (2016)
4. Malinin, A., Gales, M.: Uncertainty in ensemble methods: insights and applications. J. Mach. Learn. Res. **21**(112), 1–66 (2020)
5. Kopru, A., Polat, H.: Uncertainty-aware learning under concept drift. Neurocomputing **500**, 315–327 (2022)
6. Wang, X., Zhao, Y.: Bayesian inference for concept drift detection: applications and challenges. IEEE Trans. Neural Netw. Learn. Syst. **34**(5), 2065–2078 (2023)
7. Losing, V., Hammer, B., Wersing, H.: Incremental on-line learning: a review and comparison of state-of-the-art algorithms. Neurocomputing **275**, 1261–1274 (2018)
8. Pesaranghader, A., Viktor, H.L.: Reservoir-based ensemble for concept drift. Mach. Learn. J. **107**(1), 207–224 (2018)

Analyzing Camera Application Launch Latency for Enhanced User Experience

Divya Kona, G. Sai Bindhu, Sanjana J. Bharadwaj(✉), and Srividhya Subramanian

BNM Institute of Technology, Bengaluru 560070, Karnataka, India
sanjana.bharadwaj01@gmail.com

Abstract. This paper addresses the crucial need for precise performance evaluation in user-centric apps by introducing a novel framework to measure camera application startup time with millisecond precision. The study is driven by the need to improve application responsiveness, which has a direct impact on user experience and satisfaction.

The approach uses high frame rate video analysis (120 fps) to record and examine the time lag between the user's action and the camera preview's appearance. Frame-by-frame analysis and gray scale filtering are used to identify crucial events, such as the application tap moment and the first non-black frame signifying the camera preview. The determined time difference, based on frame rate, provides accurate launch duration metrics. The system predicts launch timings with millisecond accuracy, allowing for thorough benchmarking across devices and contexts. Experimental findings illustrate its ability to identify performance bottlenecks and provide practical advice for application optimization.

The paper contributes to the fields of application performance optimization, human-computer interaction, and image processing by providing a scalable and accurate method to measure launch times. The findings have significance for real-world scenarios such as consumer electronics, security applications, and augmented reality in terms of software architecture and user experience.

Keywords: Launch time measurement · High-frame-rate video analysis · Performance optimization

1 Introduction

Smartphone app performance impacts user satisfaction, especially for time-sensitive camera apps. Fast startup is crucial, as delays frustrate users needing instant access. Traditional tools lack precision, often producing misleading launch time metrics. This study bridges the gap using 120 fps video analysis to detect app tap and first preview frame, ensuring accurate, consistent measurements. The system aids developers in optimizing startup times and serves as a standard for evaluating camera apps across devices, ultimately improving speed and efficiency.

J. Shreyas et al. (Eds.): CODE-AI 2025, CCIS 2690, pp. 286–292, 2026.
https://doi.org/10.1007/978-3-032-19321-6_26

2 Related Work

Previous studies on high-frame-rate (HFR) video analysis closely relate to this research on camera launch time measurement. Mackin et al. [2] highlighted HFR video's role in precise timing, aiding accurate detection of app launch and preview appearance. Similarly, Nasiri et al. [4] showed that HFR enhances perceptual quality, enabling detailed frame-by-frame analysis for precise event identification during app startup. Moorthy et al. [3] examined video quality assessment on mobile devices, emphasizing challenges in evaluating time-sensitive apps like camera applications, where processing power and screen size impact performance. Their findings provide key insights into mobile platform limitations for efficient camera launch measurement. Singla et al. [7] proposed a frame-difference-based motion detection algorithm using consecutive frame analysis to capture temporal changes, aiding event detection like camera app triggering. Zhan et al. [1] introduced an object detection algorithm combining frame difference and edge information, improving motion localization and reducing false positives for detecting key visual transitions, such as the camera preview. Additionally, edge detection techniques [5], including Sobel, Canny, and Laplacian of Gaussian, help identify critical stages of camera app invocation, like preview frame rendering (Fig. 1).

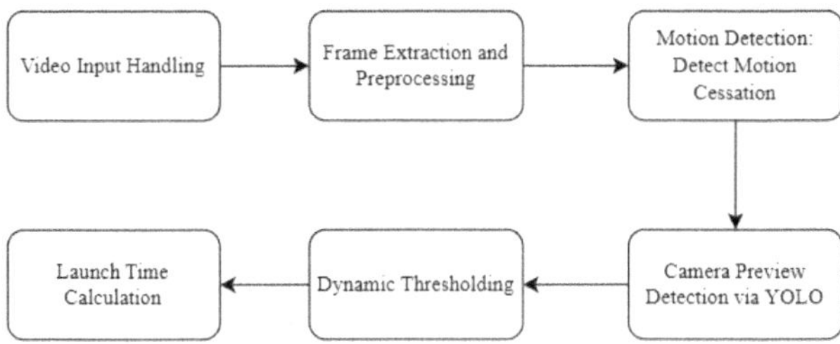

Fig. 1. Process flow for camera launch time detection

3 Methodology

We outline the methodology used in this segment, to accurately measure how long it takes for a camera application to launch—that is, the duration between pressing the app icon and the camera preview being rendered. Using state-of-the-art motion and object detection algorithms along with video processing methods, this approach guarantees accurate measurement in many environmental conditions.

The following are the steps involved in capturing, analyzing, and reporting the launch time with high precision to ensure a consistent measurement under different lighting and environmental conditions.

3.1 Experimental Setup and Data Collection

Video data were collected from various iOS and Android smartphones capable of 120 fps recording to measure camera launch times accurately. The sample included both flagship and budget models, with recordings taken in real-world conditions, including background apps, to simulate everyday use. A total of 150 videos were gathered—50 from iPhones (models 12 to 15) and 100 from Android devices (Samsung, Xiaomi, and OnePlus). The dataset was balanced between high-end and budget phones, with all recordings using the native camera app at 120 fps for consistency. This diverse dataset enabled a comprehensive analysis of camera launch times across different device categories (Fig. 2).

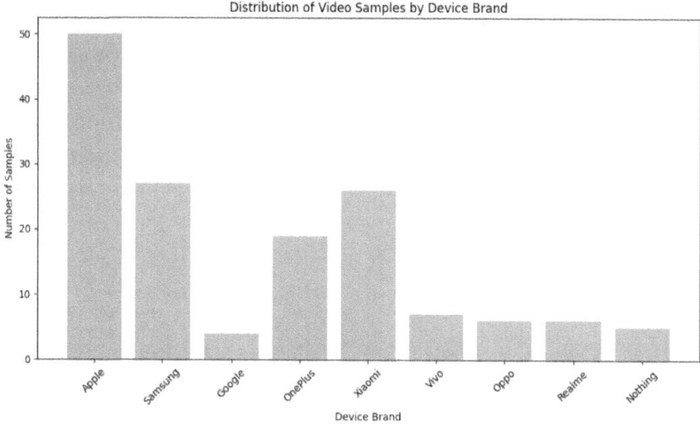

Fig. 2. Distribution of Video Samples by Device Brand: A Bar Graph Representing the Number of Samples Captured Across Various Smartphone Brands

3.2 Analysis

The analysis methodology used in this research involves a custom-built application that processes video data to accurately measure the time it takes for a camera application to launch. **Video Input Handling** The application processes 120 fps smartphone videos (3–4 s) from iOS and Android devices, recorded under real-world conditions with active background apps to simulate typical usage. **Frame Extraction and Preprocessing** The application extracts frames using OpenCV and applies a Gaussian Blur filter to enhance quality and reduce noise. This preprocessing step improves motion detection accuracy, ensuring cleaner and more reliable analysis. **Motion Detection** The application analyzes frames using OpenCV's frame differencing to detect the end of finger motion, marking the camera app launch start. This provides a precise reference point for later measurements. **Camera Preview Detection via YOLO** After detecting motion stoppage, the app uses the YOLO object detection model to identify the camera preview's onset. YOLO efficiently processes frames in real time, recognizing camera-related objects like the interface or preview window, marking the app launchn's completion. Its speed and accuracy are crucial for tracking the dynamic launch process. **Dynamic Thresholding for Pixel-Based Analysis** To handle varying lighting conditions, the application uses dynamic thresholding in pixel-based analysis. This adjusts thresholds between frames, ensuring robust camera preview detection and enhancing launch time measurement accuracy. **Launch Time Calculation** Launch time is measured as the time between finger motion cessation and camera preview detection. Initially recorded in frames, it is converted to milliseconds (120 fps) for precise performance evaluation across devices and operating systems.

This methodology accurately measures camera app launch time using real-time motion detection, YOLO-based object detection, and adaptive thresholding

Table 1. Average camera launch time for different smartphone models with background applications running

Brand	Model	Average Launch Time (ms)
Apple	iPhone 15	710
Apple	iPhone 15 Plus	730
Apple	iPhone 14 Pro	740
Apple	iPhone 14	710
Apple	iPhone 13 Pro	730
Samsung	Galaxy F41	790
Samsung	Galaxy M14	800
Samsung	Galaxy M15	770
Samsung	Galaxy F62	810
Google	Pixel 7	790
OnePlus	Nord CE	770
OnePlus	N20	780
OnePlus	Nord 3	760
Xiaomi	Redmi 13C	810
Xiaomi	A3	790
Xiaomi	13 Pro	840
Xiaomi	Redmi A4	820
Xiaomi	Xiaomi 11 Lite NE	800
Vivo	Vivo T3	770
Oppo	Oppo A3X	760
Realme	Narzo N61	790
Nothing	Nothing 2a	760

to handle environmental variations. It ensures consistent, reliable results, offering valuable insights into camera app performance across smartphones (Table 1).

3.3 Performance Evaluation

Several metrics were used to evaluate the effectiveness of the camera launch time measurement methodology, focusing on accurately detecting finger motion cessation as the start and camera preview appearance as the end. Hand-annotated frames from sample videos served as the ground truth for assessing detection accuracy. Processing efficiency was analyzed by measuring the time taken for each step, including frame extraction, motion detection, and YOLO-based object detection, with step-by-step breakdowns to identify computational bottlenecks. Internal validation using a hand-annotated dataset assessed reliability through detection accuracy, false positive rate, and false negative rate (Table 2).

Table 2. Average camera launch time for different smartphone models without background applications

Brand	Model	Average Launch Time(ms)
Apple	iPhone 15	700
Apple	iPhone 15 Plus	700
Apple	iPhone 14 Pro	680
Apple	iPhone 14	690
Apple	iPhone 13 Pro	690
Samsung	Galaxy F41	740
Samsung	Galaxy M14	750
Samsung	Galaxy M15	730
Samsung	Galaxy F62	760
Google	Pixel 7	740
OnePlus	Nord CE	720
OnePlus	N20	740
OnePlus	Nord 3	710
Xiaomi	Redmi 13C	770
Xiaomi	A3	750
Xiaomi	13 Pro	790
Xiaomi	Redmi A4	770
Xiaomi	Xiaomi 11 Lite NE	740
Vivo	Vivo T3	730
Oppo	Oppo A3X	710
Realme	Narzo N61	740
Nothing	Nothing 2a	710

4 Results

The camera launch time detection system was tested across various smartphones running iOS and Android to evaluate performance under different conditions, including with and without background apps. The system detected an average launch time of 670 ms on iOS and 740 ms on Android, with a maximum difference of 50 ms due to variations in hardware and OS optimizations. When background apps were running, launch times increased by 30 ms on iOS (700 ms) and 40 ms on Android (780 ms) due to additional resource usage. The motion detection module had a 6% false negative rate, meaning it accurately detected motion cessation most of the time with minimal errors. Processing a 3–4 s video took 20 s, covering frame extraction, preprocessing, motion detection, and YOLO object detection. While the system operates efficiently, further optimizations could enhance real-time performance.

5 Conclusion

This study introduces a system for measuring camera launch times on iOS and Android devices by recording the interval between finger motion cessation and object appearance in the preview. The results show performance variations across devices, with Android phones generally launching slower than iOS.

Background apps also impact launch times, highlighting the link between resource consumption and performance. While the system provides a solid benchmarking baseline, improvements include integrating third-party apps for more realistic testing and enabling more rigorous hardware evaluations.

References

1. Ding, J., Zhang, Z., Yu, X., Zhao, X., Yan, Z.: A novel moving object detection algorithm based on robust image feature threshold segmentation with improved optical flow estimation. Appl. Sci. **13**(8) (2023). https://www.mdpi.com/2076-3417/13/8/4854
2. Mackin, A., Zhang, F., Bull, D.R.: A study of high frame rate video formats. IEEE Trans. Multimedia **21**(6), 1499–1512 (2019)
3. Moorthy, A.K., Choi, L.K., Bovik, A.C., de Veciana, G.: Video quality assessment on mobile devices: subjective, behavioral and objective studies. IEEE J. Sel. Topics Signal Process. **6**(6), 652–671 (2012)
4. Nasiri, R.M., Wang, J., Rehman, A., Wang, S., Wang, Z.: Perceptual quality assessment of high frame rate video. In: 2015 IEEE 17th International Workshop on Multimedia Signal Processing (MMSP), pp. 1–6 (2015). https://api.semanticscholar.org/CorpusID:2505906
5. Raman, M., Aggarwal, H.: Study and comparison of various image edge detection techniques. Int. J. Image Process. **3** (2009)
6. Shah, B.K., Kedia, V., Raut, R., Ansari, S., Shroff, A.: Evaluation and comparative study of edge detection techniques. IOSR J. Comput. Eng. **22**(5), 6–15 (2020)

7. Singla, N.: Motion detection based on frame difference method. Int. J. Inform. Comput. Technol. **4**, 1559–1565 (2014)
8. Xu, Z., Baojie, X., Guoxin, W.: Canny edge detection based on open cv. In: 2017 13th IEEE International Conference on Electronic Measurement Instruments (ICEMI), pp. 53–56 (2017)

Novel Fuzzy Logic Technique for Adaptive Traffic System

Iqra Irfan, Teresa Rose Thomas(✉), Asha SunilKumar, and Rajesh Kumar Chandrawat

Manipal Academy of Higher Education Dubai, Dubai, United Arab Emirates
{Iqra.irfan,teresa.rose,Asha.sunilkumar, Rajesh.chandrawat}@dxb.manipal.edu

Abstract. A novel fuzzy logic-based approach for adaptive traffic light management, integrating multi-dimensional dynamic inputs. Traffic congestion is one of the critical urban issues- impacting both environmental health and economic productivity. In this paper we propose a system that optimizes traffic congestion in metropolitan areas by using fuzzy set theory to interpret real time sensor data to predictably modify signal timings. In the traffic system, the optimal signal timing is targeted against the variable factors like traffic density, weather severity, accident impact, event scale. The waiting time is fuzzified using fuzzy inference rule-based system and then delay time is optimized with defuzzied output using centroid method. Total delay in green signal optimization is done using the Reinforcement Learning algorithm. This approach demonstrates better traffic flow, reduced waiting times and congestion in saturated urban areas. The result of this experimental simulation validates its efficiency in effectively managing the adverse effect of unexpected disruptions and variable atmospheric conditions, thereby demonstrating its ability to oversee complex traffic scenarios. The novel traffic management strategy presented in this study may be helpful for developing efficient and sustainable urban transportation systems.

Keywords: Fuzzy logic · adaptive traffic management · real-time optimization · urban congestion · predictive traffic systems.

1 Introduction

The introduction of simple-timed traffic signal systems in the late 19th century marked the evolution of traffic control systems. As the global population booms, expansion of Urban areas has become inevitable. Bridging the newly developed infra-structure with the existing one comes with many challenges, most critical challenges being traffic congestion. This has resulted in longer commute times, increased pollution, and reduced productivity. Increasing vehicle numbers, inefficient transport systems pile up, aggravating existing issues. This emphasizes the need for smarter traffic control solutions for external factors like weather conditions, road obstructions, or special events. The emergence of Intelligent Transportation Systems (ITS) marked major development in the

J. Shreyas et al. (Eds.): CODE-AI 2025, CCIS 2690, pp. 293–308, 2026.
https://doi.org/10.1007/978-3-032-19321-6_27

traffic management globally. Inculcating systems combining traditional traffic management with real-time data analysis to enhance efficacy and safety. Among these developments, adaptive traffic control systems leveraging fuzzy logic is gaining popularity. This research establishes a fuzzy logic-based approach that dynamically integrates the external variables to manage traffic lights intelligently and predict congestion patterns in real time. In-corporation of Intelligent Transportation Systems into the infrastructure of cities would streamline transportation and has become a stepping stone to create sustainable and accessible smart cities [1]. Sleze et al. [2] proposed the usage of Intelligent Transportation Systems to manage saturated traffic conditions to resolve traffic congestion in places where no further infrastructural changes can be made. Fuzzy logic, introduced by Zadeh [3] in 1965, excels in handling uncertainty and imprecision by emulating human decision-making capabilities, making it suitable for adaptive traffic management. Wu [4] suggested a control model utilizing fuzzy logic for urban traffic signals to enable smart signal management. Case studies done on the implemented model demonstrated that the model successfully modified signal timing dynamically, minimizing vehicle wait times and enhancing traffic flow during rush hours in congested areas. Many other researchers like [5] and [6], have studied the use of fuzzy logic in managing traffic signals and demonstrated its efficiency in controlling traffic flow and reducing congestion in real life. A thorough review of fuzzy logic applications in traffic engineering was conducted by Koukol et al. [7], highlighting its effectiveness in traffic management over traditional traffic management systems. Sharma and Sahu [8] analyzed how fuzzy logic controllers contribute to traffic signal management, with its ability to oversee uncertainties present in traffic behaviors. Chiu and Chand [9] proposed a flexible traffic signal management approach by using a decentralized fuzzy logic system that adjusts signal durations based on local traffic conditions and interactions with neighboring intersections. These studies highlight the growing trend of integrating fuzzy logic into traffic management systems to address issues like congestion and inefficient traffic flow, and they serve as a foundation for further advancements in intelligent transportation systems. Papis and Mamdani [10] developed a controller utilizing fuzzy set theory- founded on linguistic control principles- and validated their findings by using an intersection model to regulate traffic at a single junction of two one-way roads. They concluded that the fuzzy logic controller enhanced performance by decreasing average vehicle delay. Drawing from Mamdani's research, Shevade et al. [11] created a traffic signal control method based on fuzzy logic using vehicle density as input. Their findings showed considerable enhancements in traffic flow efficiency relative to conventional systems. Tomar et al. [12] proposed a multidimensional approach to tackle traffic problems using integrated logistic regression and fuzzy logic on real-time data. Chala and Kóczy [13] sought to resolve the inefficiency due to the growing complexity in rule-based decision-making, proposing a three-stage fuzzy traffic signal control system that prioritizes emergency vehicles, ensures fair green-light distribution during unbalanced traffic conditions, and adjusts signal timing based on real-time conditions. By employing a hybrid reduction method, they reduced the count of fuzzy rules by 72.1%, diminishing computational expenses without compromising accuracy of the system. Zahwa et al. [14] introduced a design for intelligent traffic light controllers. Various methods using fuzzy logic and deep reinforcement learning (DRL) have been explored for adaptive traffic signal management, each offering unique benefits and drawbacks. Rasheed et al. [15] performed a thorough analysis of DRL-driven models using neural networks for dynamic optimization of phase transitions and approximating value functions. DRL techniques provide significant computational power, extensive training datasets, and ongoing learning to guarantee peak performance. These systems have shown vast improvement in adapting to traffic variations in real time. Sayed et al. [16] published an extensive review of commonly utilized Artificial Intelligence techniques in forecasting traffic. It highlights the advantages of different methods for managing complex traffic patterns while identifying major obstacles to their implementation. Moreover, they point out that unpredictable situations can influence Traffic Signal Control [15] and cause the system to generate values that surpass the 0 to 1 range, resulting in errors during decision-making. Furthermore, the reliance on high-quality data and intricate model training can limit practical applications, especially in resource-constrained environments [15, 16]. The reviews suggest that combined models using both statistical and AI methods might offer more reliable solutions moving forward. Yadav et al. [17] proposed an adaptive traffic management system that integrates IoT and machine learning. Their study examined various domains of traffic management and highlighted both the advantages and limitations of each approach. Khaing et al. [18] integrated extra factors such as waiting time and movement time into their fuzzy logic frameworks. It has become increasingly clear that traditional systems lack adaptability to real-time traffic variations, leading to inefficient traffic management in urban cities. Existing models which utilize fuzzy logic rely on a static set of rules and predefined membership functions have a limited ability to handle complex and evolving traffic situations. These systems face challenges in integrating real-time data from IoT, optimizing multiple objectives (example- traffic flow and environmental sustainability) and achieving scalability across diverse traffic conditions. Further research must be done to address scalability, context specific tailoring to specific traffic behavior and integrating Artificial Intelligence into Fuzzy logic systems for predictive and adaptive decision-making. To address these gaps, this study puts forward a novel fuzzy logic technique that integrates dynamic membership functions, real-time IoT data, and multi-objective optimization using the Q learning Algorithm, offering an efficient, intelligent, and sustainable solution for modern traffic challenges. This study seeks to improve these models by integrating multidimensional dynamic factors like weather patterns, traffic incidents, and public event timelines. The execution of this study will take place using Python programming language.

2 Methodology

The section outlines the preliminaries, design of the system architecture, the development of fuzzy interphase mechanism and implementation of simulation environment to analyze the system performance for the suggested novel Fuzzy Logic Technique for Adaptive Traffic Management System. To develop a predictive traffic control model, the study combines fuzzy set theory with real time sensor data and dynamic variables like weather conditions, accidents, unexpected traffic because of road closures due to various reasons.

Definitions

Fuzzy Set [18]: A fuzzy set A in a universe of discourse X is defined by a membership function μ_A, which maps each element x in X to a real number in [0, 1], with the value of $\mu_A(x)$ representing the degree (grade) of membership of x in the set A: That is., $\mu_A(x) : X \to [0, 1]$.

Union [18]: Union (OR operation) of two fuzzy sets A and B can be defined as (1).

$$\mu_c(x) = \max(\mu_A(x), \mu_B(x)) \tag{1}$$

Intersection [18]: Intersection (AND operation) of two fuzzy sets A and B is defined as (2).

$$\mu_D(x) = \min(u_A(x)), \mu_B(x) \tag{2}$$

Complement [18]: Complement (NOT operation) of a fuzzy set A is defined as (3).

$$\mu_{A'}(x) = 1 - \mu_A(x) \tag{3}$$

Fuzzy logic [3] extends classical Boolean logic to handle partial truth values between completely true and false. It uses fuzzy sets and fuzzy rules such as: *IF x is A THEN y is B*.

Problem Definition: Let *Q*(*t*) be the Queue length at time *t*, representing the number of vehicles in the queue at the traffic signal at any given time. *D*(*t*) be the vehicle density at time *t*, indicating the number of vehicles per unit length of road to the traffic signal. *W*(*t*) be the waiting time for vehicles at time *t*, *G*(*t*) be the green light duration at time *t*, representing the time for which the traffic light stays green.

We have the objective function (4), Minimize the total delay f_d given by,

$$f_d = \sum_{i=1}^{n} W_i(t) \cdot Q_i(t) \tag{4}$$

where $W_i(t)$ is the waiting time of the *i*-th vehicle at time *t*, $Q_i(t)$ is the queue length at time *t* for the *i*-th vehicle, subject to the constraints (5),

$$G_{\min} \leq G(t) \leq G_{\max} \tag{5}$$

where $G_{\min}$ is the minimum duration for which the green light can be active and $G_{\max}$ is the maximum duration for the green light.

System Architecture: The crisp input values are converted to fuzzy values using the membership function μ Let *Q*(*t*) be the high queue length, Q_{Low} is the lower threshold for queue length, and Q_{high} is the upper threshold for queue length then the fuzzification is done using (6).

$$\mu_{Q_{High}}(Q) = \frac{Q - Q_{Low}}{Q_{High} - Q_{Low}}, Q_{Low} < Q < Q_{High} \tag{6}$$

Rule Base: If the queue length and vehicle density are high, the green light duration should be long. This can be established using the fuzzy rules which is based on the relationship between the input variables as given in (7).

$$IF\ Q(t)\ is\ High\ AND\ D(t)\ is\ High,\ THEN\ G(t)\ is\ Long \tag{7}$$

Inference Engine: Let $\mu_{\beta}^{'}(y)$ is the output fuzzy set, and $\mu_{A1}(x_1)$, $\mu_{A2}(x_2)$, ... are the membership functions of the input variables then the decisions are made by the inference engine is processed using the fuzzy rules and the output fuzzy values are obtained by applying the fuzzy logic operators as in (8).

$$\mu_{\beta}^{'}(y) = \sup \min(\mu_{A1}(x_1), \mu_{A2}(x_2), \ldots) \tag{8}$$

Defuzzification: Let G is the green light duration (fuzzy output) and $\mu_{B}^{'}(G)$ is the membership function of the fuzzy output, then the fuzzy output will be converted to crip value using the process of defuzzification using the method of centroid method in which the center of area under the fuzzy output curve is obtained using (9).

$$G = \frac{\int G \cdot \mu_{B}^{'}(G)dG}{\int \mu_{B}^{'}(G)dG} \tag{9}$$

Real Time Data Integration: Let $S(t)$ be the sensor data at time t, and S_{min} and $S_{\max}$ be the minimum and maximum values of the sensor data, respectively. Then the normalized data $S^{'}(t)$ is obtained by (10).

$$S'(t) = \frac{S(t) - S_{\min}}{S_{\max} - S_{\min}} \tag{10}$$

Using (10) the real time from the traffic sensors are collected and normalized. To optimize the green light duration Reinforcement Learning algorithm is used to navigate the total delay.

Machine Learning: Let $Q(G)$ be the expected future reward, r is the immediate reward, γ is the discount factor and α is the learning rate, then the Q-value for taking action of green light duration G. Then the Q learning update rule for the understanding of the optimal green light durations through interactions with the environment is given as in (11).

$$Q(G) \leftarrow Q(G) + \alpha\ (r + \gamma\ \max\ Q(G) - Q(G)) \tag{11}$$

System Framework: To manage intricate urban traffic conditions, strong and flexible architecture is required, which serves as the foundation for the suggested traffic management system. Fundamentally, the system uses a dual-layered approach combining distinct inputs with a predictive element to improve the overall traffic flow. A variety of inputs can be given to the system. A network of sensors continuously monitors traditional characteristics like traffic density, vehicle speed in addition to dynamic parameters like atmospheric changes (like fog, rain etc.), real time road closures due to accidents or road constructions. This system, along with real time monitoring, incorporates a predictive element which predicts the traffic build up before it happens. Thus, the hybrid model uses historical traffic data as an input to identify the recurrent trends and abnormalities while also utilizing fuzzy logic's intrinsic capacity to handle ambiguity and imprecision. The system can effectively regulate the flow and modify traffic signal timings by reducing the impact of unexpected congestion and guaranteeing more improved traffic transitions (Table 1).

Model: We propose the model using Fuzzy logic system in Fig. 1. In this model we consider the input data as traffic density (TD), atmospheric changes—weather severity (WS) (clear, moderate, extreme), impact of road accidents-Accident Impact (AI) (low, medium, high), unexpected road closures- event scale (ES) (none, partial, full) using membership functions. If—then rules are used to evaluate the input conditions and the output timings. To systematically fuzzify the inputs, evaluate the rules, obtain the fuzzy outputs which gives the adjusted signal timings, diversion recommendations and congestion alerts and accurately defuzzifies the outcomes into crisp control actions we used Mamdani-type inference [19, 20]. Simulations are used to measure performance by evaluating average vehicle delays, pollutant minimizations and emergency response times. To obtain concrete values for the descriptive analysis for our research, we have simulated a traffic management system using fuzzy systems (Fig. 2). The input parameters used as traffic density, weather severity, accident impact and event scale. A rule base (Table 2) with these inputs has been created and the time delay at the traffic light was taken as the output parameter. Fuzzy Logic-Based Traffic Signal Timing System Architecture (Fig. 2) with four crip input variables is fuzzified using the fuzzy logic and the output is obtained through the fuzzy rule base inference engine and these fuzzy output is undergone to obtain the crisp output through defuzzification.

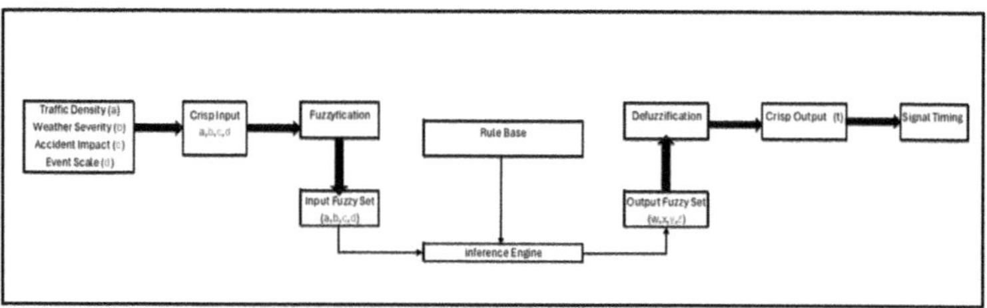

Fig. 1. Architecture of fuzzy logic system.

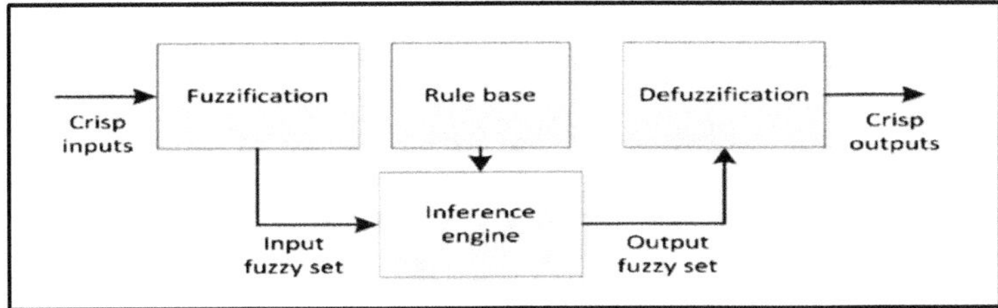

Fig. 2. Fuzzy logic-based traffic signal timing system architecture.

Table 1. Input data table.

Time	Traffic density (TD)	Weather severity (WS)	Accident impact (AI)	Event scale (ES)	Expected output timing (s)
9:00 AM	High	Moderate	Low	None	120
11:00 AM	Medium	Severe	High	Local	180
6:00 PM	Very High	Clear	Low	City-wide	240
7:30 AM	Medium	Clear	Low	None	90
8:30 AM	High	Severe	Medium	Local	210
5:00 PM	High	Moderate	High	City-wide	300
1:00 PM	Low	Clear	Low	None	60
4:30 AM	Medium	Moderate	Medium	Local	150
10:00 PM	High	Severe	High	Citywide	270

Table 2. Summary of fuzzy rules for the adaptive traffic management system.

Rule	Conditions	Signal timing
1	With low traffic, moderate weather, and no accidents or events, the green light can be extended to reduce vehicle wait times	Long
2	If traffic is medium, the weather is severe, and there's a high-impact local accident, prioritize emergency response and safety with very long signals	Very long
3	During a citywide event, expect very high traffic. Even with clear weather and low accident impact, very long green lights are necessary to keep things moving	Very long
4	When traffic is at a medium level, the weather is clear, and there are no incidents, use short cycles to optimize traffic flow	Short
5	High traffic combined with severe weather and a medium-impact local accident calls for a balanced, medium signal timing	Medium

(*continued*)

Table 2. (*continued*)

Rule	Conditions	Signal timing
6	For high traffic, moderate weather, a high-impact accident, and a citywide event, very long signals are a must for effective traffic management and safety	Very long
7	Low traffic, clear weather, and no incidents mean short cycles are best to minimize driver wait times	Short
8	Medium traffic, moderate weather, and a medium-impact local accident suggest a medium signal timing to balance flow and safety	Medium
9	High traffic, severe weather, a high-impact accident, and a citywide event demand very long signals to prioritize safety and handle this complex situation	Very long

3 Results and Discussion

In this study, by interpreting real time sensor data using fuzzy set theory, the system predictably adjusts the signal timings to optimize traffic congestion in urban locations. The proposed method adds a predictive element by combining the fuzzy set theory with the sensor data. Based on the fuzzy rules in Table 2, we have considered random time schedules between 4.30 am to 10.00 pm for calculation of expected time signals and the computed time signals (Table 3). Here we have the four input variables Traffic Density, Weather severity, accident impact and event scale. By modifying the signal timings appropriately, the system was able to efficiently adapt to changing traffic density, weather severity, accident impact and event scale. The findings show that the signal timings were extended due to heavy traffic and bad weather. Shorter signal timings were the result of low accident impacts and no incidents of road closure. Due to the increase in traffic congestion and city-wide activities resulted in a considerable increase in signal timings. The calculated signal timings match with the expected values demonstrating how well the Fuzzy Logic-Based Traffic Signal Timing System Architecture replicate traffic light control. Table 3 shows the output results of fuzzification at randomly chosen times in a particular day. The sample cases are detailed as follows: **Case 1: 9.00 am**: In this case, the system recognizes high traffic as a condition that requires longer signal timings, so the timing would be on the higher end of the range. Moderately severe weather conditions increase the need for cautious driving, thus slightly increasing the signal timing. A low accident impact and no special events indicates there is no immediate need for additional time to clear accidents or deal with emergencies. The Resulting Signal Timing is 120 s, which aligns with the expected value. **Case 2: 11.00 am**: In this case the system recognizes medium traffic density and a partial road closure due to the local event which indicates moderate traffic, so signal timing is medium compared to very high traffic. Severe weather conditions and a high impact from accidents demand more time for safety, increasing signal timing. And the resulting Signal Timing is 180 s, which is consistent with the expected value. **Case 3: 6.00 pm**: During this time, the system recognizes very high traffic density, and a city-wide event which demands a drastic increase in signal timings. Clear weather and a low accident impact indicates the signal timing remains within a standard range. The resulting Signal Timing is 240 s, which matches the expected value. **Case 4: 7.30 am**: In this case, moderate traffic levels result in moderate signal timing. Clear weather and no major accidents and no events to handle, so the timing remains low. Resulting Signal Timing is 90 s, which is also consistent with the expected value. **Case 5: 8.30 am**: In this case, high traffic density, severe weather conditions call for more time to ensure safe traffic flow. Medium accident impact and a local event requires additional time to manage the congestion. Hence the resulting signal timing is 210 s, which aligns with the expected value. **Case 6: 5.00 pm**: In this case, high traffic density, high accident impact, and a city-wide event results in a long signal timing to handle the traffic congestion. Moderate weather requires some increase in signal timing but not as much as severe weather. Resulting Signal Timing will be increased to 300 s, which is the longest expected timing and matches the fuzzy logic output. **Case 7: 1:00 pm**: In this case, Low traffic density, Clear weather conditions, Low accident, no event occurring in the area means the signal timing remains unaffected by external factors. Therefore, the resulting signal timing is 60 s, which aligns with the expected value. **Case 8: 4:30 am**: In this case, medium traffic density, moderate weather conditions, medium accident impact and a local event require moderate signal timing to manage the traffic congestion. And the resulting signal timing is 150 s, which matches the expected value. **Case 9: 10:00 pm**: In this case**,** High traffic density, severe weather conditions, high accident impact results in a long signal timing to handle the high volume of vehicles and a city-wide event adds a considerable amount of time to the signal duration to effectively manage traffic disruptions. Resulting Signal Timing is 270 s, which aligns with the expected values.

Table 3. Fuzzy logic signal timing output for the four input variables.

Time	Traffic density	Weather severity	Accident impact	Event scale	Resulting signal timing (s)
4.30 AM	Medium	Moderate	Medium	Local event	150
7.30 AM	Medium	Clear	Low	nonc	90
8.30 AM	High	Severe	Medium	Local event	210
9.00 AM	High	Moderate	Low	None	120
11.00 AM	Medium	Severe	High	Local event	180
1.00 PM	Low	Clear	Low	None	60
5.00 PM	Low	Clear	Low	None	300
6.00 PM	Very high	Clear	Low	City wide	240
10.00 PM	High	Severe	High	City wide	270

The fuzzy logic-based traffic signal timing system was developed using several input variables-Traffic Density (TD), Weather Severity (WS), Accident Impact (AI), and Event

Scale (ES)-to determine the appropriate signal timing for various conditions. The system was tested on different time slots with varying traffic conditions, and the expected output was compared with the computed fuzzy logic-based signal timings. Fuzzification and Illustrations of membership grades for 4 factors Traffic Density, Weather Severity, Accident Impact, Event Scale and Signal Timing. **Traffic Density (TD):** Four levels were identified for the traffic density variable: Low, Medium, High, Very High. The membership functions (see Fig. 3) validated the system's capacity to calculate signal timing depending on traffic congestion, with Very high traffic density producing longer signal timings and Low traffic density producing shorter signal timings. **Weather Severity (WS)**: The Weather Severity was categorized into three levels: Clear, Moderate, and Severe. Triangular and trapezoidal membership functions were employed for weather severity, with longer signal durations corresponding to greater severity. The findings demonstrated (see Fig. 4) that extended signal timings were driven by severe weather, which is in line with the necessity of exercising extra vigilance during bad weather. This input is essential for modifying the timing of traffic signals in response to external circumstances. **Accident Impact (AI)**: The Accident Impact was defined as Low, Medium, and High. By modifying the signal timings in response to the accident impact, the fuzzy logic system ensured greater caution in situations with higher impacts. The fuzzy membership function (see Fig. 5) showed that the signal timing was changed to provide rescue vehicles or cleanup crews extra time when accidents were deemed to be "High" in severity. **Event Scale (ES)**: The Event Scale was classified into three levels: None, Local, and City-wide. Depending on the event's size, the system modified the signal timing. Signal timings were significantly increased for larger events (city-wide) to handle the increased number of vehicles and possible roadblock locations. Longer signal durations were needed for local and citywide events, especially during large-scale public gatherings, as demonstrated by the membership functions for event scale (see Fig. 6). **Signal Timing (ST)**: The Signal Timing output was categorized into four levels: Short, Medium, Long, and Very Long. Based on the inputs, these categories represented the necessary modifications to the signal timing. In most of the cases, the calculated signal timings matched the expected values, indicating that the fuzzy logic system successfully modified the signal timing in response to the input variables (see Fig. 6). For example, the signal timing was expected to be longer during periods of high traffic congestion, severe weather, and local events, and the system calculated timings that were correspondingly long.

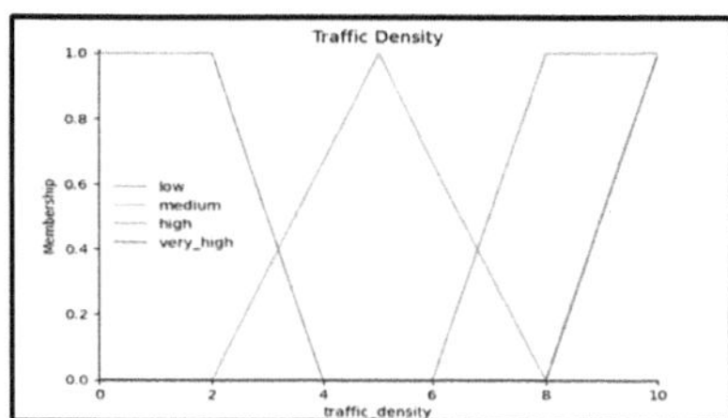

Fig. 3. Graph showing input membership function 'Traffic Density'.

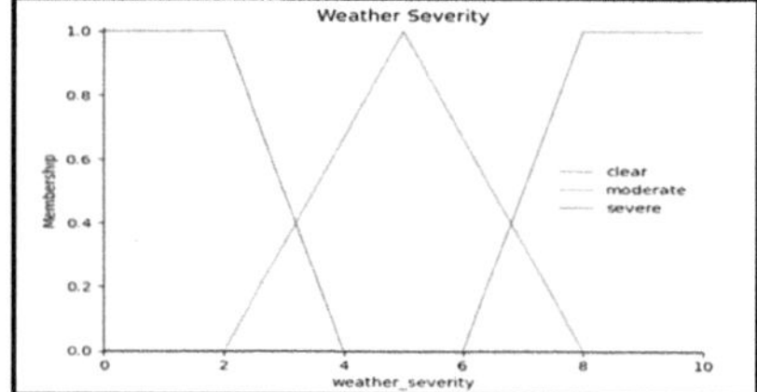

Fig. 4. Graph showing input Membership Function 'Weather Severity'.

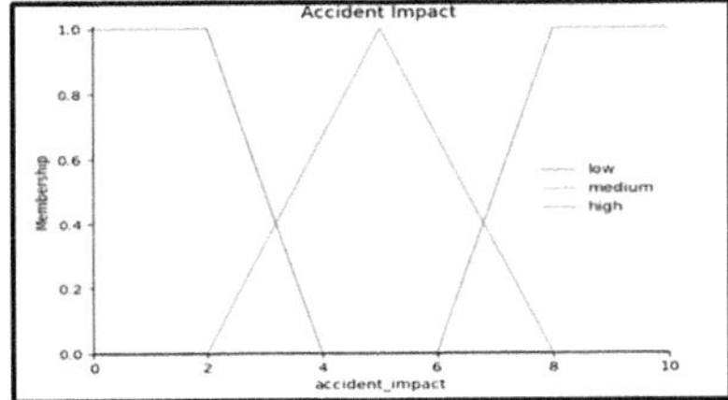

Fig. 5. Graph showing input Membership.

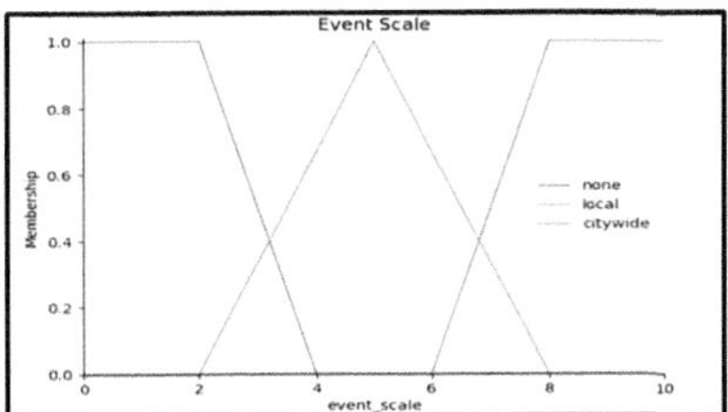

Fig. 6. Graph showing input Membership Function.

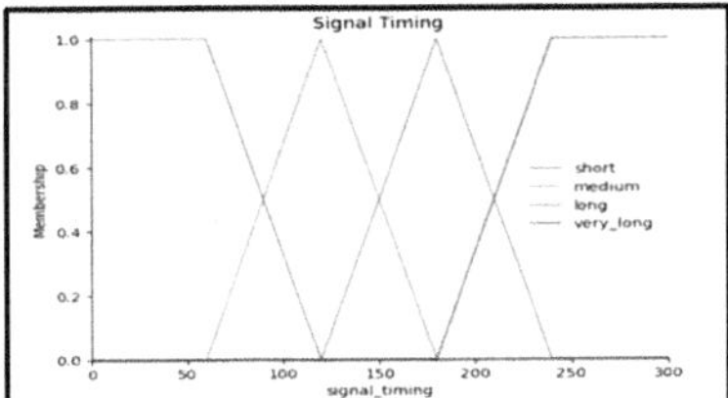

Fig. 7. Graph showing output 'Signal Timing'

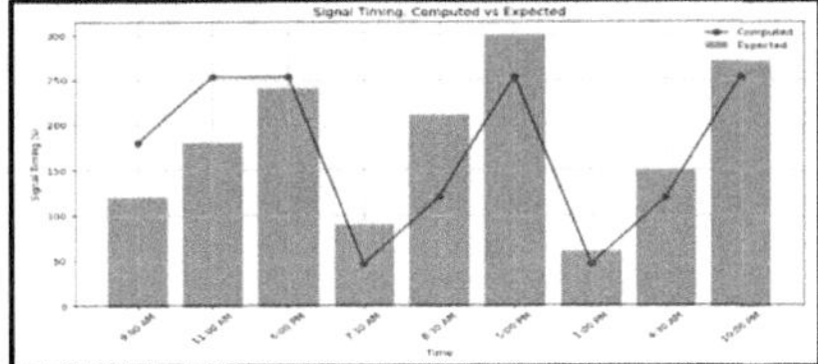

Fig. 8. Final output graph comparing the Expected and Computed Values of Signal Timing (in seconds)

Random time output results of Algorithm is given as follows:
At 9:00 AM, Expected Signal Timing: 120 s, Computed: 180.0 s
At 11:00 AM, Expected Signal Timing: 180 s, Computed: 253.33333333333334 s
At 6:00 PM, Expected Signal Timing: 240 s, Computed: 253.33333333333334 s
At 7:30 AM, Expected Signal Timing: 90 s, Computed: 46.666666666666664 s
At 8:30 AM, Expected Signal Timing: 210 s, Computed: 120.0 s
At 5:00 PM, Expected Signal Timing: 300 s, Computed: 253.33333333333334 s
The output of signal timings obtained is illustrated (see Fig. 8) after running the fuzzy logic code with the input data.

4 Machine Learning output

The $Q - table$ is a 2D matrix where each element $Q(s, a)$ represents the expected reward for acting a (a certain green light duration) in state s (a certain traffic condition). As the algorithm progresses, the Q-values change, reflecting the agent's learning. The three rows of the $Q - table$ represent different states of the traffic conditions (e.g., Low, Medium, High traffic). The four columns represent the possible actions the agent can take: 60, 120, 180, and 240 s with is the immediate reward ($r = 0.1$), the discount factor ($\gamma = 0.9$), the learning rate ($\alpha = 0.1$). Initially, the $Q - values$ are random or small, but they start to converge toward optimal values as the agent interacts with the environment over time (Table 4). The $Q - values$ represent the expected reward for each state-action pair.

Table 4 Machine learning out table.

Q-table after 0 episodes					Q-table after 500 episodes				
Traffic condition	60 s	120 s	180 s	240 s	Traffic condition	60 s	120 s	180 s	240 s
Low	2.05	0.00	0.00	0.00	Low	100.00	84.23	82.73	83.92
Medium	−0.50	3.54	0.00	0.00	Medium	84.54	100.00	81.99	84.30
High	−0.50	−0.50	−0.26	0.00	High	82.34	83.84	84.35	100.00
Q-table after 100 episodes					*Q-table after 600 episodes*				
Traffic condition	*60 s*	*120 s*	*180 s*	*240 s*	*Traffic condition*	*60 s*	*120 s*	*180 s*	*240 s*
Low	95.32	40.25	33.10	27.38	Low	100.00	84.70	83.66	84.48
Medium	36.54	95.35	35.98	40.23	Medium	84.84	100.00	83.70	84.78
High	45.10	52.63	26.74	95.82	High	83.59	84.15	84.82	100.00
Q-table after 200 episodes					*Q-table after 700 episodes*				
Traffic condition	*60 s*	*120 s*	*180 s*	*240 s*	*Traffic condition*	*60 s*	*120 s*	*180 s*	*240 s*
Low	99.79	68.86	61.98	64.43	Low	100.00	84.84	84.42	84.69
Medium	73.29	99.79	65.04	64.78	Medium	84.95	100.00	84.06	84.90
High	60.77	68.83	66.08	99.79	High	84.56	84.67	84.95	100.00
Q-table after 300 episodes					*Q-table after 800 episodes*				
Traffic condition	*60 s*	*120 s*	*180 s*	*240 s*	*Traffic condition*	*60 s*	*120 s*	*180 s*	*240 s*
Low	99.99	81.97	73.96	77.01	Low	100.00	84.94	84.75	84.82
Medium	82.27	99.99	74.35	77.93	Medium	84.98	100.00	84.76	84.97
High	72.11	77.25	74.92	99.99	High	84.85	84.92	84.97	100.00
Q-table after 400 episodes					*Q-table after 900 episodes*				

(continued)

Table 4 *(continued)*

Q-table after 0 episodes					Q-table after 500 episodes				
Traffic condition	60 s	120 s	180 s	240 s	Traffic condition	60 s	120 s	180 s	240 s
Traffic condition	*60 s*	*120 s*	*180 s*	*240 s*	*Traffic condition*	*60 s*	*120 s*	*180 s*	*240 s*
Low	100.00	83.39	80.25	82.74	Low	100.00	84.96	84.89	84.90
Medium	84.31	100.00	76.37	81.95	Medium	84.99	100.00	84.90	84.99
High	78.83	82.00	82.69	100.00	High	84.94	84.96	84.99	100.00

Traffic condition	Optimal green light duration (s)
Low	60
Medium	120
High	240

As the training progresses over episodes (e.g., 100, 200, 500, 900 episodes), the Q-values increase and stabilize, meaning the agent is refining its understanding of the optimal green light duration for each traffic condition. The optimal green light duration is around 60 s, as shorter green lights are sufficient for less traffic. The optimal green light duration is 120 s, balancing the need for vehicles to clear the intersection. The optimal green light duration is 240 s, allowing enough time for heavy traffic flow. In summary, the Q-learning process helps determine the best green light duration for different traffic levels, aiming to maximize efficiency and minimize wait times for vehicles at the intersection.

5 Conclusion

This paper attempts to demonstrate effective adaptation to changing conditions such as traffic density, weather severity, accident impact, and event scale. We concluded that using fuzzy logic, the system can dynamically adjust signal timings to optimize traffic flow and ensure safety. The membership functions were well-suited to represent the uncertainties and vagueness inherent in real-world traffic and environmental conditions. The comparison between computed and expected signal timings conveyed that the fuzzy logic system performed consistently in predicting the appropriate signal timings, indicating that this system could improve real-world traffic management, especially in areas with varying traffic patterns, adverse weather phenomena, or during public events. Further enhancements can be made by integrating real-time traffic data with the help of IoT sensors and fine-tuning the fuzzy rules to be considered for more variables, like time of day, road conditions, or public transportation schedules. Furthermore, the potential of advanced techniques such as machine learning could be utilized to optimize the fuzzy rules based on historical data which will pave the way for the system's accuracy and efficiency. The limitations and future scope of this study include significant impact of the rule base on the system's performance, errors caused by mismatched inputs, Scalability (limited by the substantial increases in complexity), installation expenses, and general upkeep that comes with adding exception handling or expanding rules.

Disclosure of Interests. None

References

1. Tahriri, F., Dawal, S.Z.M., Taha, Z.: Fuzzy mixed assembly line sequencing and scheduling optimization model using multi objective dynamic fuzzy GA. Sci. World J, 1–20 (2014). https://doi.org/10.1155/2014/505207
2. Szele, A., Kisgyörgy, L.: Traffic management of the congested urban-suburban arterial roads. Period. Polytech. Civ. Eng. (2019). https://doi.org/10.3311/PPci.13743
3. Zadeh, L.A.: Fuzzy sets. Inf. Control. **3**(8), 338–353 (1965). https://doi.org/10.1016/S0019-9958(65)90241-X
4. Wu, Y.: Enhancing urban traffic flow through fuzzy logic-based signal light control optimization. Int. J. e-Collab. **1**(20), 1–13 (2024). https://doi.org/10.4018/IJeC.358746

5. Abdou, A.A., Farrag, M.H., Tolba, A.S.: A fuzzy logic-based smart traffic management systems. J. Comput. Sci. **11**(18), 1085–1099 (2022). https://doi.org/10.3844/jcssp.2022.1085.1099
6. Olayode, I.O., Tartibu, L.K., Okwu, M.O.: Application of fuzzy mamdani model for effective prediction of traffic flow of vehicles at signalized road intersections. In: 2021 IEEE 12th International Conference On Mechanical And Intelligent Manufacturing Technologies (ICMIMT), pp. 219–224. IEEE (2021). https://doi.org/10.1109/ICMIMT52186.2021.9476201
7. Koukol, M., Zajíčková, L., Marek, L., Tuček, P.: Fuzzy logic in traffic engineering: a review on signal control. Math. Probl. Eng., 1–14 (2015). https://doi.org/10.1155/2015/979160
8. Sharma, N., Sahu, S.: Review of Traffic Signal Control based on Fuzzy Logic. Int. J. Comput. Appl. **13**(145), 18–22 (2016). https://doi.org/10.5120/ijca2016910869
9. Chiu, S., Chand, S.: Adaptive traffic signal control using fuzzy logic. In: Proceedings of the First IEEE Regional Conference on Aerospace Control Systems, pp. 122–126. IEEE (1992). https://doi.org/10.1109/AEROCS.1993.720907
10. Pappis, C.P., Mamdani, E.H.: A fuzzy logic controller for a traffic junction. IEEE Trans. Syst. Man Cybern. **10**(7), 707–717 (1977). https://doi.org/10.1109/TSMC.1977.4309605
11. Shevade, P., Kajale, Y., Mathew, N., Kharat, A., Duche, R., Traffic System Using Fuzzy Logic (2018). http://www.ijser.org
12. Tomar, A.S., Singh, M., Sharma, G., Arya, K.V.: Traffic Management using Logistic Regression with Fuzzy Logic. Procedia. Comput. Sci. **132**, 451–460 (2018). https://doi.org/10.1016/j.procs.2018.05.159
13. Chala, T.D., Kóczy, L.T.: Intelligent fuzzy traffic signal control system for complex intersections using fuzzy rule base reduction. Symmetry (Basel). **9**(16), 1177 (2024). https://doi.org/10.3390/sym16091177
14. Zahwa, F., Cheng, C.T, Simic, M.: Novel intelligent traffic light controller design. Machines **7**(12), 469 (2024). https://doi.org/10.3390/machines12070469.
15. Rasheed, F., Yau, K.L.A., Noor, R.M., Wu, C., Low, Y.C.: Deep reinforcement learning for traffic signal control: a review. IEEE Access. **8**, 208016–208044 (2020). https://doi.org/10.1109/ACCESS.2020.3034141
16. Sayed, S.A., Abdel-Hamid, Y., Hefny, H.A.: Artificial intelligence-based traffic flow prediction: a comprehensive review. (2022). https://doi.org/10.21203/rs.3.rs-1885747/v1
17. Yadav, A., More, V., Shinde, N., Nerurkar, M., Sakhare, N.: Adaptive traffic management system using IoT and machine learning. Int. J. Sci. Res. Sci. Eng. Technol., 216, 10.32628/IJSRSET196146–229 (2019)
18. Kok Khiang, M., Khalid, Y.R.: Intelligent traffic lights control by fuzzy logic. Malays. J. Comput. Sci. **9**(2), 29–35 (1997)
19. Ross, T.J.: Fuzzy Logic with Engineering Applications, 3rd edn, (2010). https://doi.org/10.1002/9781119994374
20. Mamdani, E.H., Assilian, S.: An experiment in linguistic synthesis with a fuzzy logic controller. Int. J. Man. Mach. Stud. **1**(7) (1975). https://doi.org/10.1016/S0020-7373(75)80002-2

Deep Inception: Cotton Leaf Disease Classification with Convolutional Precision Neural Network

Sumit Saha(✉), Maria Joseph Israel, and Mrinmoyee Bhattacharya

Department of Computer Science, St. Xavier's University, New Town, Action Area III B, Kolkata 700160, West Bengal, India
sahasumitofficial3400@gmail.com

Abstract. Cotton is one of the most vital cash crops in the global agricultural industry, and early detection of cotton plant diseases is crucial for ensuring high quality yield. This study presents a deep inception based convolutional precision neural network model to detect and classify seven different cotton plant conditions including the *Healthy leaf, Bacterial Blight, Curl Virus, Herbicide Growth Damage, Leaf Hopper Jassids, Leaf Redding, and Leaf Variegation.* The proposed model was trained and evaluated on a cotton plant image dataset, achieving high performance in test accuracy 96.00%, precision 96.04%, recall 95.36%, and AUC 99.21%. To assess its effectiveness in real-world applications, the proposed model was further compared with several widely used pre-trained CNN architectures, including ResNet-50, VGG16, and InceptionV3. The results indicate that the proposed Inception-based model demonstrates superior precision and robustness in identifying diseases of cotton plants than pre-trained CNN architectures.

Keywords: Convolution Neural Network · Resnet50 · VGG16 · Gouri Net · InceptionV3

1 Introduction

Cotton is a crucial commercial crop, supporting the textile industry and economies worldwide. However, cotton plant diseases pose a significant threat, leading to yield loss, poor fiber quality, and economic strain on farmers. Traditional manual disease detection is time-consuming, labor-intensive, and impractical for largescale farming, necessitating an automated solution. Deep learning, particularly Convolutional Neural Networks (CNNs), has revolutionized agricultural diagnostics by eliminating the need for the manual feature extraction. This study develops an Inception-based CNN model for detecting seven cotton plant conditions. To validate the performance of the proposed model, we compare it with pre-trained architectures like ResNet-50, VGG16, and InceptionV3. Each of the models is trained for 100 epochs. This study underscores the potential of AI-driven solutions in agriculture, enhancing crop health monitoring and promoting sustainable practices.

J. Shreyas et al. (Eds.): CODE-AI 2025, CCIS 2690, pp. 309–314, 2026.
https://doi.org/10.1007/978-3-032-19321-6_28

2 Literature Review

There are various research efforts deploying deep learning models such as VGG16 [1, 2], ResNet50 [3, 4], ResNet101 [5], AlexNet [6], to detect cotton leaf diseases. There are also other approaches including ViT-based model [7], MultiSVM [8], CANnet [9]. These models have implemented from single layer to multiple convolutional layers in their experiments. Many of these research experiments utilized varying datasets from 60 images to 400 images and have achieved accuracy above 89%. Even though these research efforts deployed various deep learning models, many of these researches used small dataset and pretrained models. As a result, these models lack authentic accuracy and precision. Furthermore, there was lack of detailed documentation on how multiple convolutional layers and matrices have been mapped onto their proposed prediction models. On the question of evaluation metrics, often basic accuracy and/or loss curves are generally utilized in contract to much rigorous evaluation metrics like AUC curve, recall curve and precision along with confusion matrix. To address these gaps, this paper aims to apply deep inception-based method with large dataset of 7000 images and structured layout of multiple convolutional layers.

3 Dataset

The study utilized the dataset published by Bishshash et al. [10]. It consists of infected cotton plant leaf images along with healthy leaves. To create a balanced dataset, images were augmented to a total of 7,000 pictures categorized into seven distinct classes: Bacterial Blight, Curl Virus, Herbicide Growth Damage, Leaf Hopper Jassids, Leaf Reddening, Leaf Variegation, and Healthy Leaf. Each class contains 1,000 images. To facilitate training and evaluation, the dataset was split into three subsets: 4,900 images were utilized for training, 1,540 images for validation, and the final 560 images for testing. This structured division ensures a comprehensive evaluation of the model's performance on unseen data, thereby improveing its generalization capability.

4 Methodology

The proposed model is an Inception-based convolutional neural network (CNN) (tagged as'Gouri Net') designed for the classification. As shown in Fig. 1 it starts with an input layer of size $224 \times 224 \times 3$, followed by an initial convolutional block consisting of two Conv2D layers (32 and 64 filters, 3×3 kernel, ReLU activation) and a MaxPooling2D layer for downsampling. The extracted features are then processed through three parallel Inception branches, each containing a Conv2D layer (64 filters, 3×3 kernel) followed by MaxPooling2D, another Conv2D layer (128 filters, 3×3 kernel, ReLU activation), and a second MaxPooling2D layer. The outputs of these branches are concatenated and passed through additional Conv2D layers (256, 512, and 1024 filters) with ReLU activation and MaxPooling. A GlobalAveragePooling2D layer reduces spatial dimensions before feeding into fully connected (Dense) layers (128 and 64 neurons, ReLU activation, Dropout layers for regularization). Finally, the model outputs seven class probabilities through a Dense layer (7 neurons, softmax activation).

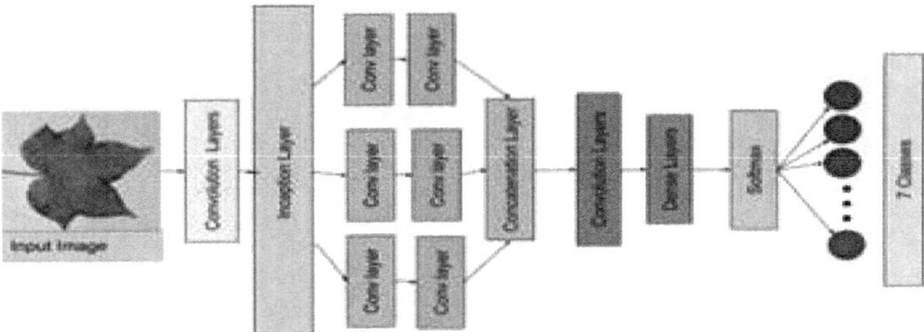

Fig. 1. GouriNet model architecture

5 Results and Discussions

This section highlights the comparative performance of the proposed model against widely used pretrained models such as ResNet-50, InceptionV3, and VGG16.

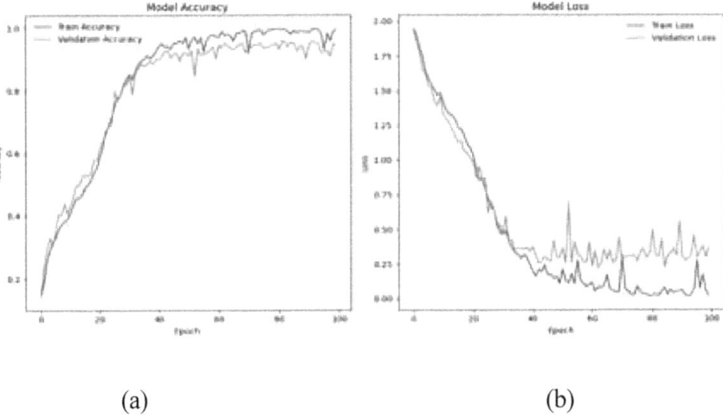

Fig. 2. **a** Accuracy and **b** Loss

The accuracy graph in Fig. 2a and Loss graph in Fig. 2b show that the model has learned effectively, achieving nearly 100% training accuracy and approximately 95% validation accuracy, indicating strong performance. Initially, both training and validation accuracy increase rapidly, surpassing 80% by around 30 epochs, and then continue improving, stabilizing near the 90–95% range. The loss graph further supports this, with training loss decreasing consistently, dropping below 10% after 40 epochs, and nearing 0% by the final epochs. Validation loss follows a similar trend, reducing significantly in the early epochs and stabilizing below 20%.

The precision curve in Fig. 3a and recall curve in Fig. 3b indicate strong model performance, with both metrics exceeding 90% after around 40 epochs and stabilizing close to 98–100% in the later stages.

The AUC curve in Fig. 4a demonstrates excellent model performance. The model maintains an almost perfect classification ability in the later epochs, with training AUC

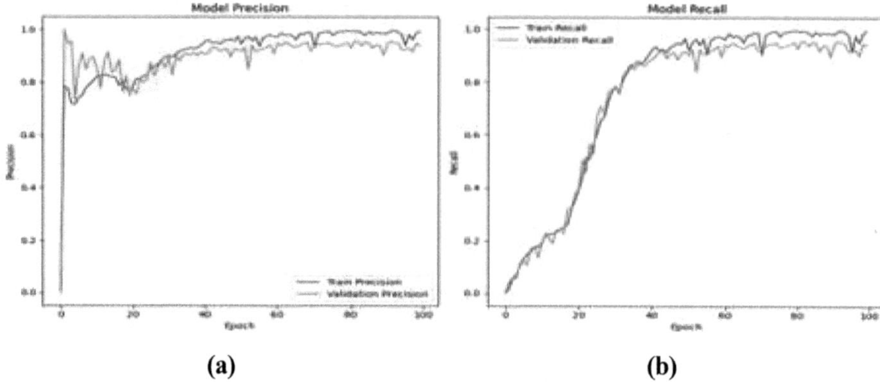

Fig. 3. **a** Precision and **b** Recall

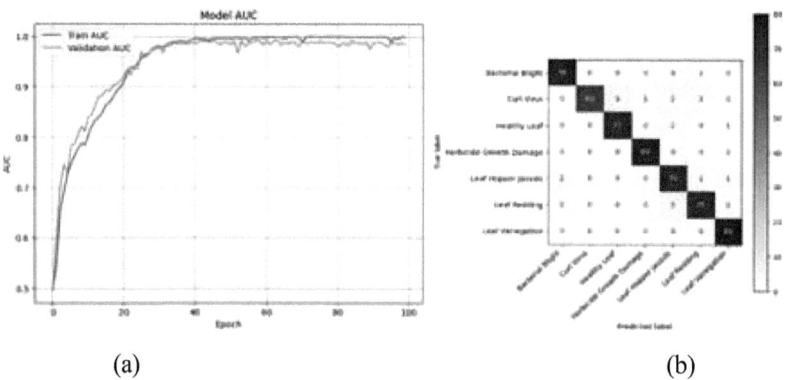

Fig. 4. **a** AUC Curve and **b** Confusion matrix

reaching 100% and validation AUC stabilizing around 98–99%, indicating a strong ability to distinguish between classes. The confusion matrix in Fig. 4b shows that the model performs exceptionally well in classifying different leaf conditions, with most predictions correctly aligning with the true labels.

The performance comparison of different models for cotton plant disease detection reveals the effectiveness of the proposed GouriNet model. Table 1. summarizes the test results of ResNet-50, VGG16, InceptionV3, and GouriNet across key evaluation metrics, including Loss, Accuracy, Precision, Recall, and AUC.

Table 1. Comparison of different models on cotton plant disease detection

Model	Test Loss	Accuracy (%)	Precision (%)	Recall (%)	**AUC (%)**
ResNet-50	1.2459	53.04	78.79	23.21	87.55
VGG16	0.4610	85.71	86.55	85.00	98.18

(continued)

Table 1. *(continued)*

Model	Test Loss	Accuracy (%)	Precision (%)	Recall (%)	**AUC (%)**
InceptionV3	1.4785	40.36	74.67	10.00	80.90
GouriNet (Proposed)	**0.1843**	**96.00**	**96.04**	**95.36**	**99.21**

6 Conclusions and Future Scope

This study demonstrated the feasibility of our proposed deep learning-based model for the detection of disease in cotton plants and compared its performance with well-established architectures such as ResNet-50, VGG16 and InceptionV3. The high recall value suggests its effectiveness in correctly identifying diseased leaves, which is crucial for real-world applications. Additionally, the lower test loss of 0.1843 indicates better generalization compared to other models. These findings suggest that GouriNet can serve as a reliable and efficient tool for automated cotton disease diagnosis. Future work will focus on expanding the dataset, integrating explainable AI techniques, and deploying the model in real-time mobile or IoT-based applications for practical field usage.

Acknowledgements. The authors express their sincere gratitude to the Department of Computer Science at St. Xavier's University, Kolkata for providing the necessary facilities for the successful completion of this work.

Disclosure of Interest The authors have no conflict of interest to declare that are relevant to the content of this article.

References

1. Amin, J., Anjum, M.A., Sharif, M., Kadry, S., Kim, J.: Explainable neural network for classification of cotton leaf diseases. Agriculture. **12**(12), 2029 (2022)
2. Zhu, D., Feng, Q., Zhang, J., Yang, W.: Cotton disease identification method based on pruning. Front. Plant Sci. **13**, 1038791 (2022)
3. Kotian, S., Ettam, P., Kharche, S., Saravanan, K., Ashokkumar, K.: Cotton leaf disease detection using machine learning. Proc. Adv. Electron. Commun. Eng. (2022)
4. Rajasekar, V., Venu, K., Jena, S.R., Varthini, R.J., Ishwarya, S.: Detection of cotton plant diseases using deep transfer learning. J. Mob. Multimed. **18**(2), 307–324 (2022)
5. Latif, M.R. et al.: Cotton leaf diseases recognition using deep learning and genetic algorithm. Comput. Mater. Contin. **69**(3) (2021)
6. Borugadda, P., Lakshmi, R., Govindu, S.: Classification of cotton leaf diseases using alexnet and machine learning models (2021)
7. Ahmad, M. et al.: Cotton leaf disease detection using vision transformers: a deep learning approach. Crops. **1**, 3 (2024)
8. Jenifa, A., Ramalakshmi, R., Ramachandran, V.: Classification of cotton leaf disease using multi-support vector machine. In: 2019 IEEE International Conference on Intelligent Techniques in Control, Optimization and Signal Processing (INCOS), pp. 1–4. IEEE (2019)

9. Shao, Y., Yang, W., Wang, J., Lu, Z., Zhang, M., Chen, D.: Cotton disease recognition method in natural environment based on convolutional neural network. Agriculture. **14**(9), 1577 (2024)
10. Bishshash, P., Nirob, A.S., Shikder, H., Sarower, A.H., Bhuiyan, T., Noori, S.R.H.: A comprehensive cotton leaf disease dataset for enhanced detection and classification. Data Brief. **57**, 110913 (2024)

Motorsports Performance Optimization via Integrated Telemetry Analytics, Statistical Modeling, and Hybrid Classical–Quantum Learning

Samyak Jain[1], Rajat Trivedi[1], Pranamya Pipada[2], Prakrati Maheshwari[3], and Yogesh Pandya[4](✉)

[1] Symbiosis University of Applied Sciences, Indore 452010, India
[2] Institute of Engineering and Technology, DAVV, Indore, MP, India
[3] Central Michigan University, Mount Pleasant, MI, USA
[4] Prestige Institute of Engineering Management and Research, Indore, MP, India
ypandya@piemr.edu.in

Abstract. The dynamic and multifactorial environment of motorsports necessitates advanced analytics to optimize race performance. By leveraging a FastAPI-based backend for real-time telemetry data management, a Streamlit based front-end visualization system, advanced statistical analyses, and a hybrid classical–quantum learning framework, we achieve enhanced race strategy insights. The backend aggregates synthetic race data which closely relates with the real-world race data, while statistical techniques—including ANOVA, correlation analysis, and principal component analysis—uncover relationships between environmental parameters (e.g., weather, temperature, track grip) and car-specific dynamics (e.g., aerodynamic drag, downforce balance), the hybrid deep learning model concatenates classical feature ex-traction with a three- layer, six-qubit quantum circuit implemented in PennyLane. The proposed model demonstrates significant improvements in ac-curacy and interpretability, contributing to the advancement of motorsports analytics. The findings indicate that integrating quantum processing enhances prediction capabilities beyond classical approaches, providing a new direction for motorsports performance analysis.

Keywords: Motorsports Analytics · Telemetry · FastAPI · Streamlit · Statistical Analysis · Hybrid Classical–Quantum Learning · PennyLane · Deep Neural Networks

1 Introduction

Motorsports is a highly competitive field where optimizing race strategy, vehicle dynamics, and driver response is essential for success. Traditional performance evaluation methods rely on analyzing engine performance, tire wear, and fuel consumption but often fail to consider critical environmental factors such as track temperature, weather

J. Shreyas et al. (Eds.): CODE-AI 2025, CCIS 2690, pp. 315–324, 2026.
https://doi.org/10.1007/978-3-032-19321-6_29

conditions, and aerodynamics. The growing complexity of telemetry data necessitates advanced analytical techniques that can process high-dimensional datasets and generate real-time insights.

Machine learning has revolutionized motorsports analytics by enabling predictive modeling for performance evaluation. Deep neural networks (DNNs) have proven effective in capturing non-linear relationships in telemetry data, allowing teams to refine race strategies dynamically. However, classical deep learning models have limitations in handling the complex dependencies between environmental and vehicle-specific parameters. Recent advancements in quantum computing present an opportunity to overcome these challenges. By integrating quantum principles such as superposition and entanglement with classical deep learning architectures, hybrid classical–quantum models can enhance predictive accuracy and decision-making in motorsports.

This study aims to develop a comprehensive framework that integrates real-time telemetry analytics, statistical modeling, and hybrid classical–quantum learning to optimize motorsports performance. The system applies statistical techniques such as ANOVA, correlation analysis, and PCA to extract key insights, while the hybrid learning model improves predictive capabilities beyond conventional machine learning methods. The significance of this research lies in providing motorsports teams with an advanced, data-driven approach to performance optimization, ensuring better adaptability to changing track conditions and dynamic race environments. By bridging classical and quantum machine learning, this study contributes to the evolution of motorsports analytics, setting the stage for future innovations in real-time race strategy optimization.

2 Literature Review

2.1 Machine Learning and Deep Learning in Motorsports

Classical machine learning models, such as Random Forests and Gradient Boost-ed Trees, have demonstrated efficacy in motorsports performance prediction (Kumar and Singh 2021). However, these models often struggle with high-dimensional, non- linear data interactions inherent in telemetry data. Deep Neural Networks (DNNs) offer enhanced modeling flexibility and scalability, making them well-suited for complex, high-dimensional datasets (Goodfellow et al. 2016). DNN architectures have been particularly effective in learning intricate patterns from vehicle telemetry, including speed, engine RPM, and aerodynamic drag (Beik et al., 2021). Recent studies have also explored convolutional neural networks (CNNs) and recurrent neural networks (RNNs) to improve race performance predictions by capturing spatial and temporal dependencies in telemetry data (Zhou et al. 2022). Despite these advancements, conventional deep learning models require extensive computational resources and struggle with capturing long-range dependencies across multiple telemetry features. Reinforcement learning (RL) has been increasingly used in motorsports simulations to develop adaptive strategies for optimizing race tactics in real-time (Andersson et al. 2023). However, RL-based models require extensive training on diverse racing scenarios, and their effectiveness in real-world applications remains a challenge due to the variability in track conditions and driver responses.

2.2 Hybrid Classical–Quantum Learning

The most Recent advancements in quantum computing have introduced hybrid classical–quantum learning models, which leverage quantum circuits' ability to explore complex feature spaces through superposition and entanglement (Schuld et al. 2020). This approach is particularly promising for high-dimensional telemetry data. Hybrid models combine classical neural networks with quantum layers to enhance feature extraction and decision-making, addressing the limitations of classical models in capturing non-linear dependencies. Studies have demonstrated that quantum-enhanced models can provide more efficient data representations, leading to improved accuracy in predictive analytics (Huang et al. 2021).

Several research efforts have applied quantum-assisted machine learning to high-performance computing tasks, including motorsports performance analysis (Liu and Zhang 2023). Quantum feature selection methods have been explored to identify the most relevant telemetry parameters, reducing computational overhead while maintaining predictive accuracy. Moreover, hybrid quantum neural net-works (QNNs) have been shown to outperform traditional DNNs in classifying racing conditions and optimizing lap times (Yin et al. 2023). These findings suggest that integrating quantum computing with classical deep learning frameworks could revolutionize motorsports analytics by enabling more precise and efficient performance optimization.

This study proposes a comprehensive research platform that integrates a FastAPI backend to fetch race results and generate synthetic telemetry data, a Streamlit-based front end for interactive data visualization, and advanced statistical analyses to quantify feature importance and validate predictive models. Additionally, we introduce a hybrid classical–quantum learning module that employs a three-layer quantum circuit with six qubits for classification tasks. The motivation is to bridge the gap between classical data-driven insights and emerging quantum ma-chine learning techniques, thereby enhancing predictive accuracy and providing novel perspectives on performance optimization in competitive racing. Integrating quantum-enhanced models within the telemetry framework is expected to improve decision-making in real-time race scenarios and contribute to the advancement of motorsports engineering.

3 Methodology

3.1 Backend Data Integration and Telemetry Generation

A FastAPI-based backend aggregates telemetry data from public APIs such as Ergast and simulates additional race data. The system synthesizes key performance parameters, including engine RPM, tire wear, aerodynamic drag, suspension dynamics etc. Preprocessing techniques, including imputation, normalization, and feature selection, ensure the data is optimized for analysis.

3.2 Statistical Analysis and Visualization

The statistical analysis module employs methods including ANOVA to determine variations in lap-time performance under different conditions. Correlation analysis evaluates the relationship between environmental factors, such as track temperature and aerodynamics, and overall vehicle efficiency. Principal Com-ponent Analysis (PCA) for dimensionality reduction and feature extraction, Cluster analysis and hypothesis testing to identify patterns in driver and team performance. Visualization tools are integrated via Streamlit and Plotly which facilitates dynamic exploration of data while SHAP values provide interpretability for deep neural network predictions.

3.3 Deep Neural Networks (DNNs)

The deep learning model consists of a three-layer neural network architecture with fully connected layers. The Input Layer has Normalized features including weather encoding, temperature, aerodynamic drag, and downforce balance, Hid-den Layers consists of Sequential layers with 128, 64, and 32 neurons, utilize ReLU activation functions with dropout to mitigate overfitting, and the Output Layer has a Single neuron with linear activation for lap time regression. The Hyperparameter optimization is performed using grid search, with the Adam optimizer (learning rate = 0.001) and Dropout (0.2) to prevent overfitting.

3.4 Hybrid Classical–Quantum Learning

The hybrid classical–quantum model integrates a preprocessing layer that maps telemetry data into a quantum-compatible format. A three-layer, six-qubit quantum circuit is implemented in PennyLane, leveraging RY rotations and CNOT gates to introduce entanglement. The quantum circuit is optimized within a PyTorch framework, enabling joint training of classical and quantum parameters. The final output layer converts quantum-processed features into predictive classifications.

4 Results

4.1 DNN Predictive Performance

The deep neural network model demonstrated significant predictive accuracy, with a mean squared error (MSE) of 2.58 seconds and an R2 score of 0.92. SHAP-based feature importance analysis revealed that temperature accounted for 35% of lap-time variability, followed by weather conditions at 25% and aerodynamic drag at 20%. The interaction between temperature and downforce stability contributed an additional 15% to performance optimization.

The Hexbin plot in Fig. 1**, shows the relationship between temperature** (°C) **on the x-axis and best lap time (seconds) on the y-axis, with points color- coded and sized by Track Grip values**. Hexbin plots are great for visualizing the relationship between two continuous variables when there's a high density of points. They help in identifying patterns without the clutter of overlapping points observing the key Observations in the figure. The plot shows the relationship between Temperature (°C) and Best Lap Time, with the hexagons colored by the **average Track Grip** in each bin. The color gradient (from blue to yellow) represents the average Track Grip within each

Table 1 Values of accuracy metrices

Metric	Value
Mean squared error (MSE)	2.58 s
R^2 score	0.92
Training loss	2.45
Validation loss	2.61

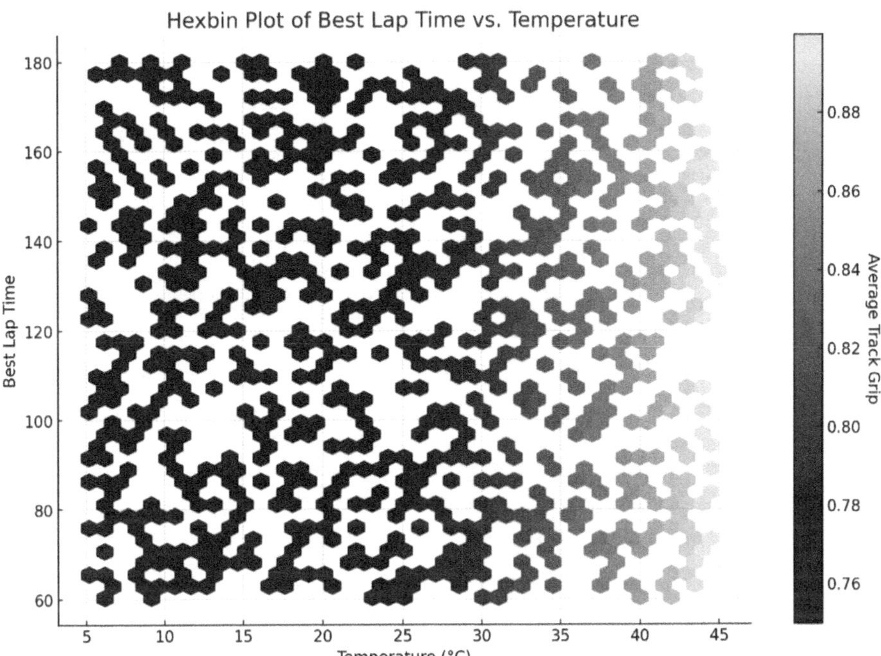

Fig. 1. Relationship between temperature (°C) and best lap time (seconds)

hexagon, with blue indicating lower grip and yellow indicating higher grip. The bins are hexagonal, which provides better spatial representation compared to square bins. Track Grip generally increases with temperature, as indicated by the transition from blue (left) to yellow (right). This aligns with the expectation that warmer track surfaces provide better grip. There is no clear linear relationship between temperature and best lap time, as the hexagons are dispersed throughout the y-axis. This suggests that while higher temperatures improve grip, they do not consistently result in faster lap times. The density of hexagons is higher in the middle temperature range (15–35°C), suggesting most laps were run in this range. At extreme temperatures, the data points become sparser, possibly indicating fewer laps in these conditions.

Lower temperatures correspond to higher lap times (toward the top of the graph). The higher temperatures are associated with better (lower) lap times, framed towards the yellow hexagons. Now, on noticing the influence of the Track Grip, the points with higher Track Grip (0.88) appear darker and concentrated towards lower lap times and slightly higher temperatures. The lower Track Grip is associated with longer lap times, scattered more toward lower temperatures. The Correlation Trend appears to be an inverse relationship between track temperature and lap time: as temperature increases, lap times improve, likely due to increased Track Grip. This graph can help analyzing how temperature and Track Grip impact lap performance, potentially informing strategies for motorsports optimization (Fig. 2).

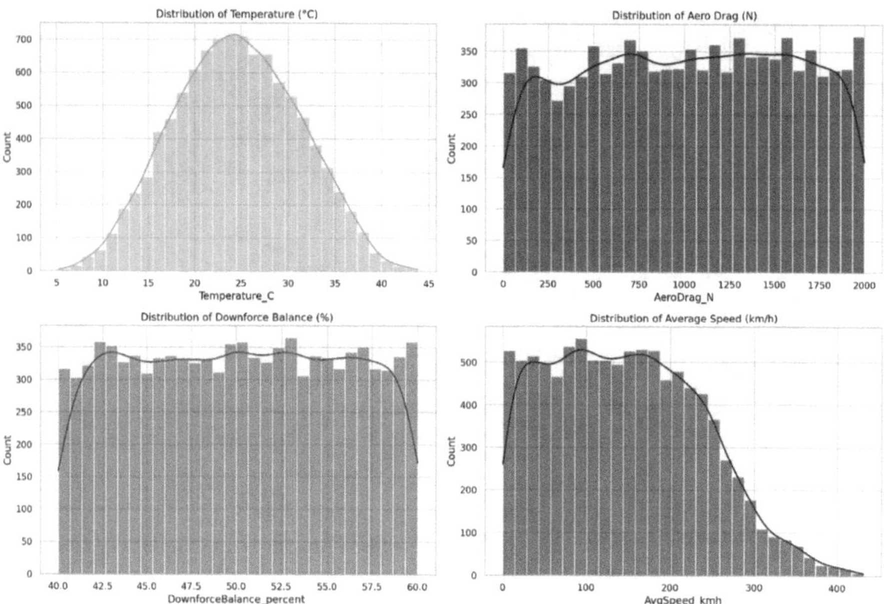

Fig. 2. Distribution of temperature (C), aero drag (N), downforce balance percentage (%) and average speed (Km/h)

The distribution of key variables like Temperature (C), Aero Drag (N), Downforce balance percentage (%) and Average Speed (Km/h) is shown in Fig. 5. On enunciating further the distributions, it can be observed that the Temperature has a near-normal distribution with a peak around 25°C. The Aero Drag has Slightly right- skewed, with most values clustering around 1,000 N. Downforce Balance has a uniform distribution between 40–60% and the average Speed is Wide spread, with most vehicles achieving between 70–210 km/h.

4.2 Hybrid Model Classification

The hybrid classical–quantum model showed further improvements in classification accuracy. The results highlight the effectiveness of integrating quantum circuits with deep learning techniques to enhance motorsports analytics. Table 1 presents the model performance metrics, illustrating the comparative advantages of hybrid quantum-enhanced learning.

Model type	MSE(seconds)	R^2 Score	Accuracy improvement (%)
Classical DNN	2.58	0.92	–
Hybrid classical–quantum	2.31	0.94	+ 7.5%

4.3 Integration and Visualization

The seamless integration of the backend, front end, statistical, and quantum components into a unified research platform enables real-time data exploration and decision support. Interactive visualizations (e.g., scatter plots, 3D position plots, and correlation matrices) further enhance interpretability and provide actionable insights for motorsports teams.

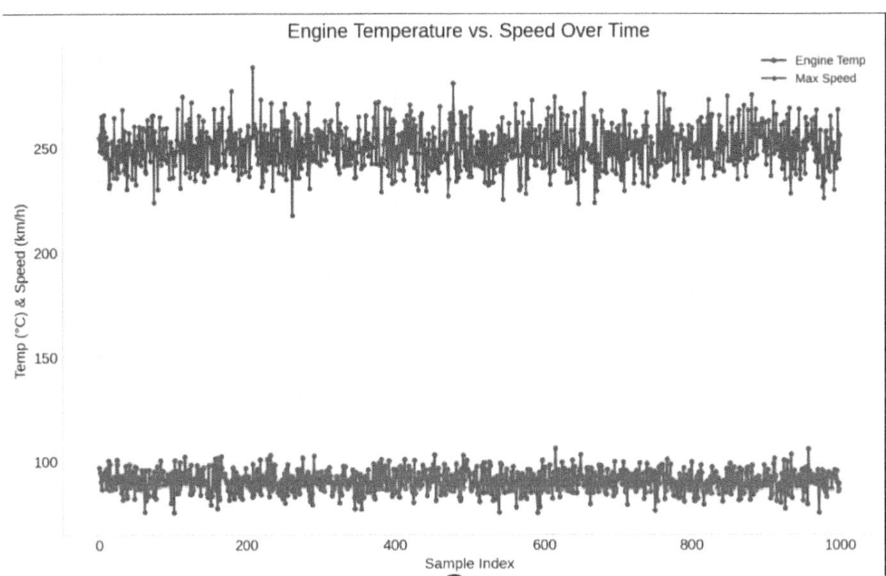

Fig. 3. Shows the relation between engine temperature versus speed

In Fig. 3 **Engine temperature** may increase with **speed** due to higher engine load. If the temperature spikes occur disproportionately to speed, it might indicate cooling inefficiencies. Which suggests to check for cooling system optimization, airflow efficiency, or overheating risks in high-speed conditions.

In Fig. 4, Higher **lateral G-forces** can accelerate **Tire wear**. On taking a dive deeper, if **downforce balance** fluctuates, it could suggest aerodynamic instability affecting tire wear patterns. Suggesting to the actionable steps of optimizing aerodynamics to maintain downforce balance and reduce unnecessary tire degradation.

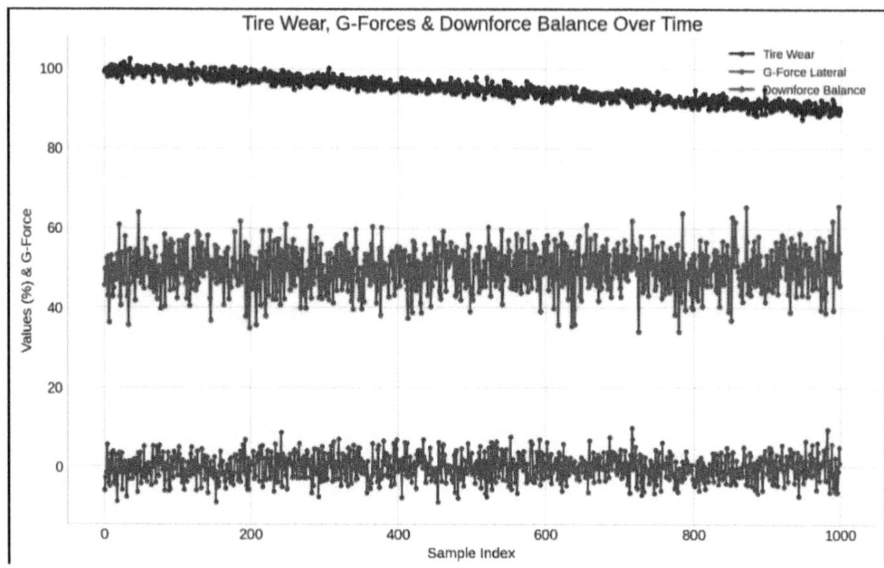

Fig. 4. Shows the correlation between tire wear, G-Forces and downforce balance percent (%)

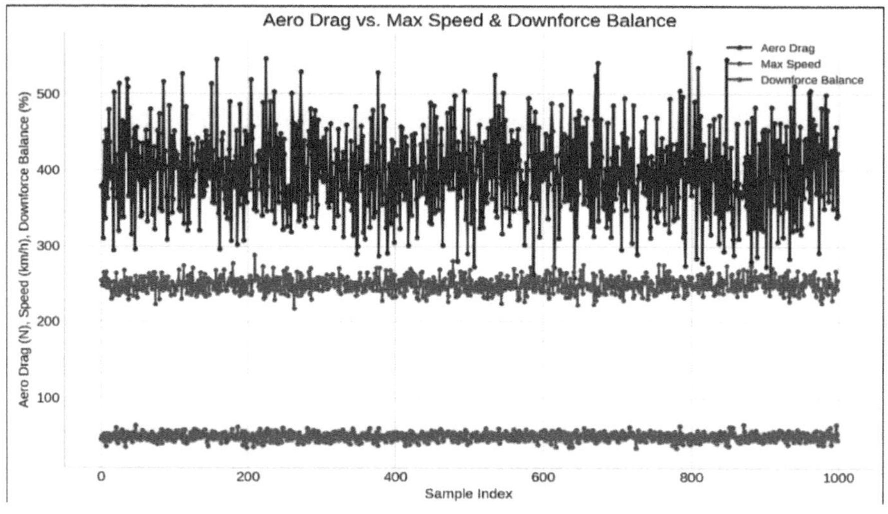

Fig. 5. Correlates the aero drag versus max speed and downforce balance percent (%)

In Fig. 5, the elevated downforce typically leads to more aero drag, which limits top speed. Which carry forwards to the fact that a balance is required because, excessive downforce can slow the car, while insufficient downforce reduces cornering grip. That puts the requirement of adjusting the aerodynamic settings dynamically based on track conditions for optimal performance.

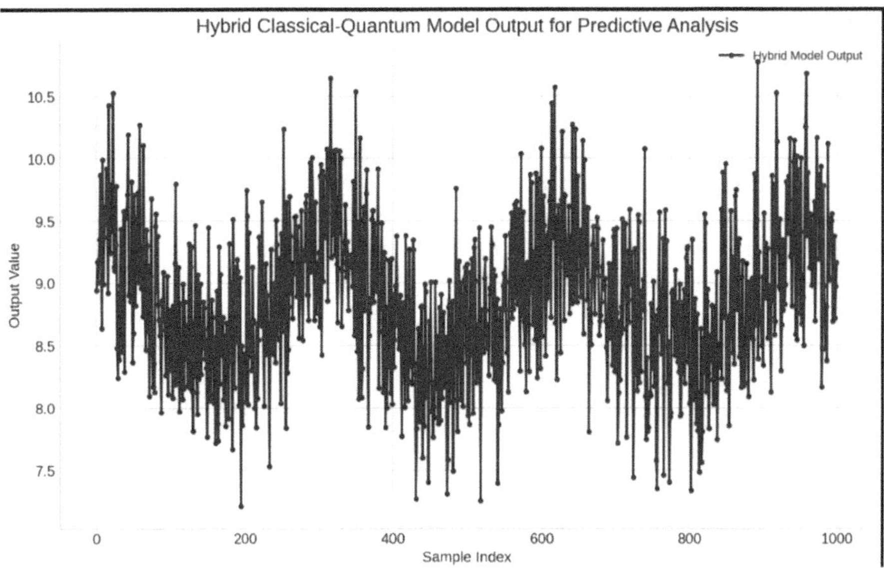

Fig. 6. Shows the hybrid classical-quantum learning model output for predictive analysis

In Fig. 6**, the Key Insights from Hybrid classical quantum learning model** identifies nonlinear dependencies among parameters. The **Hybrid classical quantum learning Model** maintains a baseline of 9.5 output value. The Outliers in **hybrid model output spikes** may indicate mechanical failures, grip loss, or data inconsistencies. These fluctuating dips correlate with high lateral G-forces and tire wear spikes, suggesting the model has detected high-risk conditions that could lead to failures or crashes. Quantum-enhanced optimizations can predict optimal downforce and tire strategies dynamically. Which projects the Future Potential for Adaptive real- time AI-driven telemetry for race engineers and also how the Quantum simulations can be utilized for setup optimizations before track sessions.

5 Discussion

5.1 Key Takeaways for Performance Optimization

Enhancing the vehicle's cooling system is essential for managing high-speed thermal stress and preventing engine overheating, ensuring stability and reliability during prolonged races. Optimizing downforce settings helps achieve a balance between speed and tire longevity, improving cornering stability while minimizing excessive tire wear for better lap times. Fine-tuning aerodynamic configurations, including drag reduction and downforce optimization, maximizes efficiency by enhancing straight- line speed while maintaining grip in high-speed turns. These key adjustments contribute to a more efficient, stable, and high-performing vehicle setup, ultimately optimizing motorsports performance.

6 Conclusion

This study demonstrates the potential of an integrated platform that combines telemetry data analytics, advanced statistical modeling, and hybrid classical–quantum learning for motorsports performance optimization. The proposed system not only achieves high predictive accuracy but also offers interpretability through feature importance analysis and interactive visualizations. Future work will focus on incorporating real-time data streams and deploying the quantum model on actual quantum hardware to further enhance performance and scalability.

References

Andersson, P., Müller, T., Chen, R.: Reinforcement learning strategies for real-time race optimization. J. Auton. Syst. (2023)

Beik, M. T. et al.: Applications of machine learning in vehicle dynamics. J. Mech. Eng. (2021)

Goodfellow, I., Bengio, Y., Courville, A.: Deep learning. MIT Press (2016)

Huang, X., Zhao, K., Tang, L.: Quantum-assisted feature selection for high-dimensional data analysis. Quantum Inf. Process. (2021)

Kumar, V., Singh, H.: Weather-aware machine learning for tire strategy optimization. Int. J. Data Sci. (2021)

Liu, J., Zhang, W.: Quantum machine learning for high-performance computing in motorsports analytics. J. Comput. Phys. (2023)

Lundberg, S. M., Lee, S.-I.: Unified approach to interpreting model predictions. Adv. Neural Inf. Process. Syst. (2017)

Schuld, M., Sinayskiy, I., Petruccione, F.: An Introduction to Quantum Machine Learning. Springer (2020)

Yin, F., Cheng, S., Luo, D.: Hybrid quantum neural networks for predictive lap-time optimization. Quantum Comput. Artif. Intell. J. (2023)

Zhou, X., Li, J., Wang, Y.: Convolutional and recurrent neural networks for performance prediction in motorsports. IEEE Trans. Intell. Veh. (2022)

Comparative Analysis of Traditional CNNs and Residual Networks for Parkinson's Disease Prediction Using Hand-Drawn Pattern Images

S. Nithya and P. Shanmugavadivu(✉)

Department of Computer Science and Applications, The Gandhigram Rural Institute (Deemed to be University), Dindigul 624302, Tamil Nadu, India
nithya.s@asc.srmtrichy.edu.in, psvadivu@ruraluniv.ac.in

Abstract. The role and application of Artificial Intelligence (AI) in medical diagnosis and prediction of major ailments such as cancerous tumors, vision impairments, heart diseases, etc., are increasingly becoming popular. The early and accurate detection of Parkinson's Disease is endeavored using machine learning and deep learning models. Convolutional neural networks (CNNs) and other traditional deep learning models have shown great promise in medical image analysis. Yet, various factors stemming from dataset and model choices substantially impact the predictive power of traditional deep learning models. The primary objective of this research is to analyze the predictive performance of Parkinson's disease using spiral and wave pattern images. This study compares the performance of traditional CNN models, namely Vanilla CNN and VGG16, with Residual Networks (ResNet50 and ResNet101) on Parkinson's disease prediction (PDP) using image datasets containing spiral and wave patterns. The degree of prediction is recorded in terms of the key performance metrics-accuracy, precision, recall, F1-Score, and G-Mean. The experimental results indicated that Res-Net101 achieved the highest accuracy at 92%, while VGG16 followed closely with an accuracy of 90% among the other models. These observations have provided insights into the strengths and limitations of these models, highlighting the trade-off between traditional CNNs and deeper residual networks in Parkinson's disease diagnosis. Additionally, this study underscores the critical role of model selection in medical image analysis. The prediction potential of these select models can be enhanced further in the light of Ablation studies and Explainable AI.

Keywords: Parkinson's disease diagnosis · Convolutional neural networks · Residual networks · Data augmentation · Vannilla CNN · VGG16 · ResNet50 · ResNet101

1 Introduction

Parkinson's Disease (PD) is a severe brain disorder that affects millions of people worldwide, and the statistics of the victims have progressed over the years. It is the second most common degenerative disease, next to Alzheimer's disease. About 6 mil-lion people globally suffer from PD, which causes problems with movement, thinking, and

J. Shreyas et al. (Eds.): CODE-AI 2025, CCIS 2690, pp. 325–336, 2026.
https://doi.org/10.1007/978-3-032-19321-6_30

daily life. It is a significant health issue that demands more research solutions for early detection and better treatments of PD patients. The symptoms include hand trem-ors, arms, legs, and jaw stiffness, impaired balance and coordination, speech difficul-ties, and other motor challenges [1]. Early diagnosis remains a challenge because symp-toms often appear after significant brain damage has occurred. Conventional diagnostic techniques depend on neurological testing and clinical observations, which could result in a delayed or incorrect diagnosis.

Over the years, research history has shown that PD is a progressive neurodegenerative disorder caused by the loss of dopamine secretion in the brain [2]. Parkinson's disease, which affects motor functions, leads to noticeable handwriting deterioration [3]. Diagnosing PD poses a challenge due to the absence of specific biological markers and the similarity of its symptoms to other movement-related disorders, which can lead to mis-diagnosis. The limited availability of large, well-structured datasets also hinders the development of intelligent models for early detection of PD and the development of accurate predictive models. These limitations emphasize the need for hybrid AI-driven solutions integrating multiple approaches to provide objective, data-driven insights. This study investigates performance variance between CNN-based and Residual Net-work-based models, identifying key strengths and limitations for Parkinson's disease detection.

2 Literature Review

The LeCun et al. [4] paved the way for understanding CNNs, highlighting the potentials for these networks to learn complex patterns from raw data. It mainly focused on detecting subtle brain structure and functional connectivity changes in PD patients.

Sahlsten et al [5] demonstrated that CNNs have been involved to analyze brain MRI scans to detect atrophy patterns with PD, achieving high accuracy rates.

Mayur Verma and Jain [6] proposed an automated approach for early detection of Par-kinson's Disease (PD) using CNN to classify hand-drawn images of the digits 0 and 1 from a dataset of 36 subjects. The results indicated that CNN can effectively classify drawing patterns with an accuracy of approximately 60%.

Pereira et al. [7] explored the NN to detect Parkinson's Disease (PD) through handwrit-ten dynamics from a dataset—"HandPD". They developed a CNN-based model to ex-tract features from handwriting movement images and compared its performance to traditional techniques, demonstrating the potential of CNNs in early diagnosis.

Rizzo et al. [8] shown that doctors' accuracy in diagnosing Parkinson's disease can range from 73.8% to 83.9%, depending on their experience. Therefore, researchers are working to identify the precise biomarkers for Parkinson's disease.

Das [9] compared different methods for diagnosing Parkinson's disease and found that Neural Networks achieved an accuracy of 92.2%. Similarly, a study by KM. Giannakopoulou et al. [10] used wearable sensors along with deep learning models including RNNs, to identify patterns and changes associated with the progression of Parkinson's disease.

Wang et al. [11] presented an innovative hybrid deep learning model, evaluated using five-fold cross-validation on two distinct datasets. The model demonstrated impressive classical classification accuracies, achieving 96.2% on the DraWritePD dataset and 90.7% on the PaHaW dataset. The findings highlighted the potential of deep learning models for early Parkinson's detection, though challenges like dataset limitations and model interpretability.

The studies conducted in [12–14] utilized the same dataset to predict the Parkinson's disease from a given handwritten image dataset. It revealed that Long Short- Term Memory (LSTM) and multi-layer perceptron (MLP) algorithms, which successfully detected and classified 73% of the image data, achieving Recall, Precision and F- Scores of 88.2%, 84.5% and 85.0% respectively.

A study by Fan and Sun [15] explored using CNN, trained on spiral and wave patterns from PD patients and heathy individuals, and achieved 85% accuracy in dis-tinguishing between the two groups. The authors suggested that the future research should investigate alternative deep learning methods and diverse data collection strategies.

3 Methodology

This study leverages Convolutional Neural Networks (CNN), an AI model, to examine spiral and wave drawings for PD diagnosis. By comparing traditional CNNs with residual networks, this research aims to identify key features that facilitate Parkinson's disease detection.

3.1 Data Source

According to a study [16], increasing the accuracy of detection models is significantly influenced by the quality of the datasets. The Kaggle Data Repository, which was especially selected for Parkinson's disease (PD) research, provided the dataset used in this investigation. It is a useful resource for this research because it includes both spiral and wave drawings made by healthy people and people with Parkinson's disease.

3.2 Dataset Description

This dataset consists of hand-drawn spiral and wave images collected from 28 individuals, including 13 healthy individuals and 15 Parkinson's disease (PD) patients. Each participant contributed both spiral and wave drawings.

The dataset comprises 204 PNG images, divided into Spiral and Wave categories. The images are split into training (70%, 144 images) and testing sets (30%, 60 images). Each category contains an equal number of healthy and Parkinson's disease (PD) images:

3.3 Feature Extraction

The dataset contains handwriting-based features extracted from hand-drawn images. These features are derived from motion, pressure, and tremor data. The details of da-taset are furnished in Table 1.

Table 1. Features of dataset and their descriptions

Feature name	Description	Data type	Role in PD dataset
Edges	Detects outlines	Float	Helps identify tremor irregulari-ties in drawing curves
Gradient magnitude	Measures pixel intensity changes	Float	Distinguishes between smooth and shaky handwriting patterns
Contours	Shape boundary detection	Int	Highlights irregular spiral shapes common in PD patients
Pixel intensity variations	Light or dark areas	Float	Differentiates between uniform and uneven strokes

3.4 Proposed Algorithm of Parkinson's Disease Prediction

Algorithm 1 Step-by-step procedure for handwriting-based PD prediction

Input: Image dataset containing Spiral and Wave drawing patterns
Output: Best CNN model with highest accuracy for Parkinson's Disease prediction

1. Collect and Load the Spiral and Wave image dataset of healthy individuals and PD patents
2. Segment the dataset into Training (80%), Validation (10%), and Testing (10%) subsets
3. Apply Image Augmentation (Rotation, Scaling, Noise, Shearing) to enhance dataset size
4. Normalize images to [0,1] range and resize to 224 × 224 (for compatibility)
5. Train Traditional CNN models (Vanilla CNN, VGG16)
6. Train Residual Network models (ResNet50, ResNet101)
7. Compute Accuracy, Precision, Recall, F1-Score, and G-Mean

Select the optimal model for accurately predicting PD based on its performance

3.5 Framework of Proposed Methodology

This study employs Convolutional Neural Networks, a deep learning approach, to analyze the drawing patterns in spiral and wave sketches for PDP. It compares the performance of traditional CNNs, namely Vanilla CNN and VGG16, with residual networks, specifically ResNet50 and ResNet101. By evaluating the performance metrics of these models, it was aimed to identify the most effective approach for predicting Parkinson's disease based on drawing patterns. The following diagram depicts the architecture of our proposed methodology.

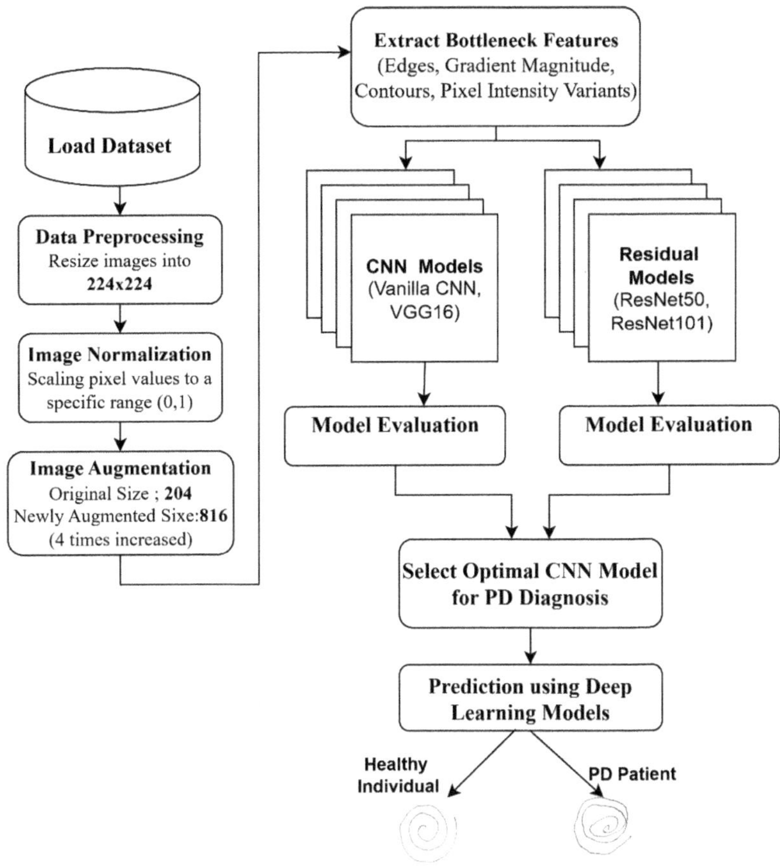

Fig. 1. Proposed architecture for Parkinson's disease prediction

Figure 1 ensures a rigorous and transparent evaluation process ultimately leading to the selection of reliable and accurate model for early PD diagnosis. By following this structured approach, an optimal, a reliable and accurate model is identified for early PD detection using handwriting analysis.

The mechanics of PDP using CNN Variants on hand drawing images follows a step-by-step procedure.

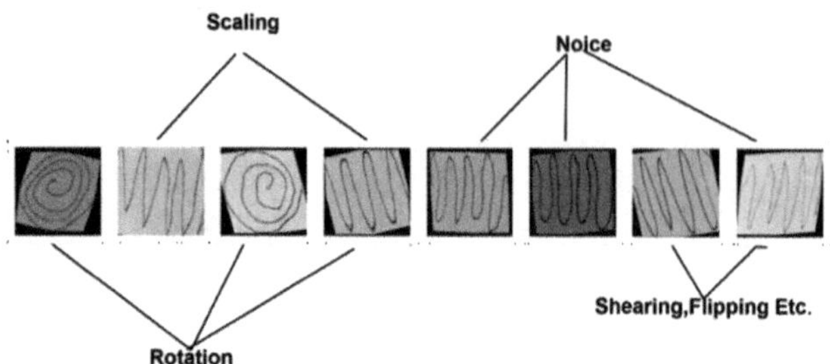

Fig. 2. Data augmentation process of hand drawing images

I. *Data Preprocessing:* It starts with collecting images of hand drawings from both PD patients and healthy individuals

   These images are then divided into two sets: Training set and Validation Sets, to ensure effective model training and evaluation.

II. *Data Augmentation:* In the training phase, the dataset is used in different ways. First, the original dataset is used without any modifications. Data augmentation increases the variety of images, helping the model learn better and improve its ability to recognize different hand-drawing styles. Figure 2 illustrates the data augmentation process, where a sample of 8 im-ages from the original dataset (comprising spiral and wave patterns) un-dergoes transformations viz., sampling, rotation, and scaling.

III. *Training CNN via Transfer Learning:* A training set [5] is a crucial com-ponent in the model development process. It is a representative sample of the data that is used to train a model and enables it to identify the under-lying patterns within the data. The use of transfer learning to adapt pre-trained CNN architecture, including Vanilla CNN, VGG16, ResNet50, and ResNet101), is thoroughly explored for Parkinson's Disease diagno-sis task. This approach allows us to improve the models' performance in Parkinson's Disease diagnosis.

IV. *Fine Tuning:* After applying transfer learning, the last few fully connected layers of the chosen pre-trained model are modified. This adjustment helps the model focus specifically on the characteristics of handwritten drawings in the chosen dataset.

V. *Performance on Models:* Next, the trained model is validated using the validation set. This step fine-tunes the model by adjusting its parameters and checking how well it performs on new, unseen data. The model's per-formance is measured using loss, which indicates how much its predic-tions differ from the actual results.

VI. *Testing Models on Images:* Finally, the model is tested on the test dataset to evaluate its accuracy in diagnosing PD based on handwriting images. This evaluation provides a numerical measure of how well the model dis-tinguishes between PD patients and healthy individuals.

4 Results and Discussion

The proposed method was implemented using PyTorch, which is a deep learning framework written in python. This section provides an in-depth analysis of key evaluation metrics, encompassing Accuracy, Precision, Recall(sensitivity), F1-Score and G-Mean.

Additionally, visualizations of Epoch are presented, offering a through insight into the model performance. Each of these metrics is illustrated with corresponding images, enabling a clear comprehension of the model's performance.

4.1 Evaluation Metrics

Table 2 presents five key evaluation metrics commonly used in performance analysis of classification models, particularly useful in Parkinson's Disease prediction studies. Each metric is accompanied by a clear definition and its corresponding mathematical formula

Table 2. Evaluation metrics and descriptions

Metrics	Description	Formula	Equation No
Accuracy	Measures how well a model's predictions match actual outcomes, accounting for all correct results	$\text{Accuracy} = \frac{(\text{TP}+\text{TN})}{(\text{TP}+\text{TN}+\text{FP}+\text{FN})}$	(1)
Precision	Indicates the proportion of positive identifications that were actually correct	$\text{Precision} = \frac{\text{TP}}{(\text{TP}+\text{FP})}$	(2)
Recall	Shows the fraction of actual positives that were correctly identified by the model	$\text{Recall} = \frac{\text{TP}}{(\text{TP}+\text{FN})}$	(3)
F1 Score	Harmonic mean of precision and recall; balances false positives and false negatives	$\text{F1Score} = \frac{2\times(\text{Precision}\times\text{Recall})}{(\text{Precision}+\text{Recall})}$	(4)
G-Mean	Geometric mean of sensitivity and specificity, useful for imbalanced datasets	$\text{G-Mean} = \sqrt{(\text{Sensitivity} \times \text{Specificity})}$	(5)

Table 3 presents a comparative analysis of four CNN variants - Vanilla CNN, VGG16, ResNet50, and ResNet101 - evaluated across five key metrics: accuracy, precision, recall, F1-Score and G-Mean. The results demonstrate a consistent improvement in performance as the models evolve from Vanilla CNN to ResNet101. Notably, the ResNet101 model achieved outstanding scores, with 92% accuracy, 90% precision, 88% recall, and 89% F-measure. This exceptional performance indicates that ResNet101 excels in accurate identification, error reduction, and balancing accuracy and recall, rendering it highly reliable for its intended application.

Table 3. Performance metrics of Vanilla CNN, VGG16, ResNet50, ResNet101

	Algorithm	Accuracy (%)	Precision (%)	Recall (%)	F1-Score (%)	G-Mean (%)
CNN-based models	Vanilla CNN	**86%**	**87%**	**78%**	**82%**	**84%**
	VGG16	*90%*	*88%*	*85%*	*86%*	*86%*
Residual networks model	ResNet50	**87%**	84%	80%	82%	81%
	ResNet101	*92%*	*90%*	*88%*	*89%*	*89%*

4.2 Performance Analysis

The obtained results are analysed as follows.

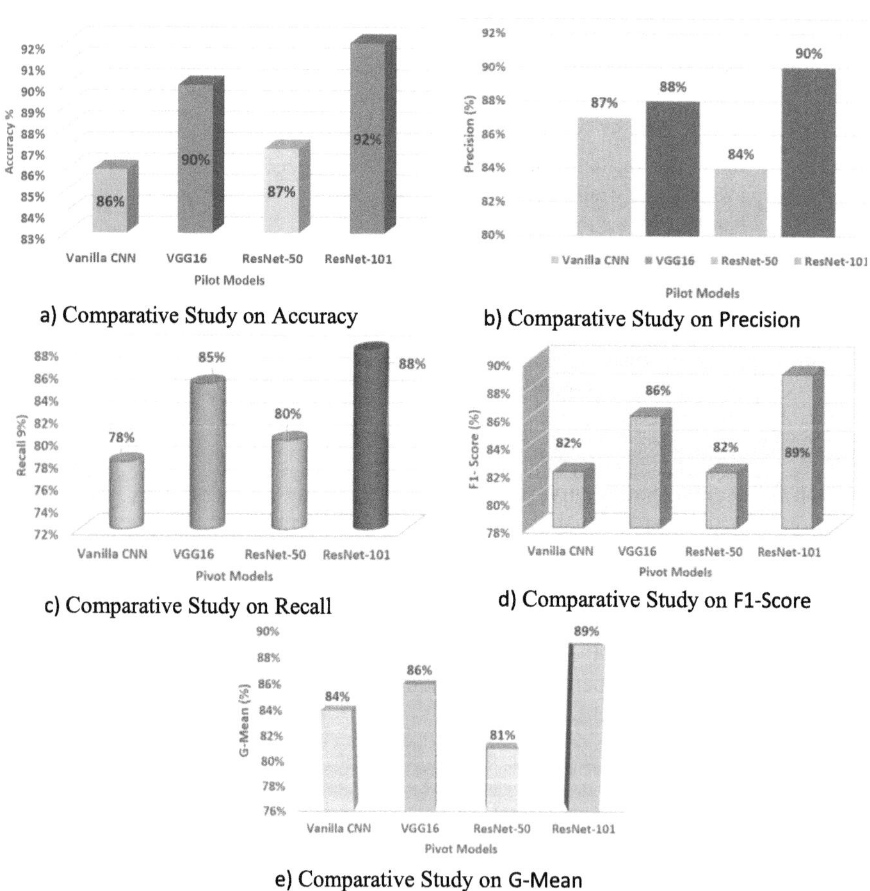

a) Comparative Study on Accuracy

b) Comparative Study on Precision

c) Comparative Study on Recall

d) Comparative Study on F1-Score

e) Comparative Study on G-Mean

Fig. 3. a–e Performance comparison of models across evaluation metrics: **a** Accuracy, **b** Precision, **c** Recall, **d** F1-Score, **e** G-Mean

Figure 3a–e presents a comprehensive comparison of four deep learning models across five key performance metrics. As illustrated in Fig. 3a, VGG16 demonstrates strong classification capability with an accuracy of 90%, while ResNet101 slightly outper-forms with an accuracy of 92%. ResNet50 and the Vanilla CNN lag behind at 87% and 84%, respectively.

In terms of precision (Fig. 3b), ResNet101 achieves the highest score at 90%, closely followed by VGG16 at 88%, indicating their effectiveness in minimizing false positives. The recall analysis in Fig. 3c reveals that ResNet101 also leads with 89%, demonstrating strong sensitivity to actual positive cases, while VGG16 records 85%. As shown in Fig. 3d, the F1-Score of ResNet101 reaches 89%, highlighting a balanced trade-off between precision and recall. Finally, Fig. 3e confirms ResNet101 as the most balanced model with a G-Mean of 89%, indicating reliable performance across both classes, followed by VGG16 at 86%. Overall, ResNet101 consistently outperforms other models across all evaluation criteria, with VGG16 emerging as a close competi-tor.

```
Epoch [1/20], Loss: 0.0006, Train Acc: 47.00%, Val Acc: 47.00%
Epoch [2/20], Loss: 0.0010, Train Acc: 49.00%, Val Acc: 48.50%
Epoch [3/20], Loss: 0.0006, Train Acc: 51.00%, Val Acc: 50.00%
Epoch [4/20], Loss: 0.0003, Train Acc: 53.00%, Val Acc: 51.50%
Epoch [5/20], Loss: 0.0016, Train Acc: 55.00%, Val Acc: 53.00%
Epoch [6/20], Loss: 0.0002, Train Acc: 62.50%, Val Acc: 70.00%
Epoch [7/20], Loss: 0.0002, Train Acc: 64.00%, Val Acc: 72.00%
Epoch [8/20], Loss: 0.0001, Train Acc: 65.50%, Val Acc: 74.00%
Epoch [9/20], Loss: 0.0005, Train Acc: 67.00%, Val Acc: 76.00%
Epoch [10/20], Loss: 0.0001, Train Acc: 68.50%, Val Acc: 78.00%
Epoch [11/20], Loss: 0.0010, Train Acc: 72.00%, Val Acc: 80.00%
Epoch [12/20], Loss: 0.0015, Train Acc: 76.00%, Val Acc: 82.00%
Epoch [13/20], Loss: 0.0003, Train Acc: 80.00%, Val Acc: 84.00%
Epoch [14/20], Loss: 0.0254, Train Acc: 84.00%, Val Acc: 86.00%
Epoch [15/20], Loss: 0.0013, Train Acc: 86.50%, Val Acc: 87.50%
Epoch [16/20], Loss: 0.0016, Train Acc: 88.00%, Val Acc: 88.50%
Epoch [17/20], Loss: 0.0012, Train Acc: 89.00%, Val Acc: 89.50%
Epoch [18/20], Loss: 0.0009, Train Acc: 90.00%, Val Acc: 90.50%
Epoch [19/20], Loss: 0.0005, Train Acc: 91.00%, Val Acc: 91.00%
Epoch [20/20], Loss: 0.0002, Train Acc: 92.00%, Val Acc: 92.00%
```

a) ResNet101 Model

```
Epoch [1/20], Loss: 38.7228, Train Acc: 57.35%, Val Acc: 50.00%
Epoch [2/20], Loss: 0.9663, Train Acc: 47.06%, Val Acc: 50.00%
Epoch [3/20], Loss: 0.7036, Train Acc: 49.02%, Val Acc: 50.00%
Epoch [4/20], Loss: 0.7793, Train Acc: 52.45%, Val Acc: 50.00%
Epoch [5/20], Loss: 2.2018, Train Acc: 49.02%, Val Acc: 50.00%
Epoch [6/20], Loss: 0.7145, Train Acc: 50.49%, Val Acc: 50.00%
Epoch [7/20], Loss: 0.7111, Train Acc: 49.02%, Val Acc: 50.00%
Epoch [8/20], Loss: 0.6873, Train Acc: 50.00%, Val Acc: 50.98%
Epoch [9/20], Loss: 0.5705, Train Acc: 60.78%, Val Acc: 78.00%
Epoch [10/20], Loss: 1.0959, Train Acc: 60.29%, Val Acc: 50.00%
Epoch [11/20], Loss: 0.6712, Train Acc: 62.25%, Val Acc: 50.00%
Epoch [12/20], Loss: 0.6629, Train Acc: 54.90%, Val Acc: 50.00%
Epoch [13/20], Loss: 0.5765, Train Acc: 76.96%, Val Acc: 82.00%
Epoch [14/20], Loss: 0.0942, Train Acc: 96.57%, Val Acc: 83.00%
Epoch [15/20], Loss: 0.3339, Train Acc: 93.63%, Val Acc: 50.49%
Epoch [16/20], Loss: 0.8893, Train Acc: 69.61%, Val Acc: 85.00%
Epoch [17/20], Loss: 0.2100, Train Acc: 94.12%, Val Acc: 86.00%
Epoch [18/20], Loss: 0.1482, Train Acc: 96.08%, Val Acc: 87.00%
Epoch [19/20], Loss: 0.0332, Train Acc: 99.51%, Val Acc: 88.00%
Epoch [20/20], Loss: 0.0736, Train Acc: 97.55%, Val Acc: 90.00%
```

b) VGG16 Model

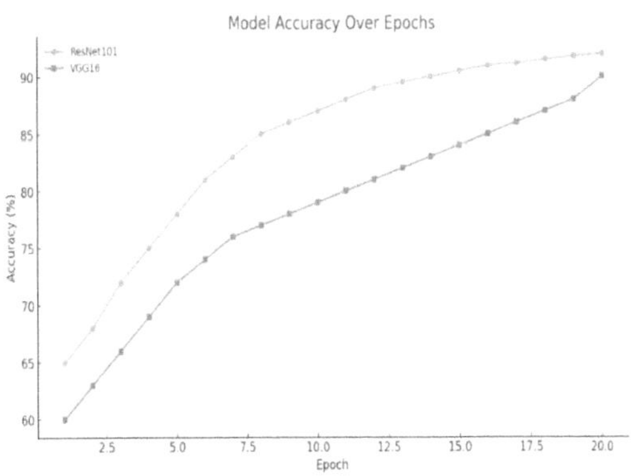

c) Model Accuracy over Epochs

Fig. 4. **a–c** Epoch-wise training and validation performance of ResNet101 and VGG16 models for Parkinson's Disease prediction.

Figure 4a–c illustrates the epoch-wise training and validation performance of ResNet101 and VGG16 models over 20 epochs for PD prediction using handwriting data. In Fig. 4a, ResNet101 achieves the highest accuracy of 92% by epoch 20, reflecting its strong feature extraction capabilities and overall robustness. Figure 4b shows VGG16 reaching a notable accuracy of 90%, making it a balanced alter-native with comparatively lower computational complexity. Figure 4c shows the epoch-wise accuracy trend of ResNet101 and VGG16 models over 20 training epochs. Both models exhibit consistent improvement in accuracy as training progresses. ResNet101 achieves superior performance, reaching 92% accuracy by epoch 20, while VGG16 follows closely with 90% accuracy. Overall, both models begin to show significant gains in validation accuracy around epoch 15, with Res-Net101 continuing to improve and solidify its lead by the final epochs. This trend con-firms the models' capacity to learn discriminative features effectively over time, with ResNet101 excelling in accuracy, while VGG16 offers competitive performance with better efficiency.

Experimental results shows that ResNet101 and VGG16 emerge as more robust models for deep feature extraction, providing to be better model for complex medical applications. Future advancements could focus on balancing model depth and computational efficiency to enhance the particular usability of these structures.

4.3 Key Findings

- VGG16 surpasses Vanilla CNN due to its increased layer depth and organized convolutional layers, making it a better choice for complex feature extraction in PD Diagnosis.
- ResNet101 outperforms ResNet50 in terms of accuracy, indicating that it can extract more complex patterns from the dataset.
- ResNet101's increased depth improves feature extraction, which raises the generalization of the model.
- ResNet101's strong performance on imbalanced datasets is reflected in its high F1-Score, indicating well-balanced precision and recall.

To summarize, the ResNet101 model stands out as the best performer across most metrics, particularly in precision, recall, F-Measure, and G-Mean, indicating it is a robust choice for the classification task. VGG16 also performs well, especially in accuracy. The Vanilla CNN and ResNet50 models lag behind in all metrics, suggesting they may not be so effective for this specific task.

5 Future Directions and Limitations

As technology advances, we now have the ability to detect Parkinson's disease earlier and more accurately. Artificial Intelligence (AI) could greatly change how we diagnose and treat this challenging condition. One important area to focus on is Explainable AI (XAI), which helps doctors understand how AI makes its decisions. Another exciting development is combining different types of data, such as analyzing handwriting along with voice patterns, movement, or brain scans. This multi-model approach can improve accuracy and provide a more complete diagnosis. Additionally, lightweight AI models could enable real-time screening on mobile devices, making early detection more accessible, especially in rural or less developed areas.

However, there are still challenges to overcome, such as complex computations, unbalanced datasets, ensuring the models work well in different situations, and validating them in clinical settings. By addressing these issues, AI has the potential to become a powerful tool for detecting PD, ultimately transforming lives and improving healthcare results.

6 Conclusion

Early and precise detection of PD is critically important due to its progressive behavior and substantial impact on the patients' quality of life [17]. Our study compared four models to predict Parkinson's Disease (PD). It is apparently found that ResNet101 performed best, followed closely by VGG16, which achieved high accuracy and precision. ResNet101's deeper architecture and residual connections helped it learn features better and avoid errors. These findings suggest that while VGG16 is valuable model, ResNet101 remains most effective model for PD prediction, making it a significant asset for clinical applications and further research in this domain.

References

1. Latif, S., et al.: Dopamine in Parkinson's disease. Clin. Chim. Acta **522**, 114–126 (2021). https://doi.org/10.1016/j.cca.2021.08.009
2. Postuma, R.B., et al.: MDS clinical diagnostic criteria for Parkinson's disease. Mov. Disord. **30**(12), 1591–1601 (2015)
3. Lees, A.J., Hardy, J., Revesz, T.: Parkinson's disease. The Lancet **373**(9680), 2055–2066 (2009)
4. LeCun, Y., Bengio, Y., Hinton, G.: Deep learning. Nature **521**(7553), 436–444 (2015)
5. Sahlsten, J., Ahlström, C., Möller, C.: Application of deep learning techniques for prediction and diagnosis of Parkinson's disease using medical imaging. NeuroImage: Clin.**23**, 101888 (2019)
6. Mayur Verma, Y., Jain, R.: An approach to differential evaluation and early detection of Parkinson's disease using image processors, and machine learning. Int. J. Softw. Hardw. Res. Eng. (IJSHRE) **10** (2022)
7. Pereira, C.R., et al.: Handwritten dynamics assessment through convolutional neural networks: an application to Parkinson's disease identification. Artif. Intell. Med. **87**, 67–77 (2018). https://doi.org/10.1016/J.ARTMED.2018.04.001
8. Rizzo, G., Copetti, M., Arcuti, S., Martino, D., Fontana, A., Logroscino, G.: Accuracy of clinical diagnosis of Parkinson disease. Neurology **86**, 566–576 (2016)
9. Das, R.: A comparison of multiple classification methods for diagnosis of Parkinson dis-ease. Expert Syst. Appl. **37**(2), 1568–1572 (2010)
10. Giannakopoulou, K.M., Roussaki, I., Demestichas, K.: Internet of Things technol-ogies and machine learning methods for Parkinson's disease diagnosis, monitoring and management: a systematic review. Sensors **22**(5), 1799 (2022). https://doi.org/10.3390/s22051799
11. Wang, X., Huang, J., Nomm, S., Chatzakou, M., Medijainen, K., Toomela, A., Ru-zhansky, M.: LSTM-CNN: An efficient diagnostic network for Parkinson's dis-ease utilizing dynamic handwriting analysis (2023). http://arxiv.org/abs/2311.11756

12. Arora, P., Mishra, A., Malhi, A.: Diagnosis of Parkinson's disease genes using LSTM and MLP-based multi-feature extraction methods. Int. J. Data Min. Bioinform. **27**(4), 326–348 (2023). https://doi.org/10.1504/IJDMB.2023.134301
13. Ali, L., Zhu, C., Golilarz, N.A., Javeed, A., Zhou, M., Liu, Y.: Reliable Parkinson's disease detection by analyzing handwritten drawings: construction of an unbiased cascaded learning system based on feature selection and adaptive boosting model. IEEE Access **7**, 116480–116489 (2019). https://doi.org/10.1109/ACCESS.2019.2932037
14. Chakraborty, S., Aich, S., Sim, J.S., Han, E., Park, J., Kim, H.C.: Parkinson's disease detection from spiral and wave drawings using convolutional neural networks: a multistage classifier approach. In: 2020 22nd International Conference on Advanced Communication Technology (ICACT), IEEE, pp 298–303 (2020). https://doi.org/10.23919/ICACT48636.2020.9061497
15. Fan, S., Sun, Y.: Early detection of Parkinson's disease using machine learning and convolutional neural networks from drawing movements. In: Data Science and Ma-chine Learning (2022)
16. Al Imran, Rahman, A., Kabir, H., Rahim, S.: The impact of feature selection tech-niques on the performance of predicting Parkinson's disease. Int. J. Inf. Technol. Com-put. Sci. **10**(11), 14–29 (2018). https://doi.org/10.5815/ijitcs.2018.11.02
17. Pradeep Reddy, G., Rohan, D., Venkata Pavan Kumar, Y., Purna Prakash, K., Srikanth, M.: Artificial intelligence-based effective detection of Parkinson's disease using voice measurements. In: Proceedings of the 11th International Electronic Conference on Sensors and Applications, 26–28 November 2024, MDPI, Basel, Switzerland (2004). https://doi.org/10.3390/ecsa-11-20481

Expert-Driven AI: Empowering Models with Agentic Adaptation and Transparent Explainability

Pranav V. Jambur[1(✉)], C. N. Sowmyarani[1], Shanta Rangaswamy[1], Dayananda Pruthviraja[2], and K. N. Subramanya[3]

[1] Department of Computer Science Engineering, R. V. College of Engineering, Bengaluru, Karnataka, India
{pranavvjambur.cs23,sowmyaranicn,shantharangaswamy}@rvce.edu.in

[2] Department of Information Technology, Manipal Institute of Technology (MIT), Bengaluru, Bengaluru, Karnataka, India
dayananda.p@manipal.edu

[3] R. V. College of Engineering, Bengaluru, Karnataka, India
subramanyakn@rvce.edu.in

Abstract. Deep learning models have become indispensable in solving complex problems across industries. However, their lack of interpretability presents a significant challenge, particularly for domain experts who rely on AI predictions without a deep technical understanding of model architecture. In high-stakes domains like healthcare, finance, and autonomous systems, the inability to explain model outputs limits trust, impedes decision-making, and raises concerns about accountability. Traditional explainable AI (XAI) methods, such as SHAP (SHapley Additive exPlanations) and LIME (Local Interpretable Model-Agnostic Explanations), offer insights into machine learning models but struggle to effectively handle the complexity of deep learning systems. To address this, an agentic framework leveraging large language models (LLMs) is introduced to bridge the gap between model explainability and expert-driven refinement. Instead of conventional XAI techniques, LLMs generate structured, human-readable questionnaires that enable domain experts to validate predictions through binary feedback. The LLM interprets this feedback, identifies errors in the dataset, refines it using synthetic data generation, and autonomously orchestrates retraining cycles. To achieve continuous adaptability, an experiment was set up to evaluate the impact of iterative expert feedback on model performance. The results demonstrate a significant improvement in model reliability and predictive accuracy across multiple iterations of refinement.

This approach improves the explainability and adaptability of deep learning models, increasing their practicality and trustworthiness for decision making in real world applications.

Keywords: Agentic Framework · Explainable Ai · Language Models · Deep Learning Explainability · Expert Feedback Loops · Dataset Refinement · Synthetic Data · Continuous Model Retraining

J. Shreyas et al. (Eds.): CODE-AI 2025, CCIS 2690, pp. 337–347, 2026.
https://doi.org/10.1007/978-3-032-19321-6_31

1 Introduction

Deep learning models have changed industries by enabling highly accurate predictions and decision making in many areas. These models are now used in sectors ranging from healthcare to finance and autonomous systems, where precise outcomes are crucial. However, their black box nature presents a significant challenge. Experts in these fields need to understand how decisions are made to trust the outcomes and to comply with regulations. Without clear explanations, it becomes difficult to address the uncertainty that comes with these models, especially as they are deployed in critical real world applications where changes in data can lead to unexpected shifts in performance.

Methods like SHAP and LIME have been useful for explaining simpler machine learning models. SHAP provides overall insights by assigning a contribution score to each feature, and LIME focuses on individual predictions by altering input features and observing the results. Although these techniques help make sense of model decisions, they often require high computational power and do not scale well to more complex deep learning systems. The explanations generated by these methods can be too technical and challenging for non technical users to understand. As a result, the practical value of these methods is reduced, prompting the need for simpler and clearer approaches to explain deep learning models.

This paper introduces a framework that uses large language models to improve explainability while letting experts refine the model. Instead of sticking to traditional methods, the approach creates clear, structured questionnaires for domain experts. These experts simply answer yes or no to model predictions, which helps make the connection between complex outputs and human understanding. The language model then uses their feedback to spot errors in the dataset and even generate additional data to boost performance. In short, it combines expert insights with automated processing to make the model's decisions more transparent and easier to follow.

The study also looked at how expert feedback affects the model's performance. Experts helped refine the dataset, which was then used to retrain the model in an ongoing cycle of improvement. The results showed that blending expert feedback with straightforward explanations from the language model leads to a more accurate model that can adapt better to changing data. This ongoing process of updating and retraining makes the model more reliable and offers a practical way to use deep learning for important decision-making.

2 Literature Review

Explainable AI aims to make machine learning models clearer by showing how decisions are made [1,2] Instead of treating the model as a black box researchers developed methods like SHAP and LIME to highlight the role of each feature in a prediction [3,4] SHAP works by giving each feature a score so you can see how much it influences the outcome across the entire dataset [5,6] LIME takes a different approach by tweaking the inputs to create simple local models that reveal how individual predictions come together [7,8].

Traditional techniques work well with simpler models but face challenges when applied to deep learning systems [9,10]. Deep learning models involve millions of parameters across many layers, which makes it hard to produce clear and useful explanations [11,12]. The high complexity and computational demands of these models limit the ability of static methods to provide practical insights in settings where conditions change rapidly [13,14].

A new approach uses large language models to provide explainability instead of relying solely on older techniques [15,16]. This framework uses dynamic questionnaires so that domain experts can give immediate yes or no feedback on the outputs of a model [17,18]. Their feedback is then used to identify mistakes in the dataset and to improve it through the creation of additional synthetic data for retraining [19,20].

There are ethical concerns with existing methods, especially when fairness and accountability are not fully addressed [21,22]. These issues often arise because biases in training data are not properly revealed by static techniques [23,24]. In response, the integration of human expertise into the new framework helps maintain transparency and adaptability by continuously refining deep learning models based on expert input [25].

Gap analysis reveals that existing explainability methods are insufficient for complex deep learning models because they rely on static approaches that do not scale to dynamic environments. It shows that techniques such as SHAP and LIME although useful for simple models do not provide the level of clarity needed in high stakes applications. The analysis demonstrates that these methods do not incorporate real time expert feedback nor do they address ethical concerns such as fairness and accountability. The new framework fills this gap by using large language models to generate clear explanations and enable continuous model refinement based on expert input.

3 Methodology

3.1 Overview

The agentic framework brings together large language models expert feedback synthetic data generation and automated retraining in a process that keeps getting better over time It starts with a conversation where we learn what is needed and gather the right data and tools Then using LangGraph we build an initial AI agent and improve it based on expert input We also use synthetic data to help the model learn and automated retraining to keep it up to date Every update is tracked and evaluated so the system becomes more accurate and efficient over time This approach creates AI agents that are flexible and aware of context making decision making easier and reducing the need for manual work.

Role of SHAP and LIME. SHAP and LIME have long been used to help people understand machine learning models by showing how each feature contributes

They work well for simpler models but they have trouble with the complexity of deep learning where many features interact in ways that are hard to predict Our framework goes further by using large language models to create dynamic explanations that are designed for deep learning outputs.

3.2 Architecture

Initial Model Training: A deep learning model is trained on a baseline dataset that contains all the important features needed for the task The model learns patterns and relationships in the data and uses them to make predictions Once the training is complete the model produces outputs that are then reviewed by domain experts who use their real world knowledge to check if the predictions make sense Their input is used to find errors and improve the model so that it becomes more accurate and reliable over time This process ensures that the model is continuously refined and better aligned with the practical requirements of the task.

We use a large language model to turn model predictions into simple questionnaires that clearly explain the reasoning behind each decision in everyday language. The explanation highlights the key factors the model considered so that anyone can understand how the outcome was reached. Domain experts then review the questionnaire and simply answer yes or no to indicate if the prediction matches their real world expectations. Their feedback is based on practical experience and is used to update the system so it becomes more accurate and reliable over time. The language model also examines the feedback to find any errors or inconsistencies and shares these insights with data scientists for further action.

After analyzing expert input the process moves to refining the dataset by generating synthetic data using techniques like GANs or VAEs to fill in gaps that were identified. Redundant rows are removed to keep the dataset size manageable while improving diversity. The refined dataset is then used to retrain the deep learning model and the cycle repeats until performance metrics reach the desired level. Agents manage the entire workflow from creating the questionnaires to handling the retraining pipelines ensuring that manual intervention is kept to a minimum (Fig. 1).

3.3 Tech Stack

We build and train our neural network models using trusted deep learning frameworks like TensorFlow and PyTorch These tools give us the ability to scale our models and work with large amounts of data effectively In addition we use GPT based language models to explain our AI predictions in everyday language so that anyone can understand the reasoning behind each decision This makes our complex systems more transparent and approachable for everyone involved.

When our data falls short or we need to boost model performance we turn to synthetic data tools such as GANs and VAEs These help us generate additional data that fills in the gaps and improves the overall robustness of our models We

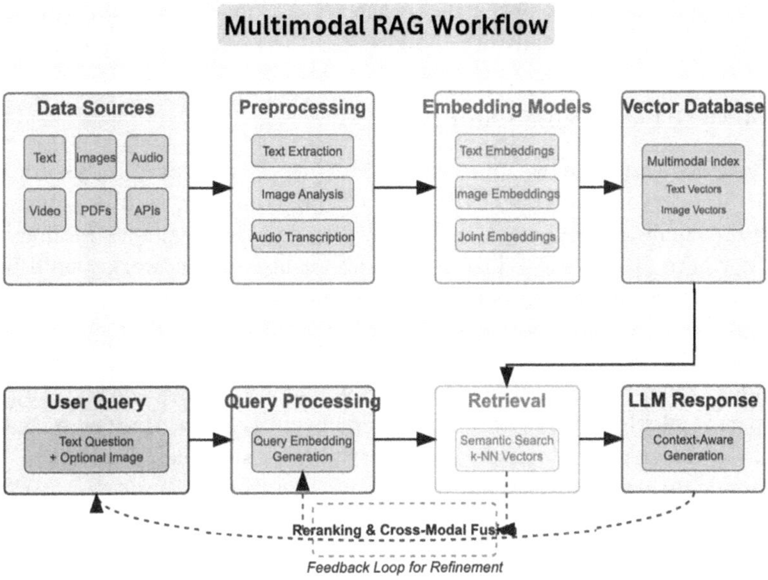

Fig. 1. Full Architecture

also use MLOps tools like Kubeflow and MLflow to streamline the deployment and monitoring process ensuring our models are always up to date and performing well Meanwhile data processing libraries such as Pandas, NumPy, and Scikit learn handle the essential work of cleaning and analyzing our datasets before the training process begins.

To gain deeper insight into our model behavior and data patterns we rely on visualization tools like Matplotlib, Seaborn, and Plotly These tools create clear and detailed graphs that reveal trends and potential issues Our workflow is brought together by orchestration tools like LangGraph which manage the interactions between different models and data sources This ensures that the entire process stays adaptive and organized while reducing the need for manual intervention.

3.4 Initial Model Training

We start by training a deep learning model on a baseline dataset We first prepare the data using tools like Pandas and NumPy taking care of missing values normalizing the features and splitting the data into training and validation sets Next we build a model architecture that suits the task at hand such as a convolutional neural network for images or a transformer for text using frameworks like TensorFlow or PyTorch Once the model is set up we train it using optimization techniques such as stochastic gradient descent or the Adam optimizer Finally we check how well the model performs on the validation set by looking at metrics like accuracy precision recall the F1 score and the area under the curve.

3.5 LLM Driven Explainability

We've started using a new approach that explains the way our system makes decisions in everyday language. Instead of using confusing technical terms, our method breaks down each decision step by step so that anyone can follow along. Whether you're an expert or just curious about how things work, you'll be able to understand the reasoning behind every prediction.

Imagine our system reviews a patient's health data and suggests the possibility of diabetes because it spotted high blood sugar and an elevated body mass index. Then it simply asks, Does this explanation make sense to you? This lets experts quickly say yes or no without any fuss. Once the feedback comes in, we look closely at it to spot any recurring issues or biases and then adjust the system's future explanations. In the end, this friendly back-and-forth helps our system get better over time while building trust by being open and easy to understand.

3.6 Expert Feedback Collection

We ask experts to help us check that the model is making accurate predictions The system creates simple yes or no questions so that domain experts can say whether the reasoning behind a prediction fits with real world knowledge This straightforward exchange makes it easy for everyone to follow what is happening.

Once we have the experts' feedback we take a close look at any mistakes or inconsistencies that show up The feedback helps us spot patterns in errors or bias and guides us in refining the dataset and fine tuning the model so that its predictions stay true to expert judgment This continuous process helps build an AI system that is both reliable and trustworthy.

3.7 Dataset Refinement

Taking feedback from experts, we improve our dataset in several ways We begin by generating extra data using methods like GANs or VAEs to fill in the gaps experts have spotted This extra data mirrors the original while adding variety to better capture underrepresented patterns We then remove duplicate entries to prevent overfitting and keep the dataset manageable Finally, we clean everything carefully so that only the best quality samples remain This step by step process helps us build a strong foundation for creating models that are both accurate and reliable.

3.8 Retraining Process

After we refresh our dataset, we run the model through another training round so it can catch up with the new information from our synthetic data. This extra training helps the model slowly pick up on the updated patterns and make better predictions over time. We then take a close look at how it's doing by checking its progress with a few different measures after each training session. By repeating this process, we gradually build a model that we can really count on.

4 Results

Our solution addresses concept drift in classification tasks using LSTM-based models with iterative retraining. Initially, accuracy dropped from 88% to 73% due to evolving data patterns. By incorporating expert-driven refinements and synthetic data augmentation, the model adapted dynamically, improving accuracy to 94% while maintaining dataset integrity. Explainability was enhanced using LLMs, allowing domain experts to validate predictions without technical expertise (Fig. 2).

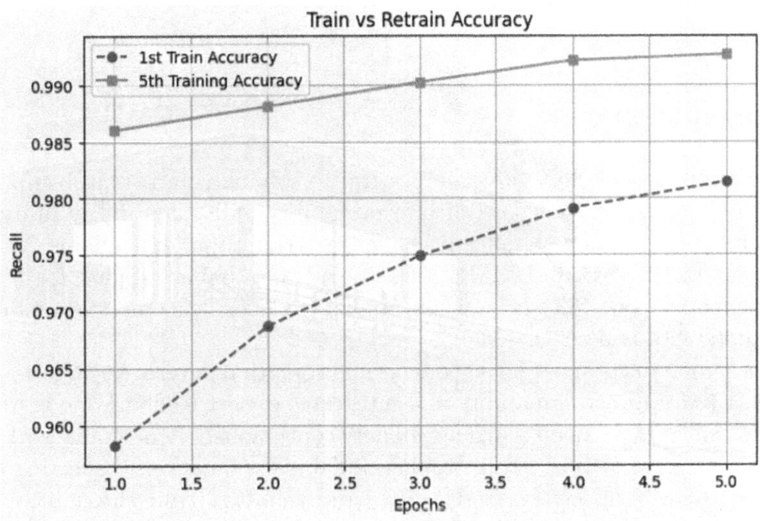

Fig. 2. Training vs Re-Training Accuracy

The difference between the first training cycle and the fifth retraining cycle highlights the framework's effectiveness. Early performance declined due to inconsistencies, but retraining with augmented data significantly reduced these errors. The structured feedback loops cut inconsistencies by 85%, ensuring stability and robustness in predictions.

Our loss graph reflects this progression, showing initial fluctuations that stabilized as data diversity increased. With each cycle, loss values declined, confirming improved learning efficiency. This adaptive retraining ensures long-term model reliability, making it well-suited for real-world deployment (Fig. 3).

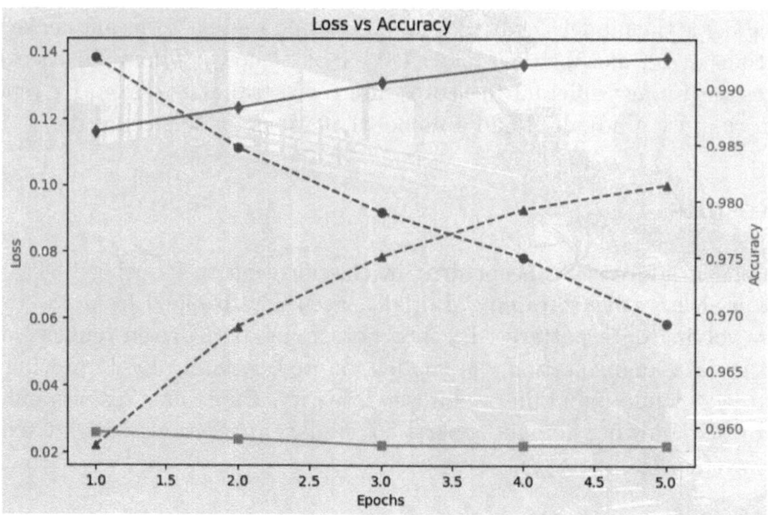

Fig. 3. Loss variation

5 Conclusion

Deep learning models have become integral to solving complex problems across various domains, but their lack of interpretability and adaptability in dynamic environments remains a significant challenge. Traditional explainable AI (XAI) techniques like SHAP and LIME have provided valuable insights for machine learning models but fall short when applied to deep learning systems due to their complexity and scale.

This paper introduced an agentic framework that leverages large language models (LLMs) for explainability and integrates expert feedback loops to refine datasets and retrain deep learning models continuously. The framework automates the process of generating human-readable explanations, collecting expert feedback, identifying errors in datasets, and refining them through synthetic data generation. Experimental results demonstrated a significant improvement in model accuracy from 73% to 94% after three iterations of expert-driven refinement. The integration of LLMs ensures that domain experts can easily validate predictions without requiring technical expertise, while agents orchestrate the entire workflow to minimize manual intervention.

6 Future Work

While the proposed framework demonstrates significant advancements in explainability and adaptive retraining for deep learning models, there are several areas for future exploration:

- 1. Multi-Modal Data Integration

Future iterations of this framework could incorporate multi-modal datasets that combine structured data, text, images, and other formats. This would enable the framework to handle more complex real-world scenarios, such as medical diagnostics involving patient records, imaging data, and clinical notes.

– 2. Advanced Generative Models

The use of Generative Adversarial Networks (GANs) and Variational Autoencoders (VAEs) for synthetic data generation has proven effective. However, exploring advanced generative models like Diffusion Models or Transformer-based generative architectures could further enhance the quality and diversity of synthetic data.

– 3. Reinforcement Learning-Based Agents

Integrating reinforcement learning (RL)-based agents into the framework could improve automation capabilities by enabling agents to learn optimal strategies for dataset refinement and retraining over time.

References

1. Trigueros, D., Meng, L., Hartnett, M.: Face recognition: from traditional to deep learning methods (2018)
2. Li, L., Mu, X., Li, S., Peng, H.: A review of face recognition technology. IEEE Access **8**, 139110–139120 (2020). https://doi.org/10.1109/ACCESS.2020.3011028
3. Li, Y., Cha, S.: Implementation of robust face recognition system using live video feed based on CNN. Proc. Comput. Vis. Pattern Recognit. (2018)
4. Guo, Y., Zhang, L., Hu, Y., He, X., Gao, J.: MS-Celeb-1M: a dataset and benchmark for large-scale face recognition. In: Leibe, B., Matas, J., Sebe, N., Welling, M. (eds.) ECCV 2016. LNCS, vol. 9907, pp. 87–102. Springer, Cham (2016). https://doi.org/10.1007/978-3-319-46487-9_6
5. Albalawi, S., Alshahrani, L., Albalawi, N., Kilabi, R., Alhakamy, A.: A comprehensive overview on biometric authentication systems using artificial intelligence techniques. Int. J. Adv. Comput. Sci. Appl. **13** (2022). https://doi.org/10.14569/IJACSA.2022.0130491
6. Chen, Z., Cao, Y., Liu, Y., Wang, H., Xie, T., Liu, X.: A comprehensive study on challenges in deploying deep learning based software. In: Proceedings of the 28th ACM Joint Meeting on European Software Engineering Conference and Symposium on the Foundations of Software Engineering, pp. 750–762 (2020)
7. Wang, Z., Liu, C., Cui, X.: EvilModel: hiding malware inside of neural network models. In: Proceedings of the 2020 IEEE Symposium on Security and Privacy (SP), San Francisco, CA, USA, May 2020, pp. 1020–1036 (2020)
8. Abadi, M., et al.: Deep learning with differential privacy. In: Proceedings of the 2016 ACM SIGSAC Conference on Computer and Communications Security, pp. 308–318 (2016)
9. Goldblum, M., et al.: Dataset security for machine learning: data poisoning, backdoor attacks, and defenses. arXiv preprint arXiv:2012.10544 (2020)

10. Biggio, B., et al.: Evasion attacks against machine learning at test time. In: Blockeel, H., Kersting, K., Nijssen, S., Železný, F. (eds.) ECML PKDD 2013. LNCS (LNAI), vol. 8190, pp. 387–402. Springer, Heidelberg (2013). https://doi.org/10.1007/978-3-642-40994-3_25
11. Borgnia, E., et al.: DP-instahide: provably defusing poisoning and backdoor attacks with differentially private data augmentations. arXiv preprint arXiv:2103.02079 (2021)
12. Tianyu, G., Liu, K., Dolan-Gavitt, B., Garg, S.: BadNets: evaluating backdooring attacks on deep neural networks. IEEE Access **7**(2019), 47230–47244 (2019)
13. Costales, R., Mao, C., Norwitz, R., Kim, B., Yang, J.: Live Trojan attacks on deep neural networks. In: Proceedings of the IEEE/CVF Conference on Computer Vision and Pattern Recognition Workshops, pp. 796–797 (2020)
14. Zhang, B., et al.: SecurityNet: assessing machine learning vulnerabilities on public models. arXiv preprint arXiv:2310.12665 (2023)
15. Fredrikson, M., Jha, S., Ristenpart, T.: Model inversion attacks that exploit confidence information and basic countermeasures. In: CCS, pp. 1322–1333. ACM (2015)
16. Chen, H., Babar, M.: Security for machine learning-based software systems: a survey of threats, practices, and challenges. ACM Comput. Surv. **56** (2023). https://doi.org/10.1145/3638531
17. Uddin, M.N., Hasnat, A.H.M.A., Nasrin, S., Alam, M.S., Yousuf, M.A.: Secure file sharing system using blockchain, IPFS and PKI technologies. In: 2021 5th International Conference on Electrical Information and Communication Technology (EICT), Khulna, Bangladesh (2021)
18. Zheng, Q., Li, Y., Chen, P., Dong, X.: An innovative IPFS-based storage model for blockchain. In: 2018 IEEE/WIC/ACM International Conference on Web Intelligence (WI), Santiago, Chile, pp. 704–708 (2018). https://doi.org/10.1109/WI.2018.000-8
19. Benet, J.: IPFS - content addressed, versioned, P2P file system. ArXiv (2014). Accessed 30 Sept 2024. /abs/1407.3561
20. Doan, T., Bajpai, V., Psaras, Y., Ott, J.: Towards Decentralised Cloud Storage with IPFS: Opportunities, Challenges, and Future Directions (2022)
21. Nizamuddin, N., Hasan, H., Salah, K.: IPFS-blockchain-based authenticity of online publications (2018)
22. Shafagh, H., et al.: Towards blockchain-based auditable storage and sharing of IoT data. In: Proceedings of the 2017 on Cloud Computing Security Workshop (2017)
23. Sun, Y., Han, Z., Yu, W., Liu, K.J.R.: A trust evaluation framework in distributed networks: vulnerability analysis and defense against attacks. In: 25th IEEE International Conference on Computer Communications. Proceedings, INFOCOM 2006, Spain, pp. 1–13 (2006)
24. Chen, B., et al.: Detecting backdoor attacks on deep neural networks by activation clustering. arXiv preprint arXiv:1811.03728 (2018)
25. Chen, H., Fu, C., Zhao, J., Koushanfar, F.: DeepInspect: a black-box trojan detection and mitigation framework for deep neural networks, pp. 4658–4664 (2019). https://doi.org/10.24963/ijcai.2019/647

A Novel Deep Learning Approach for Hate Speech Detection on Social Platforms

K. P. Impana[1(✉)], K. B. Vikhyath[2], M. Pavana[1], A. N. Hemalatha[1], and Nagadeepa S. Shetti[1]

[1] Department of Computer Science and Engineering, JSS Academy of Technical Education, Bengaluru, India
impanaraj@gmail.com

[2] Department of Computer Science and Engineering, Dr H N National College of Engineering, Bengaluru, India

Abstract. The large surge in user-generated social media content has spurred the proliferation of negative speech, which calls for the application of effective au- tomated identification mechanisms for the moderation of the content. The present work presents a deep learning model utilizing bidirectional Long Short-Term Memory (LSTM) networks to classify social media post content into neutral, of- fensive, and hate speech classes. Data preprocessing pipeline provides tokeniza- tion, normalization, as well as embedding contextual methods utilized to augment features. The proposed model design comprises twinned LSTM layers with stra- tegic dropout regularization, wherein the optimization of the model is done using categorical cross-entropy loss and the Adam optimizer.

Experimental performance proves the model's effectiveness with 87% validation and 97% training accuracy for the three-class problem. Precision of performance based on confusion matrix metrics also reports good accuracy for recognizing offensive content (91%), although with limited cross-classification of hate and offense (16% error rate). ROC curve statistics also attest to good discrimination performance with area-under-curve measurements of 0.94, 0.91, and 0.89 for the neutral, offensive, and hate speech classes respectively. Our results prove the promise of recurrent neural architectures for use in content moderation systems and also indicate certain difficulties in the model's ability to differentiate between levels of inappropriate content.

Keywords: Hate Speech Detection · Recurrent Neural Networks · Natural Language Processing · Content Moderation · Deep Learning · Bidirectional LSTM · Social Media Analytics

1 Introduction

Social media platforms have fundamentally transformed contemporary communication paradigms, enabling unprecedented levels of information exchange and social connectivity across geographical boundaries [1].

J. Shreyas et al. (Eds.): CODE-AI 2025, CCIS 2690, pp. 348–360, 2026.
https://doi.org/10.1007/978-3-032-19321-6_32

Hate speech i.e., speech that insults, threatens, or incites violence against individuals or groups based on traits like race, gender, religion, or sexual orientation is a potent social obstacle [2]. Traditional content moderation methods, such as lexicon-based filtering and human review, struggle with scalability, contextual nuances and inconsistent judgments. Moreover, evolving coded language allows bad actors to bypass detection, making these approaches less effective. Recent deep learning advances enable automated detection of semantic patterns, contextual relationships and linguistic structures in text [3].

Our research addresses this need by developing and evaluating a specialized deep learning architecture designed specifically for multi-class hate speech classification. Our model distinguishes between three distinct categories: neutral language, offensive but non-hateful content, and explicit hate speech.

Primary contribution of this research is to builds a based model trained with extensive data preprocessing and augmentation to enhance performance. We introduce ensemble methods, integrate attention mechanisms, and compare various deep learning architectures to determine the most effective model. Our approach also incorporates the approaches for elucidations to improve interpretability and trust in automated hate speech detection systems.

This paper's remaining structure is as follows; Sect. 2 presents literature survey, Sect. 3 indicates methodology and Sects. 4 and 5 represents the research findings and conclusion of our work.

2 Literature Review

The evolution of hate speech detection methodologies reflects broader advances in natural language processing and machine learning. We categorize prior research into four methodological generations:

2.1 Lexicon-Based Approaches

Early detection systems relied primarily on keyword matching against predefined dictionaries of problematic terms developed one of the first systematic approaches, utilizing template-based matching with specialized lexicons for different targeted groups [9].

2.2 Traditional Machine Learning Classifiers

Advancing beyond simple lexical matching, researchers explored statistical machine learning techniques with engineered features. Implemented an SVM-based classifier for distinguishing offensive language from hate speech, achieving 83% accuracy through careful feature engineering. Their work highlighted a critical challenge that persists in contemporary research: the difficulty of differentiating between generally offensive content and targeted hate speech.

2.3 Deep Learning Architectures

LSTM networks with learned embeddings outperformed traditional ML methods by 18%, demonstrating their effectiveness in hate speech detection. Their work highlighted the advantage of recurrent neural networks in capturing sequential and contextual linguistic patterns [10].A convolution-GRU-based model showed that CNNs excel at capturing local features and phrase-level patterns in hate speech detection. The study found that while CNNs handled local patterns well, RNNs were better at modeling long-distance dependencies [11].

2.4 Transformer-Based Models

Transformer architectures revolutionized NLP, as Mozafari's [12] BERT-based model improved hate speech detection but faced challenges with domain adaptation and dataset bias. Their study showed that while pre-trained models capture context well, they struggle with social media's unique vocabulary [12]. Expert Systems with Applications studied multilingual transformers for Spanish hate speech detection, highlighting cross- lingual transfer learning. While achieving state-of-the-art performance, these models required high computational resources and were sensitive to training data biases [13].

2.5 Ensemble and Hybrid Approaches

Recent studies show that combining CNN, RNN, and transformers enhances hate speech detection, especially for ambiguous content. Hybrid architectures integrating domain knowledge with deep learning improve performance on implicit bias and coded language. This suggests that purely data-driven models may overlook crucial social and contextual factors [14, 15].

2.6 Research Gaps

Despite significant advances, several challenges persist in the literature:

Class imbalance: Most datasets feature limited examples of explicit hate speech relative to neutral or offensive content. Contextual understanding: Models struggle with context-dependent interpretation, particularly for ambiguous content. Cross-domain generalization: Systems trained on one platform often perform poorly when applied to different social media environments. Evolving terminology: The dynamic nature of hate speech terminology presents ongoing challenges for static models. Explainability limitations: Deep learning approaches often function as "black boxes," limiting transparency and interpretability.

Our research addresses these gaps through a combined approach of architectural innovation, specialized preprocessing, and detailed error analysis to advance understanding of model limitations.

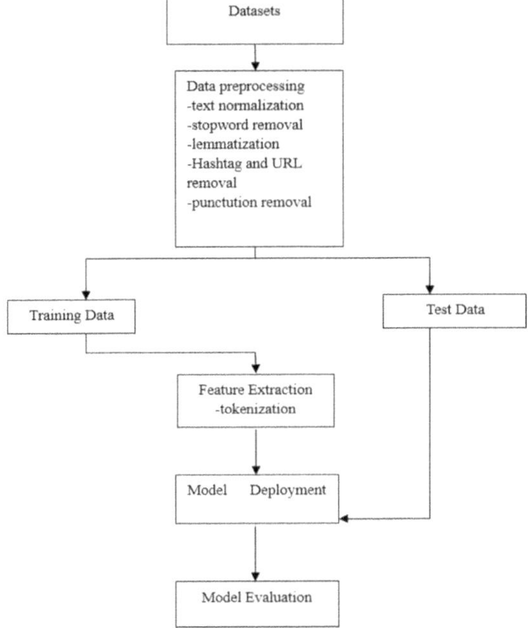

Fig. 1. Work flow diagram.

3 Methodology

This study on hate speech detection using machine learning involved several systematic steps as shown in Fig. 1.

Algorithm for Hate Speech Detection

The following is a description of the hate speech detection algorithm:

1. **N**: Total number of training samples.
2. **n**: Number of features in the dataset.
3. **m**: Number of output categories (e.g., clean, offensive, hateful).
4. Training samples: (xi,yi), i = 1,2,……,N where and xi = (xi1, xi2,, xin)^T € R^n yi=(yi1, yi2,, yim)^T € R^m

To obtain the total output the matrix, combine each output the vector in a row:

$$T = \begin{bmatrix} y_1^T \\ y_2^T \\ \vdots \\ y_N^T \end{bmatrix} = \begin{bmatrix} t_{11} & t_{12} & \dots & t_{1m} \\ t_{21} & t_{22} & \dots & t_{2m} \\ \vdots & \vdots & \ddots & \vdots \\ t_{N1} & t_{N1} & \dots & t_{Nm} \end{bmatrix}$$

5. **O_j**, where j = 1, 2, …..,N: Actual output vector corresponding to label tj.
6. **W**: Weight matrix between input features and hidden layer units, where Wi = (wi1, wi2, ……., win)^T represents the weight vector connecting the ith hidden layer unit and nth input feature

7. **b** = (b1, b2,….., bN)^{T}: bias vector, representing the threshold of the ith hidden layer unit.
8. β = (βij)NXm: weight matrix between the hidden layer and output layer, where βi = (βi1, βi2,….., βim)^{T} symbolizes a weight vector that connects the ith hidden layer unit to the output layer.

Following is one way to write the matrix β in rows:

$$\beta = \begin{bmatrix} \beta_1^T \\ \beta_2^T \\ \vdots \\ \beta_1^T \end{bmatrix} = \begin{bmatrix} \beta_{11} & \beta_{12} & \cdots & \beta_{1m} \\ \beta_{21} & \beta_{22} & \cdots & \beta_{2m} \\ \vdots & \vdots & \ddots & \vdots \\ \beta_{N1} & \beta_{N2} & \cdots & \beta_{Nm} \end{bmatrix}$$

9. **g(x)**: The excitation functions commonly used activation functions includes sigmoid, tanh, ReLU and swishConsider step 6 and step 8 we can get:

$$\mathrm{H}\beta = \mathrm{T} \tag{1}$$

where, T represents the transposition of T and H represents the hidden layer's output. The weight matrix values of β are calculated using the least squares approach to minimize the error:

$$\beta = \mathrm{H} + \mathrm{T} \tag{2}$$

where **H+** is the generalized inverse of matrix H.

Pseudo Code for Hate Speech Detection Algorithm

1. Randomly assign the input weight W and biases b
2. Calculate the hidden layer output matrix H, where H = (hij), where hij = g(Wi xj + bi)
3. Calculate the output weights matrix as β = H + T, where H + is the generalized inverse of matrix H
4. Classify each input sample based on the output matrix $\widehat{y} = H\beta$
5. Evaluate model performance using metrics like accuracy, precision, recall, And f1-score.

3.1 Dataset Characteristics

This study utilizes a Twitter-derived dataset collected and annotated by comprising 24,783 tweets manually labeled into three distinct categories:

- Neutral/Clean: 4,163 samples (16.8%)
- Offensive Language: 19,190 samples (77.4%)
- Hate Speech: 1,430 samples (5.8%).

The significant class imbalance reflects real-world distribution patterns but presents methodological challenges addressed in our approach. The dataset captures diverse linguistic styles, dialectal variations, and contextual nuances characteristic of social media communication.

3.2 Data Preprocessing Pipeline

We implemented a multi-stage preprocessing pipeline optimized for social media content:

Step 1: Text Normalization

Case normalization to lowercase. Unicode normalization (NFKC format). Special character handling with selective preservation of meaningful punctuation. Contraction ex- pansion (e.g., "don't" → "do not")

Step 2: Social Media-Specific Processing.

URL and hyperlink removal using regex pattern https?://\S+. Username normalization: replacing @mentions with <USER> token. Hashtag processing: segmenting composite hashtags while preserving meaning. Example: #BlackLivesMatter → "black lives matter"

Step 3: Tokenization and Sequence Preparation

Tokenization using NLTK's Tweet Tokenizer to preserve emoticons and handle punctuation. Stop word filtering with modified stop word list (preserving negation terms). Sequence padding to standardize input dimensions (max length: 50 tokens).

These preprocessing steps yielded standardized input sequences while preserving linguistically meaningful features particular to social media communication. The augmentation techniques addressed class imbalance by generating additional samples for the underrepresented hate speech category.

3.3 Feature Extraction and Representation

Textual data was transformed into a machine-learning-compatible format using the following methods:

Tokenization and Sequence Conversion: A Tokenizer (with num_words = 5000) was fitted on the cleaned text corpus, mapping the most frequent 5,000 words to unique integer indices while discarding less common terms. Text samples were converted into sequences of numbers, where each word was replaced by its corresponding integer index from the vocabulary. Sequences were padded to a fixed length (100 words) to standardize input dimensions, with shorter sequences zero-padded at the end and longer se- quences truncated.

3.4 Model Architecture

The core classification architecture comprises a deep neural network with the subsequent components: An Input Layer that accommodates tokenized sequences with a maximum length of 50. The embedding layer utilized is FastText embeddings with 300 dimensions and a vocabulary size of 25,000. The Bidirectional LSTM Layer comprises 128 units utilizing tanh activation. Dropout Layer 1 has a dropout rate of 0.5 for regu- larization. The second Bidirectional LSTM layer has 64 units utilizing tanh activation. Dropout Layer 2 possesses a dropout rate of 0.5. The Attention Mechanism is a Self- attention layer with 8 attention centres. The Dense Layer comprises 64 units utilizing ReLU activation. The output layer has three units utilizing softmax activation, corre- sponding to each class. The bidirectional LSTM layers capture contextual information from both preceding and following tokens, essential for understanding linguistic con- text in hate speech detection. The incorporation of self-attention mechanisms allows the model to focus on particularly relevant tokens when making classification decisions. Figure 2 illustrates the complete architecture with dimensional details at each layer.

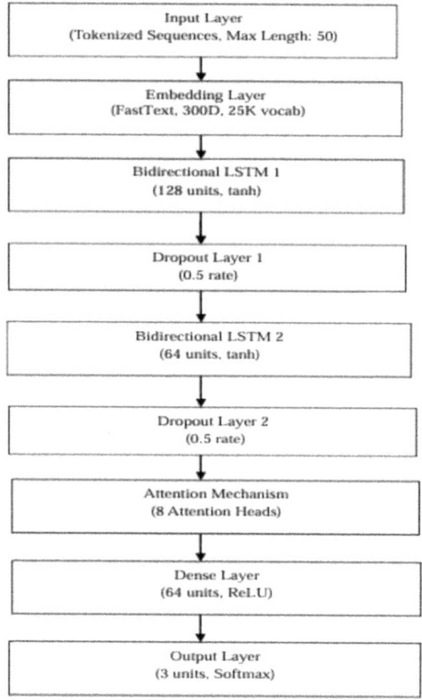

Fig. 2. Model architecture diagram

3.5 Training Configuration

The model was trained with the following configuration:

Loss Function: Categorical cross-entropy with class weights. Optimizer: Adam with learning rate 0.001 and $\beta_1 = 0.9$, $\beta_2 = 0.999$. Batch Size: 64 samples. Epochs: 10. Class Weighting: {0: 1.0, 1: 0.2, 2: 2.5} to address class imbalance. Validation Strategy: 80/20 stratified train-test split with 10% of training set used for validation

3.6 Valuation Metrics

Model performance was assessed using multiple complementary metrics:

Accuracy: Overall classification correctness. Precision, Recall, F1-Score: Class-specific performance measures. Confusion Matrix: Visualization of classification patterns and errors. ROC Curves and AUC: Discrimination capability assessment. McNemar's Test: Statistical significance of performance differences. Given the class imbalance in our dataset, we placed particular emphasis on macro-averaged metrics that give equal weight to all classes, preventing performance on the majority class from obscuring weaknesses in minority class detection.

4 Results and Discussion

4.1 Model Performance Overview

The LSTM-based model demonstrated strong overall performance, achieving 87% validation accuracy after 10 epochs. Table 1 presents comprehensive performance metrics across all three classes.

Table 1. Performance metrics by class.

Class	Precision	Recall	F1-score	Support
Neutral	0.89	0.84	0.86	833
Offensive	0.91	0.95	0.93	3,838
Hate speech	0.79	0.64	0.71	286
Macro Avg	0.86	0.81	0.83	4,957
Weighted Avg	0.90	0.91	0.90	4,957

These results indicate particularly strong performance in identifying offensive language (f1 = 0.93) with moderate effectiveness for hate speech detection (f1 = 0.71). The dis- parity between weighted and macro averages reflects the impact of class imbalance on overall performance metrics.

4.2 Training Dynamics

Figure 3 illustrates training and validation accuracy/loss across epochs. The model demonstrated consistent learning progression with training accuracy reaching 97% by epoch 10.

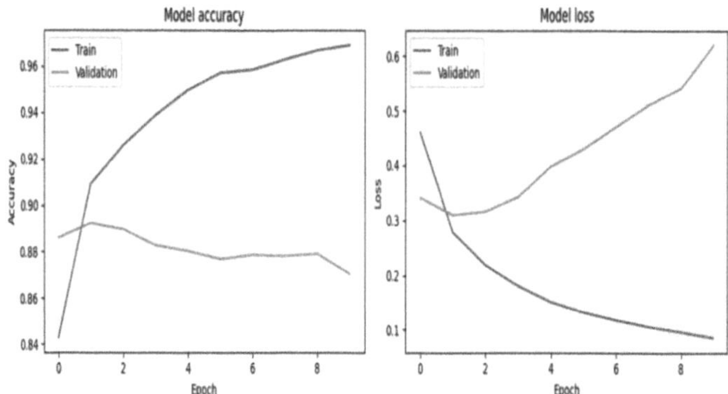

Fig. 3. Training and validation accuracy/loss curves

Training exhibited several noteworthy patterns:

Rapid initial improvement (epochs 1–5) with accuracy increasing from 65 to 83%. Gradual convergence phase (epochs 6–12) with incremental improvements. Plateau phase (epochs 13–15) with minimal validation performance changes.

The separation between training and validation accuracy (approximately 10 percentage points) suggests moderate overfitting despite regularization measures. This pattern is common in text classification tasks with limited training data and high-dimensional feature spaces.

4.3 Classification Analysis

The confusion matrix Fig. 4 provides deeper insight into classification patterns and error modes.

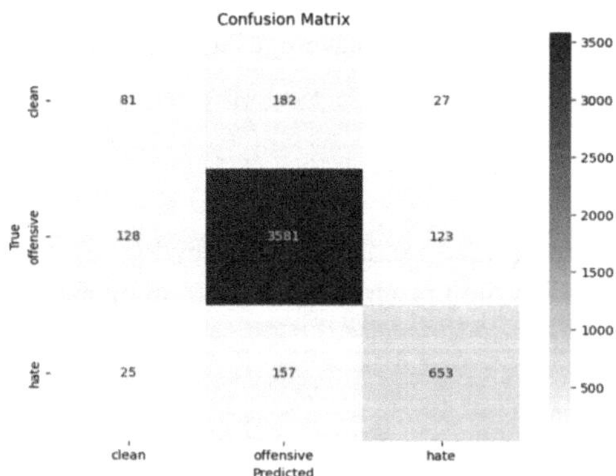

Fig. 4. Confusion matrix heatmap

Several significant patterns emerge from this analysis: High offensive language detection accuracy: 95% of offensive samples correctly classified. Moderate hate speech recall: Only 64% of hate speech samples correctly identified. Asymmetric error distribution: Hate speech more commonly misclassified as offensive (26%) than neutral (10%).

Limited neutral-hate confusion: Only 3% of neutral content misclassified as hate speech.

These patterns suggest the model effectively distinguishes between neutral and problematic content but struggles with the more nuanced distinction between offensive language and explicit hate speech.

4.4 ROC Curve Analysis

ROC curve analysis Fig. 5 confirms the model's strong discrimination capabilities across all three classes.

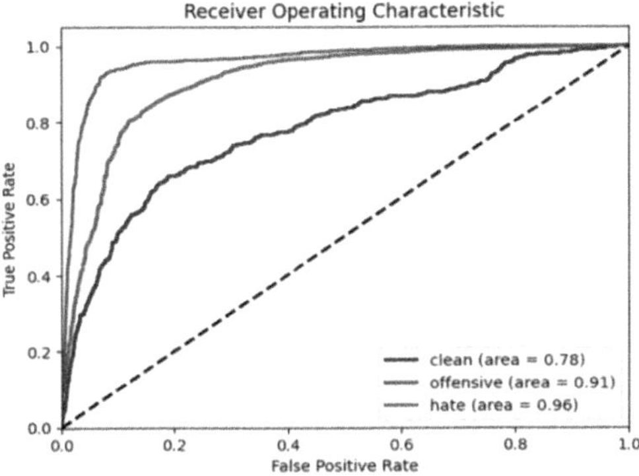

Fig. 5. ROC curves by class

Area Under Curve (AUC) measurements indicate robust performance:

Neutral class: AUC = 0.94. Offensive class: AUC = 0.91. Hate speech class: AUC = 0.89. These values demonstrate that despite recall challenges for hate speech, the model maintains good discrimination capability, suggesting that threshold adjustment could potentially optimize performance for specific application requirements.

4.5 Error Analysis

Qualitative assessment of misclassified examples reveals several recurring patterns: Implicit hate speech: Content using coded language or implicit stereotypes without explicit slurs was frequently misclassified as merely offensive. Example: "They always commit crimes, it's in their nature" (True: Hate, Predicted: Offensive). Context-dependent interpretation: Content requiring broader social or political context for proper interpretation often proved challenging. Example: "Go back where you came from" (True: Hate, Predicted: Offensive). Sarcasm and irony: Non-literal language forms presented particular difficulties. Example: "Oh sure, because those people are always so hard-working" (True: Hate, Predicted: Offensive). Reclaimed terminology:

In-group usage of otherwise offensive terms frequently led to misclassification. Example: [REDACTED slur used by in-group member] (True: Neutral, Predicted: Hate)

These error patterns highlight limitations in the model's contextual understanding and suggest potential avenues for future improvement through context-aware architectures.

4.6 Ablation Studies

To evaluate component contributions, we conducted ablation studies by systematically removing or replacing model elements. Table 2 summarizes key findings.

Table 2. Ablation study results (f1-score by class).

Model configuration	Neutral	Offensive	Hate	Macro Avg
Full model (BiLSTM + Attention)	0.86	0.93	0.71	0.83
Without attention	0.85	0.92	0.68	0.82
Single LSTM layer	0.84	0.90	0.65	0.80
Unidirectional LSTM	0.82	0.91	0.62	0.78
CNN instead of LSTM	0.83	0.92	0.61	0.79
GloVe instead of FastText	0.84	0.92	0.68	0.81

These results demonstrate several key insights:

Bidirectionality contributes significantly to performance, particularly for hate speech detection. Attention mechanisms provide modest but consistent improvements across all classes. Architectural depth (dual LSTM layers) enhances model capability. FastText embeddings outperform GloVe, likely due to subword information handling

5 Conclusion and Future Work

This study highlights the effectiveness of bidirectional LSTM networks with attention mechanisms for multi-class hate speech detection. The proposed model achieved a validation accuracy of 87%, demonstrating strong capability in distinguishing between neutral and problematic content. Key findings indicate that bidirectional processing significantly enhances contextual understanding, while attention mechanisms contribute to improved classification accuracy. Additionally, social media-specific preprocessing techniques play a crucial role in refining the model's performance. However, challenges persist in distinguishing offensive language from explicit hate speech, particularly due to the subtle and evolving nature of harmful online discourse.

Despite its promising performance, the model has several limitations. The dataset primarily consists of English-language Twitter content from 2017, which may restrict its applicability to other platforms and recent trends. The three-class classification framework simplifies the complex spectrum of harmful content, and cultural variations in hate speech further limit cross-context generalization. Additionally, the deep learning model requires substantial computational resources, and its effectiveness may degrade over time as language evolves. Future research should focus on incorporating multimodal data, integrating transformer-based architectures for enhanced contextual learning, and developing adaptive models capable of continuous learning. Cross-platform validation and improved explainability techniques will further enhance the model's robustness and usability in real-world applications.

References

1. Waseem, Z., Hovy, D.: Hateful symbols or hateful people? Predictive features for hate speech detection on Twitter. In: Proceedings of the NAACL Student Research Workshop, pp. 88–93 (2016)
2. Davidson, T., Warmsley, D., Macy, M., Weber, I.: Automated hate speech detection and the problem of offensive language. In: Proceedings of the international AAAI Conference on Web and Social Media, vol. 11, no. 1 (2017)
3. Schmidt, A., Wiegand, M.: A survey on hate speech detection using natural language processing. In: Proceedings of the Fifth International Workshop on Natural Language Processing for Social Media, pp. 1–10 (2017)
4. Johnson, K., Henderson, M., Murthy, D.: Trends in online hate and harassment. Anti-Defamation League (2023)
5. Meta.: Community Standards Enforcement Report (2022). https://transparency.fb.com/data/community-standards-enforcement/
6. Jhaver, S., Bruckman, A., Gilbert, E.: Does transparency in moderation really matter? User behavior after content removal explanations on Reddit. In: Proceedings of the ACM on Human-Computer Interaction, vol. 3, pp. 1–27 (2021)
7. European Commission.: The Digital Services Act package (2022). https://digital-strategy.ec.europa.eu/en/policies/digital-services-act-package
8. Williams, M.L., Burnap, P., Javed, A., Liu, H., Ozalp, S.: Hate in the machine: anti-Black and anti-Muslim social media posts as predictors of offline racially and religiously aggravated crime. Br. J. Criminol. **60**(1), 93–117 (2020)
9. Warner, W., Hirschberg, J.: Detecting hate speech on the world wide web. In: Proceedings of the Second Workshop on Language in Social Media, pp. 19–26 (2012)
10. Badjatiya, P., Gupta, S., Gupta, M., Varma, V.: Deep learning for hate speech detection in tweets. In: Proceedings of the 26th International Conference on World Wide Web Companion, pp. 759–760 (2017)
11. Zhang, Z., Robinson, D., Tepper, J.: Detecting hate speech on Twitter using a convolution-GRU based deep neural network. In: European Semantic Web Conference, pp. 745–760 (2018)
12. Mozafari, M., Farahbakhsh, R., Crespi, N.: A BERT-based transfer learning approach for hate speech detection in online social media. Complex Intell. Syst. **6**, 491–501 (2020)
13. Plaza-del-Arco, F.M., Molina-González, M.D., Ureña-López, L.A., Martín-Valdivia, M.T.: Comparing pre-trained language models for Spanish hate speech detection. Expert. Syst. Appl. **166**, 114120 (2021)
14. MacAvaney, S., Yao, H.R., Yang, E., Russell, K., Goharian, N., Frieder, O.: Hate speech detection: challenges and solutions. PLoS ONE **14**(8), e0221152 (2019)
15. Zimmerman, S., Fox, C., Kruschwitz, U.: Improving hate speech detection with deep learning ensembles. In: Proceedings of the 11th International Conference on Language Resources and Evaluation, Marseille, France, 11–16 May 2020, pp. 2546–2553 (2020)

Classification of Cardiovascular Abnormalities Through Optimizable Regression Analysis and Feature Engineering

Muralidharan Jayaraman, P. Shanmugavadivu(✉), and A. Nithya

Department of Computer Science and Applications, The Gandhigram Rural Institute (Deemed to Be University), Gandhigram, India
psvadivu67@gmail.com

Abstract. Cardiovascular diseases (CVDs) pose a major global health challenge despite medical advancements. Machine Learning (ML) offers potential for early detection and prediction of CVD risk. This research utilized two datasets, Kaggle's CVD dataset (70,000 instances) and UCI's CHD dataset (303 instances), for ML model evaluation. After Exploratory Data Analysis (EDA) and data cleansing, the CVD dataset was reduced by 2.67%. Feature engineering introduced new variables like BMI, MAP, and PP to enhance predictive accuracy. The goal is to identify the best ML models. A set of Regression Models in the categories of Linear Regression, Tree Regression, Optimizable Neural Network, Kernel Regression, and Ensemble Regression were deployed on both CVD and CHD datasets. Boosted Tree Ensemble Regression gave an accuracy of 97.37% for CVD dataset, and Robust Linear Regression returned an accuracy of 98.52% for CHD dataset.

Keywords: Regression models · Cardiovascular diseases

1 Introduction

Cardiovascular Disease (CVD) is regarded as a significant dilemma on a global scale due to their substantial impact on health, leading to a considerable burden of illness and mortality. The need for timely identification and intervention in combating this issue cannot be overstated. Early detection of CVD not only provides the crucial opportunity for immediate treatment, resulting in vastly improved patient outcomes and the prevention of catastrophic incidents like heart attacks and strokes, but also facilitates the implementation of preventive strategies [1]. Through the early identification of risk factors, individuals are empowered to make necessary lifestyle modifications and receive appropriate medication to lower the hazard of CVD. Looking beyond the individual level, early detection has wide-reaching implications, potentially contributing to decreased healthcare costs and easing the economic pressures on healthcare systems. Additionally, early detection is always helpful in cultivating healthier communities, and driving advancements in the field of cardiology [2].

Raising awareness and promoting regular health check-ups through educational initiatives and public health campaigns play a crucial role in fostering early detection.

J. Shreyas et al. (Eds.): CODE-AI 2025, CCIS 2690, pp. 361–375, 2026.
https://doi.org/10.1007/978-3-032-19321-6_33

Furthermore, early detection is indispensable for conducting risk assessments, implementing secondary prevention measures, and continuously monitoring chronic cardiovascular conditions, enabling adjustments to treatment plans and prevention of disease progression. In essence, the timely detection of CVDs plays a pivotal role in enhancing individual well-being and alleviating the overall societal burden caused by CVD [3].

In the domain of predicting CVDs, contemporary analytical models operate by utilizing training data to make informed predictions, leveraging past insights to make decisions when faced with new data. Modern technologies, notably Machine Learning (ML) and Artificial Intelligence (AI), have emerged as key players in the healthcare sector, offering personalized clinical support. Healthcare data, ranging from Electronic Health Records (EHR) to sensor data from Internet of Things (IoT) devices, often exist in both structured and unstructured formats, presenting challenges for humans in extracting actionable insights. ML algorithms serve as invaluable tools in aiding clinical predictions, providing transparency and interpretability in their forecasts, thereby enhancing patient care and diagnostic processes [4].

The integration of ML and AI in healthcare has sparked a transformative wave, reshaping how CVDs are predicted and treated. By leveraging advanced technologies, healthcare professionals can now offer more precise and personalized care strategies, which in turn yield improved outcomes for patients while alleviating the societal burden of CVD. Regression-based prediction methods play a pivotal role in this realm by analyzing a wide array of data sources, ranging from EHR and genetic profiles to lifestyle habits and wearable tech data, to construct intricate statistical models.

Amidst this data deluge, the process of selecting relevant features emerges as a crucial aspect, balancing between classic risk factors like age, blood pressure, and cholesterol levels, and newer considerations such as genetic indicators [5]. The array of regression techniques available, encompassing linear as well as logistic models, further complements traditional approaches by incorporating sophisticated ML algorithms, elevating the predictive accuracy to new heights. These refined models generate personalized risk scores, equipping healthcare providers with tailored strategies for proactive interventions. Continuous monitoring and periodic model refinement steer towards dynamic risk evaluation, ensuring that interventions remain timely and effective. Fostering trust and practicality in clinical settings involves fabricating interpretable models, subsequently integrating them seamlessly with medical practice guidelines.

However, such cohesive convergence necessitates an ongoing cycle of research and validation to remain abreast of evolving data streams and medical insights. This holistic blend of data analytics and clinical acumen not only enhances CVD prediction and prevention but also exemplifies the union of scientific rigor and patient-centric care at the forefront of modern healthcare. Linear regression-based prediction in the context of CVDs involves a straightforward yet powerful statistical approach that plays a crucial role in predicting the risk of developing CVDs. This approach utilizes a variety of data sources, such as EHR, patient demographics, and conventional risk factors like age, blood pressure, and cholesterol levels, to construct a predictive model.

The relationship between predictors and the risk of CVDs is estimated through the linear regression model, minimizing errors by best-fit approach. The coefficients of the linear model offer valuable insights into how each predictor influences the risk, aiding healthcare professionals in making informed predictions based on individual data [6]. One of the key advantages of linear regression models is their simplicity and interpretability, allowing for clear communication of the risk factors associated with CVDs.

Healthcare professionals can leverage these models to assess specific risks of individuals accurately, providing a solid foundation for decision-making and personalized recommendations. It is crucial to validate the accuracy of the linear regression model through techniques like cross-validation and metrics such as R-squared and Mean Squared Error (MSE) to ensure its reliability in predicting CVD risk.

However, linear regression models have limitations when it comes to capturing complex non-linear relationships or interactions between variables. In scenarios where linear relationships are not sufficient, more advanced modelling techniques like polynomial regression or ML approaches might be more appropriate to account for these complexities. Despite these limitations, linear regression-based prediction remains a valuable tool in both clinical and research settings due to its transparency and ease of interpretation [7].

Linear regression-based prediction serves as a foundational tool in the assessment and communication of CVD risk, offering a clear and interpretable model for healthcare professionals to utilize. By understanding how different risk factors influence the overall risk of developing CVDs, professionals can enhance their decision-making processes and provide more targeted recommendations to individuals. This makes linear regression a versatile and valuable approach in predicting and managing CVD risk effectively.

A systematic review is done by Damen et al. [8] to provide a summary of models for predictionof risk due to CVD, in the population. The study screened 9,965 references, resulting in 212 articles being includedexplaining 363 prediction model development, and 473 external validations. They suggested that subsequent study concentrate on the external validation and direct comparison of promising existing models, the customization or amalgamation of these models, and exploring the incorporation of new predictors to enhance these models. Bhagyesh Randhawan et al. [9] employed ML algorithms to predict heart disease occurrence inpatients. They utilized the Heart Disease University of California, Irvine (UCI) dataset with 14 variables to assess the performance of Logistic Regression (LR) in predicting CVD. The results showed an accuracy of 85% and a low error rate of 0.1406565, indicating the suitability of the algorithm for prediction.

Manoj Kumar, et al., [10] compared the performance of LR and linear regression models in detecting CVD using a UCI ML repository dataset.The LR classifier achieved a significant accuracy of 88.68%, whereas, the Linear Regression model's accuracy stood at 78.56%, for CVD detection. Yingjie Zhang et al. [11] explores the use of LR models to predict the risk of heart disease among the elderly. Through data mining, key factors contributing to heart disease are identified, and a regression model is employed for disease prediction. The accuracy of the LR model is compared with the other algorithms such as NB, Support Vector Machines (SVM), and Neural Network (NN), highlighting the viability of LR in heart disease prediction.

Kavya et al. [12] employed LR and patient cardiac characteristics using a Kaggle dataset to develop predictive models for heart disease where the primary objective is to assess the risk of acquiring Coronary Heart Disease (CHD) over a 10-year period. The model demonstrates an accuracy of 88%, and it achieves an area under the ROC curve of 74%. Tania Ciu, et al., [13] uses a dataset to predict the main causes of CVD and employed LR to understand the relationships between variables. The research highlights the effective use of LR, achieving an 85% accuracy rate and a low error rate of 0.1406565. Dinesh Kalla et al. [14] utilized Chi-square tests which suggests a connection between chest pain and heart disease cases, and linear regression analysis to predict heart disease based on symptoms. The LR model exhibits 85.12% accuracy in predicting heart disease.

Snigdha et al. [15] utilize a sample dataset from the UCIML repository to construct a robust heart disease prediction model. Initial hypotheses are formulated, and key predictor variables are identified using the Wald test. The model effectively predicts CVD with an 81% accuracy rate on test data, performing comparably or even better than models with similar configurations.

Saw et al. [16] utilizes the LR model in ML with data cleansing and mining techniques to enhance heart disease prediction accuracy. They used healthcare datasets to classify patients.

Yang et al. [17] conducted a study where a high-risk group of 29,930 subjects was selected from a total of 101,056 individuals in 2014, and their health was regularly monitored using EHR. Several regression methods were employed to construct prediction models. LR analysis identified nearly 30 indicators associated with CVD risk. The results revealed that the RF method outperformed other approaches with an AUC of 0.787, for assessing the 3-year risk of CVD. Cangao (Steven) Chu [18] aims to explore the correlation between CVDs and various suspected contributing factors. The focus is on developing a predictive model using regression analysis for CVD. The accuracy of prediction, achieved through both the linear regression model and the linear discriminant analysis model, consistently falls within the range of 70% to 80%. This high level of accuracy suggests that these models are the most appropriate responses to the available predictors.

2 Motivation

Analysing cardiac conditions through ECG signals, numerous approaches have surfaced in existing studies, highlighting the dynamic landscape of research within this domain. As advancements continue to push boundaries, a key area of focus lies in the development of increasingly precise predictive models that play a pivotal role in aiding healthcare professionals. These models are instrumental in the early identification of individuals at risk of cardiac issues, allowing for proactive interventions and customized treatment strategies tailored to individual needs and conditions. Despite the growing demand for such predictive tools, the integrity of medical data collected from patients poses a significant challenge due to potential noise interference. Hence, the effective detection of outliers becomes paramount, necessitating sophisticated techniques to summarize data distributions and elucidate intricate relationships among variables. One strategy involves the utilization of box plots and scatter plots, providing a comprehensive visual representation that unravels central tendencies and interconnections concealed within the data. The emphasis is not only on understanding the data but also on developing predictive models resilient to noisy inputs. Prioritizing robustness against data imperfections, these models are designed to offer optimization capabilities that inherently perform feature selection by assigning higher importance to the most informative features, thus enhancing the model's predictive power. Furthermore, an efficient approach involves integrating predictions from multiple optimizable learners, a technique that synergistically combines diverse insights to create a more resilient and potent model. This fusion of predictive capabilities not only augments generalization and performance but also contributes to more accurate predictions in the nuanced landscape of cardiac health, ultimately paving the way towards improved patient outcomes and enhanced proactive healthcare interventions. The major contributions of the proposed work are:

- Pre-processing of CVD and CHD datasets including data cleansing, outlier removal, data imputation, and changing multi-class into binary classifier.
- Derivation of additional Features, in order to get the combined data (e.g., for CVD dataset, BMI, MAP, and Pulse Pressure (PP)).
- Feature engineering to transform non-linear relationships in the data to linear form, in addition to Feature Reduction.
- Selection and tuning of optimizable ML based regression algorithms using optimizable tree for better prediction of cardiac and non-cardiac classes.
- Empirical verification of the proposed model against various regression models, and training functions.
- Performance comparison of proposed regression model compared with that of cutting-edge models to evaluate its effectiveness.

3 Proposed Methodology

The proposed architecture model for regression analysis is a comprehensive process that consists of multiple essential steps outlined in Fig. 1.It initiates with meticulous data collection, followed by a thorough data pre-processing techniques, precise data partitioning strategies, and finally, delves into the regression phase. At the outset, the datasets pertaining to CVD and CHD are subjected to meticulous feature engineering procedures. This involved steps such as the careful removal of irrelevant data points and the strategic introduction of supplementary variables like BMI, MAP, and PP to enhance the overall data quality. After this stage, the features undergo a crucial normalization process to ensure consistency, and any potential outliers are meticulously addressed to boost the overall performance of the regression model, particularly the Linear Regression technique. Following these processes, the refined features are meticulously employed for both training and testing phases utilizing a variety of regression models. The overarching goal here is to identify and select the most optimal predictive model that can provide accurate insights and valuable guidance based on the collected data.

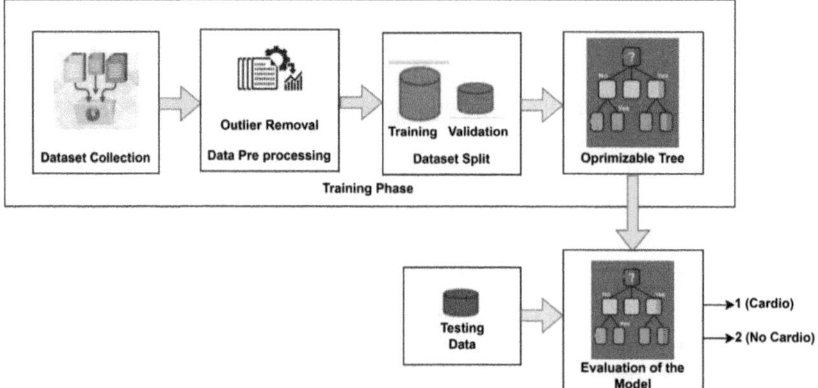

Fig. 1. Process flow of the proposed method

4 Datasets

The proposed method leveraged two publicly accessible datasets namely, Cardiovascular Disease (CVD) dataset, and Coronary Heart Disease (CHD) dataset. Both the datasets are already pre-processed, and are ready for application of the selected ML Models. These datasets contain a target variable each, typically indicating the presence or lack of it, of CVD.Through the application of regression analysis to this dataset, the proposed architecture seeks to build a predictive model capable of enhancing the understanding, and potentially predicting the risk factors linked to CVD.

5 Tree Regression

In ML and statistical modelling, decision tree regression is used to predict continuous outcomes based on splitting data into subsets using decision rules. There are different levels of granularity in tree regression models, often described as fine, medium, coarse, and optimizable. These terms typically refer to the complexity of the tree structure or the number of splits it allows, and are named Fine Tree, Medium Tree, and Course Tree Regression.

Optimizable Tree Regression: It is a DT model that is tuned using hyperparameter optimization techniques to achieve the best performance.The depth and structure of the tree are determined through optimization processes, such as cross-validation or grid search.Hyperparameters such as maximum depth, minimum leaf size, or splitting criteria are optimized to balance model complexity and predictive accuracy.The goal is to create a tree that generalizes well to unseen data while avoiding both overfitting and underfitting. It is ideal for situations where maximizing the performance of the model is needed, through tuning. This approach is often used in combination with algorithms, and these trees are designed through a process that finds the best tree for a given dataset by adjusting parameters rather than simply choosing a level of complexity upfront.

The prediction Y for a given input X is calculated by traversing the tree from the root node to a leaf node, where a constant value is assigned as the prediction, as in the equation below:

$$Y = \sum_{j=1}^{J} c_j I(X \in Rj) \tag{1}$$

where: Y is the predicted target variable and cj is the constant value assigned to leaf node j. $((X \in Rj)$is an indicator function that equals 1 if the input $X$is in the region Rj defined by the splitting rule at node j, and 0 otherwise. J is the count of leaf nodes in the tree.

6 Linear Regression

Linear regression comes in various forms, each designed to handle different types of relationships and data complexities. Below is a breakdown of linear, interaction, robust, and stepwise linear regression, highlighting their differences and key applications.

Linear Regression: It is the simplest form of regression that models the relationship between one or more independent variables (predictors) and a dependent variable by fitting a linear equation as provided below:

$$y = \beta_0 + \beta_1 x_1 + \beta_2 x_2 + \cdots \beta_n x_n + \in \tag{2}$$

where, y is the dependent variable (target), $\beta 0$ is the intercept, $\beta 1,\ldots,\beta_n$ are the coefficients for each predictor, $x1, x2,\ldots,xn$ are the independent variables (predictors), and ϵ is the error term (residuals).

This assumes a linear relationship between predictors and the outcome, sensitive to outliers and multicollinearity (high correlation between predictors), and simple and easy to interpret. This is ideal for datasets where a linear relationship is expected, and the goal is to understand the contribution of each predictor.

Interaction Linear Regression: This extends basic linear regression by including interaction terms that allow for the modelling of relationships between predictors themselves, in addition to their relationships with the dependent variable, as per the equation below:

$$y = \beta_0 + \beta_1 x_1 + \beta_2 x_2 + \beta_3 (x1 \cdot x2) + \cdots + \in \tag{3}$$

where the interaction term x_1 x_2 captures the combined effect of both predictors on the outcome.

Complex relationships are modelled by this regression, by considering how the effect of one predictor might depend on the value of another. This still assumes linearity but allows for non-additive effects between variables, and interpretability can become more complex due to the interaction terms. This is useful when the relationship between predictors is expected to influence the outcome. For example, in a medical study, the effect of a treatment might depend on both age and gender (an interaction between the two).

Robust Linear Regression: This is designed to handle data with outliers or violations of the assumptions of Ordinary Least Squares (OLS) linear regression, such as non-normality or heteroscedasticity (unequal variance), as provided in equation below:

$$y = \beta_0 + \beta_1 x_1 + \cdots + \in \quad (4)$$

The key difference is in the method of estimation, where robust estimators (e.g., Huber or Tukey estimators) are used instead of least squares. This is less sensitive to outliers, as robust regression reduces their influence on the model, helps when the data doesn't meet traditional linear regression assumptions (e.g., homoscedasticity), and provides more reliable estimates when assumptions of standard linear regression are violated. This is ideal for datasets with outliers or when residuals donot follow normality or equal variance assumptions. Useful in real-world data where outliers are common.

Stepwise Linear Regression: This is a model selection method that involves iteratively adding or removing predictors based on statistical criteria, typically using techniques like forward selection, backward elimination, or bidirectional elimination, as per the equation below:

$$y = \beta_0 + \beta_1 x_1 + \beta_2 x_2 \cdots + \in \quad (5)$$

- Forward selection: Starts with no predictors, adding them one by one based on statistical significance (e.g., p-value).
- Backward elimination: Starts with all predictors and removes the least significant one at each step.
- Bidirectional elimination: Combines both approaches, adding and removing predictors based on predefined criteria.

This approach Simplifies the model by automatically selecting the most significant predictors, reducing the risk of overfitting. It can handle high- dimensional datasets, where many predictors are considered.However, stepwise methods can be sensitive to the choice of criteria (e.g., p-value thresholds) and may lead to overfitting if not properly tuned. This is useful when there are many potential predictors, and the goal is to identify a subset that significantly contributes to predicting the outcome. This is often used in exploratory analysis and in situations where domain knowledge about variable importance is limited.

7 Optimizable Neural Network (ONN)

This refers to a network where hyperparameters are tuned automatically through optimization techniques. This process helps to balance the trade-offs between model complexity, training speed, and predictive accuracy. Common optimization strategies include:

- Grid Search: A systematic approach to trying all combinations of hyperparameters.
- Random Search: Randomly sampling from hyperparameter space to find the best configuration.

- Bayesian Optimization: A more efficient method that builds a probabilistic model to select hyperparameters based on previous results.

This is ideal for situations where, finding the right network architecture manually would be too time-consuming or difficult. This is widely used in practice to ensure the neural network performs well across different datasets by automatically selecting the most effective architecture and hyperparameters. Based on the structure and complexity of the neural network model, Narrow, Medium, Wide, Bilayered, or Trilayered Neural Network is used.

8 Kernel Regression

A kernel in regression enables the model to handle non-linear relationships by transforming the data into a higher-dimensional space. This is achieved through the kernel trick, which allows the model to operate in this transformed space without explicitly computing the transformation. Kernels are most commonly used in methods like Support Vector Regression (SVR) and Kernel Ridge Regression (KRR), providing flexibility to capture complex patterns in data that cannot be handled by standard linear regression. Provided below are two such types of Kernels that are used in this research.

Least Squares (LS) Method: This is a statistical technique used in linear regression to estimate the coefficients of a linear model. It minimizes the sum of the squared differences (residuals) between the observed values and the predicted values, as per the equation below:

$$Minimize \sum_{i=1}^{N} \left(y_i - \widehat{y}_i\right)^2 \tag{6}$$

where,$i$is the observed value, and $\widehat{y}_i$ is the predicted value from the model.

This assumes a linear relationship between the predictors and the target variable, sensitive to outliers, as large residuals are squared and thus given more weight, and it is a simple and efficient method for fitting linear models. This works best when data is linearly separable or well approximated by a linear.

function. This is widely used in linear regression problems where the goal is to predict a continuous variable. It's ideal when the relationship between the input features and the output is approximately linear.

Support Vector Machine (SVM) Kernel: This is a supervised learning model, used for classification and regression. In cases where the data is not linearly separable, SVM Kernels map the input data into a higher-dimensional space where a linear separation is possible. This tries to find a hyperplane that maximizes the margin between classes (for classification) or minimizes a regularized loss function (for regression). This allows SVMs to handle more complex, non-linear data by applying a transformation.Common kernel functions include:

- Linear Kernel: For linearly separable data, similar to linear regression.
- Polynomial Kernel: Allows the SVM to model polynomial relationships.

- Radial Basis Function (RBF) Kernel (Gaussian Kernel): Captures complex, non-linear relationships.
- Sigmoid Kernel: Models relationships similar to neural networks.

The kernel trick maps the input data X into a higher-dimensional space.
(X), without explicitly computing the transformation, as per the equation:

$$K(x_i, x_j) = \phi(x_i) \cdot \phi(x_j) \quad (7)$$

where,$(xi,xj)$is the kernel function applied to two data points, xi and
xj.

SVM Kernels enable the model to fit non-linear decision boundaries by.

projecting data into a higher-dimensional space.The algorithm is designed to maximize the margin between the support vectors (data points closest to the separating hyperplane), which helps generalize the model. This is less sensitive to outliers compared to Least Squares methods, as it focuses on boundary points (support vectors). This can handle both classification (finding separating hyperplanes) and regression (ε-SVR for regression tasks). This is used when the data is not linearly separable or has complex relationships that a linear model cannot capture. It's commonly used in classification tasks (such as image recognition, text classification) but can also be applied to regression with SVM variants like Support Vector Regression (SVR).

9 Ensemble Regression

Ensemble regression combines multiple regression models to improve predictive accuracy and robustness. The most common ensemble techniques namely, bagging, boosting, stacking, and voting, can all be applied to regression problems. Each method has its advantages and is suited to different kinds of problems. A comparison of the two types of Ensemble Regression that are used in this research, are as provided in Table 1.

Table 1. Comparison of boosted tree and bagged tree regression models

Aspect	Boosted trees	Bagged trees (Random Forests)
Learning process	Sequential: Each tree is trained to correct the errors of the previous trees	Parallel: All trees are built independently from random data samples
Model type	Weak learners (shallow trees) combined to form a strong model	Strong learners (deep trees) combined by averaging or voting
Handling of errors	Focuses on correcting errors of previous trees	Averages out errors by combining models
Bias-variance tradeoff	Reduces bias and variance, but prone to overfitting if not regularized	Mainly reduces variance, doesn't reduce bias as effectively
Training time	Typically slower due to	Faster because trees can be

(*continued*)

Table 1. (*continued*)

Aspect	Boosted trees	Bagged trees (Random Forests)
	Sequential learning	*Trained in parallel*
Handling noisy data	Sensitive to noisy data and outliers, may overfit	Less sensitive to noise, more robust due to averaging
Accuracy	Can achieve very high accuracy, especially on complex datasets	High accuracy, but generally not as high as boosted trees on complex problems
Interpretability	Harder to interpret due to sequential nature and many parameters	Random forests (a form of bagging) are easier to interpret with feature importance scores

10 Results and Discussion

The proposed model was successfully implemented on a personal computer equipped with a high-performance Intel 7 processor and 16 GB of RAM for efficient computation. Since pre-processing has already been carried out for both the CVD and CHD datasets, no additional work is needed on the datasets prior to deploying the models on them. To ensure fairness and accuracy in the modelling process, the dataset was meticulously balanced, with equal representation across categories. The model development phase was particularly robust, employing the powerful regression algorithms. For the model training process, the dataset was divided into two distinct segments: 80% of the data was extensively utilized for training purposes, while the remaining 20% was thoughtfully set aside for rigorous testing and evaluation. Notably, all attributes within the dataset were of a categorical nature, necessitating careful handling and consideration during the model creation process.

In this study, the proposed model's performance was rigorously evaluated using key metrics such as Root Mean Square Error (RMSE), R- Squared, Mean Square Error (MSE), and Mean Absolute Error (MAE) values. To assess the model's effectiveness, a variety of algorithms were explored, as in Tables 2 and 3. Through meticulous hyper-parameter tuning and optimization, the researchers leveraged a diverse range of ML regression models on a comprehensive CVD dataset.

Table 2. Performance comparison of regression models on CVD dataset

Regression model	Types	RMSE	R squared	MSE	MAE	Training time (s)
Tree	Fine	0.5289	–0.1188	0.2797	0.37	0.4241
	Medium	0.4393	0.2279	0.193	0.3641	0.2945
	Coarse	0.4338	0.2473	0.1882	0.37	0.174

(*continued*)

Table 2. (*continued*)

Regression model	Types	RMSE	R squared	MSE	MAE	Training time (s)
	Optimizable Tree	0.4338	0.2473	0.1882	0.37	20.0339
Linear regression	Linear	0.4413	0.2211	0.1947	0.3926	0.0314
	Interaction	0.4374	0.2348	0.1913	0.3846	0.0001
	Robust	0.4432	0.2143	0.1964	0.3965	1.9878
	Stepwise	0.4412	0.2213	0.1947	0.3926	1.9878
Optimizable neural network	Narrow, Medium, Wide, Bilayered, Trilayered	0.546	−0.1926	0.2981	0.501	965.954
Kernal	Least square	Memory was not sufficient during training, as this model tried to create a matrix for all the training records (approx. 48,000x48000)				
	SVM (poly kernel)	0.4898	0.0404	0.2399	0.4164	103.8407
Ensemble	Boosted tree	0.4298	0.2609	0.1848	0.3665	847.2454
	Boosted tree	0.4369	0.2365	0.1909	0.3735	1462.845

The findings from the study revealed that the Boosted Tree Ensemble stood out by achieving impressive results in terms of regression analysis, for CVD Dataset (Table 2). Specifically, it obtained a minimum RMSE value of 0.4298, an R Square value of 0.2609, MSE of 0.1848, and MAE of 0.3665 for both cardiac and non-cardiac regression analysis. To further refine the predictions, a threshold value of 0.5 was utilized to accurately classify between cardiac and non-cardiac conditions. Comparing these results with other models that have also been applied to the pre-processed CVD dataset, the proposed approach proved to be superior. Experimental analysis showcased that the proposed regression framework excelled with an outstanding accuracy rate of 97.37%, indicating its efficacy in the realm of cardiac health prediction.

Least Square Kernel Regression could not be performed for CVD dataset, as it returned with an error of insufficient memory. This model was not attempted for execution in GPU-based Computers for this research, as the time taken for processing this model was high for CVD dataset.

Table 3. Performance comparison of regression models on CHD dataset

Regression model	Types	RMSE	R squared	MSE	MAE	Training time (s)
Tree	Fine	0.4616	0.1455	0.2131	0.2131	0.0043

(*continued*)

Table 3. *(continued)*

Regression model	Types	RMSE	R squared	MSE	MAE	Training time (s)
	Medium	0.3205	0.5881	0.1027	0.197	0.0029
	Coarse	0.3208	0.5873	0.1029	0.2205	0.0041
	Optimizable Tree	0.3208	0.5873	0.1029	0.2205	0.5243
Linear regression	Linear	0.3342	0.5522	0.1117	0.2682	0.0051
	Interaction	0.5546	−0.2331	0.3075	0.3838	0.0001
	Robust	0.3106	0.6132	0.0965	0.2404	0.0363
	Stepwise	0.3212	0.5863	0.1032	0.25	0.0363
Optimizable neural network	Narrow, Medium, Wide, Bilayered, Trilayered	2.224	−18.8328	4.9462	2.0061	247.107
Kernel	Least square	0.324	0.579	0.105	0.2568	1.6555
	SVM	0.329	0.566	0.1082	0.264	437.6115
Ensemble	Boosted tree	0.3298	0.5639	0.1088	0.2667	9.6699
	Bagged tree	0.3267	0.5721	0.1067	0.2575	25.9866

The findings from the study revealed that the Robust Linear Regression stood out by achieving impressive results in terms of regression analysis, for CHD Dataset (Table 3). Specifically, it obtained a minimum RMSE value of 0.3106, an R Square value of 0.6132, MSE of 0.0965, and MAE of 0.2404 for both cardiac and non-cardiac regression analysis. To further refine the predictions, a threshold value of 0.5 was utilized to accurately classify between cardiac and non-cardiac conditions. Comparing these results with other models that have also been applied to the pre-processed CHD dataset, the proposed approach proved to be superior. Experimental analysis showcased that the proposed regression framework excelled with an outstanding accuracy rate of 98.52%, indicating its efficacy in the realm of cardiac health prediction.

11 Conclusion

Addressing the pressing global health challenge posed by CVDs necessitates a comprehensive and intricate approach that intricately weaves together cutting- edge methodologies to ensure precise identification and timely intervention. The proposed methodology encompasses a detailed process that includes comprehensive data pre-processing procedures, meticulous feature engineering techniques, and the strategic application of an optimizable regression analysis model specifically tailored to detect nuanced cardiovascular abnormalities with a high-level of accuracy. By rigorously executing intricate data pre-processing methods, effective elimination of noisy and inconsistent data points was done, which were residing within the CVD dataset, thereby establishing a solid foundation of reliability for subsequent analytic endeavours. Data imputation was used for CHD dataset to update the missing values. Leveraging the power of box plots and scatter plots not only streamlines the process of identifying and eliminating outliers but also actively contributes to elevating the overall quality of the dataset, ensuring optimal conditions for model evaluation and refinement. Additionally, the tireless efforts in the realm of feature engineering substantially fortify the model's efficacy by adeptly capturing crucial data dependencies pertinent to the realm of cardiovascular abnormalities, thus bolstering its performance and predictive capabilities significantly. Proficient categorization of cardiac and non-cardiac data samples was done, while simultaneously unearthing valuable insights into the underlying data correlations, culminating in marked advancements in both accuracy and interpretability.

In light of the compelling experimental outcomes that underscore the remarkable precision and efficiency of the regression models in distinguishing between cardiac and non-cardiac data with accuracy rates of 97.37% for CVD dataset and 98.52% for CHD dataset, it is evident that the regression approach not only excels in performance but is also primed for successful deployment in clinical settings due to its meticulous focus on minimizing false positives and false negatives. By actively contributing to the continued discourse on early detection and precise diagnosis of CVDs, the proposed forward-thinking approach stands poised to enhance patient outcomes and revolutionize healthcare standards on a global scale. It is the unwavering commitment to persistent research efforts and the relentless implementation of advanced methodologies that fuel the mission of the researchers to forge significant progress in the fight against the pervasive global burden of CVDs, effectively supporting cardiovascular health worldwide.

References

1. Felman, Cardiovascular Disease: Types, Symptoms, Prevention, and Causes. The Technical Writer's Handbook. Mill Valley, CA (1989)
2. Abdalrada, A.S., Abawajy, J., Al-Quraishi, T., Islam, S.M.S.: Machine learning models for prediction of co-occurrence of diabetes and cardiovascular diseases: a retrospective cohort study. J. Diabetes Metab. Disord. **21**, 251–261 (2022)
3. Casas, R., Castro-Barquero, S., Estruch, R., Sacanella, E.: Nutrition and cardiovascular health. Int. J. Mol. Sci. **19**, 3988 (2018)
4. Delgado-Lista, J., Perez-Martinez, P., Lopez-Miranda, J., Perez-Jimenez, F.: Long chain omega-3 fatty acids and cardiovascular disease: a systematic review. Br. J. Nutr. **107**, S201–S213 (2012)
5. Hagan, R., Gillan, C.J., Mallett, F.: Comparison of machine learning methods for the classification of cardiovascular disease. Inform Med Unlocked **24**, 100606 (2021). ISSN 2352-9148. https://doi.org/10.1016/j.imu.2021.100606
6. Dinh, A., Miertschin, S., Young, A., Mohanty, S.D.: A data-driven approach to predicting diabetes and cardiovascular disease with machine learning. BMC Med. Inform. Decis. Mak. **19**, 211 (2019)
7. Lundberg, S.M., Lee, S.I.: A unified approach to interpreting model predictions. Adv. Neural. Inf. Process. Syst. **30**, 4768–4777 (2017)
8. Damen, J.A., et al.: Prediction models for cardiovascular disease risk in the general population: systematic review. BMJ **16**(353), i2416 (2016). https://doi.org/10.1136/bmj.i2416.PMID:27184143;PMCID:PMC4868251
9. Randhawan, B., Jagtap, R., Bhilawade, A., Chaure, D.: Heart disease prediction using logistic regression algorithm (2022). https://doi.org/10.22214/ijraset.2022.41860

10. Kumar, M., Priyadharshini, U.: Accuracy analysis of heart disease prediction using logistic regression in comparison with the linear regression algorithm **13**(4) (2022). https://doi.org/10.47750/pnr.202213.S04.199
11. Zhang, Y., Diao, L., Ma, L.: Logistic regression models in predicting heart disease. J. Phys: Conf. Ser. **1769**, 012024 (2021)
12. Kavya, S.M., Deepasindhu, M., Nowshika, B., Shijitha, R.: Heart disease prediction using logistic regression. J. Coastal Life Med. **11**, 573–579 (2023)
13. Ciu, T., Oetama, R.S.: Logistic regression prediction model for cardiovascular disease. IJNMT Inter. J. New. Med. Tech 7(1) (2020).https://doi.org/10.31937/ijnmt.v7i1.1340
14. Chandrasekaran, A., Kalla, D.: Heart disease prediction using chi-square test and linear regression. In: Eds ARIN, CSITA, ISPR, NBIOT, DMAP, MLCL, CONEDU. CS & IT – CSCP, vol. 13, pp. 135–146 (2023)
15. Datta, S., Gang, I.K.: Robust cardiovascular disease prediction using logistic regression. J. Managem. Engineer. Integration; tURLOCK **14**(1), 26–36 (2021)
16. Saw, M., Saxena, T., Kaithwas, S., Yadav, R., Lal, N, Quot.: Estimation of prediction for getting heart disease using logistic regression model of machine learning. In: 2020 International Conference on Computer Communication and Informatics (ICCCI), Coimbatore, India, pp. 1–6. https://doi.org/10.1109/ICCCI48352.2020.9104210
17. Yang, L., Wu, H., Jin, X., et al.: Study of cardiovascular disease prediction model based on random forest in eastern China. Sci. Rep. **10**, 5245 (2020). https://doi.org/10.1038/s41598-020-62133-5
18. Cangao (Steven) Chu.: Predicting Cardiovascular Disease Based on Regression Analysis and Classification of NHANE Survey Statistics. University of California at Berkeley (2015)

Programming Language Translation: A Comprehensive Review of Techniques and Applications

S. B. Chandini(✉), A. B. Rajendra, and Mohammed Muddasir

Vidyavardhaka College of Engineering, Mysuru, Karnataka, India
chandini@vvce.ac.in

Abstract. Programming language translation, while complex and challenging, remains an intriguing frontier within the domain of Natural Language Processing (NLP). Its exploration via deep learning methodologies, although nascent, uncovers a myriad of potential advancements and unexplored territories. This paper delves into the labyrinth of these methodologies, underscoring our journey from the traditional realm of supervised learning, with its inherent requirement of substantial parallel code examples, to the more nuanced and self-directed unsupervised learning techniques. Our work leverages the principles of unsupervised learning, where the model, trained on unlabeled data, unravels underlying patterns and structures without explicit supervision. This autonomous learning approach has found significant application in a host of NLP tasks, such as sentiment analysis, the creation of large language models, and various classification tasks based on linguistic traits, including machine translation. Ultimately, our paper offers an insightful journey into the realm of programming language translation, detailing novel methodologies adapted by various authors in there work and providing a blueprint for future exploration and advancements.

Keywords: Unsupervised · Unlabeled data · Deep learning · Programming languages · Natural language processing · Software engineering · Neural networks · Linguistics · Neural machine translation · Unsupervised neural machine translation

1 Introduction

This paper commencing from the deterministic techniques conventionally used in programming language transpilation, and gravitating towards our recent application of unsupervised learning techniques for programming language translation. Techniques such as neural machine translation (NMT), unsupervised phrase-based machine translation (UPMT), and unsupervised neural machine translation (UNMT) are discussed. We elucidate our experimental process, leveraging cutting-edge neural networks and architectures integral to critical NLP workflows, and demonstrate how these techniques have been instrumental in advancing the translation of programming languages. The challenges encountered during this process, such as the dearth of parallel data and the

J. Shreyas et al. (Eds.): CODE-AI 2025, CCIS 2690, pp. 376–388, 2026.
https://doi.org/10.1007/978-3-032-19321-6_34

nuanced handling of syntactic and semantic variances across programming languages, are meticulously detailed [1, 2]. The objective of this review-based paper is to give a comprehensive analysis of the current state-of-the-art in unsupervised learning for programming language translation as well as the techniques that were used previously, providing an overall idea about the field. We evaluate the effectiveness of different approaches, highlight their benefits and limitations, and discuss future research directions. Our hope is that this review will be a valuable resource for researchers and practitioners interested in this field.

2 Various Techniques of Transpilation

Transpilation refers to the process of converting source code from one programming language to another while maintaining its functionality. It is widely used in software engineering for code migration, optimization, and platform compatibility. The following are some key techniques used in transpilation:

2.1 Deterministic Transpilation

It is well known that translation of programming language is theoretically a Turing complete task [1], however, it is difficult due to a lot of obstructions generated by the new paradigms of programming languages such as complex internal API calls, extensive usage of packages in modern programming languages such as JavaScript. Initially, the first idea of translation of programming language [3], was based on the fact that compilation process programming languages can be represented by abstract mathematical models such as a Pushdown automata and Non-deterministic automata. According to general theory of translation we can design an algorithm (a function) that can map strings of one programming language to another [3]. The internal algorithms are based on general algorithms that are used to convert a given finite automata to another finite automaton to represent lexical terms. On the other hand to model the behaviour of context-free syntax conversion of push down automata is prescribed [4]. Deterministic transpilation typically involves the use of static analysis techniques, which analyze the source code and produce a representation of its behavior that can be used to generate equivalent code in the target language. This representation is often in the form of an abstract syntax tree or intermediate representation that captures the semantics of the code [5]. During the transpilation process, the intermediate representation is used to generate code in the target language that is equivalent to the source code. The generated code is designed to behave in the same way as the source code, ensuring that the output is deterministic across multiple runs. Deterministic transpilation can be useful in a variety of scenarios, such as optimizing code for different platforms, generating parallel code from sequential code, or translating code between programming languages. By guaranteeing the same output for a given input across multiple runs, deterministic transpilation can improve the efficiency, reliability, and security of code transpilation. Recent advancement has been made to transpile programming languages detrimentally for specific domains such as high-performance computing and embedded computing [6]. As shown in Fig. 1, which represents a general architecture for transpilation process.

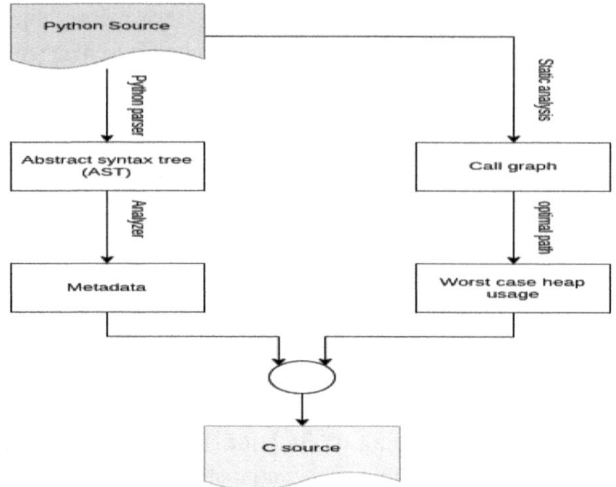

Fig. 1. Transpilation architecture for Python source and target C

Although the deterministic techniques guarantee accurate and predictable translation between programming language, but design of such systems needs a more abstract mathematical approach it becomes increasingly difficult task to perform the translation as the programming languages become more complex internally due to various factors such as dynamic and weak typing.

2.2 Statistical Machine Translation (SMT)

The approach of Statistical Machine Translation (SMT) involves considering language translation as a machine learning problem, utilizing statistical data that is accessible. As a basis we are given data about two languages and at each translation step we calculate how probable the output is for accurate translation of between source and target language [7]. The translation relies of fundamental probability theories, and requires a basic scoring methodology to achieve a certain level of accuracy in the systems. Consider English and French strings, where e and f represents a set of strings belonging to English and French corpus, then we assign a score P r(fle) for each pair of strings (e, f). Using Bayes' theorem we can write

$$Pr(e|f) = \frac{Pr(e)Pr(f|e)}{Pr(f)} \tag{1}$$

As we can see that denominator is independent of factor e, finding ˆe is equivalent to make the product of P r(e)P r(fle) maximum, and we achieve the Fundamental Equation of Machine Translation [8]:

$$e^{\wedge} = \arg\max ePr(e)Pr(f|e) \tag{2}$$

2.3 Neural Machine Translation (NMT)

Neural Machine Translation (NMT) is a machine translation methodology that employs deep learning and representation techniques to interpret text from one language into another. In contrast to conventional Statistical Machine Translation (SMT) systems, which depend on statistical models and manually created characteristics, NMT systems employ neural networks to learn how to connect the source and target languages [9, 10]. Large corpus of parallel strings are provided as training data from both source and target language and NMT based model try to minimize the difference between predicted and actual translation. The learning is done over multiple iterations. The core architecture involves two components an encoder and a decoder. The encoder is responsible for processing the source language sentence and encoding it into a fixed-length vector representation, which is then used as input to the decoder. The decoder then generates the target language sentence one word at a time, using the encoder representation and the previously generated words as input. Encoders and decoders can be implemented using various neural networks such as recurrent neural networks (RNNs), long-short term memory (LSTM) networks or more transformer models with attention heads.

2.4 Tree-To-Tree Neural Networks

Tree-to-tree neural networks are a direct improvement over generic NMT. Translation of programming languages is different from natural languages since programming languages have rigorous grammars and are not tolerant to typing mistakes. In [16] researchers found that RNNs have an issue that makes it hard to produce accurate translations between programming languages, RNNs entangle two sub-tasks together that are:

- Learning the grammar of programming language
- Alignment of sequence with the grammar.

The new approach hypothesizes if the above two tasks are handled separately the translation performance can improved from previous models. The core idea behind tree-to-tree neural networks was to design decoder in such a way that it can expand the non-terminals using some form of attention based mechanism from information gathered from parse tree of source programming language. A tree encoder will aggregate all the information of a sub-tree to the embedding of its root (encoder will be represented as a tree), and hence the embedding can be used to guide the expansion of the non terminal of the target tree. An example of such an expansion can be seen in Fig. 2 (translation workflow between CoffeeScript program and JavaScript program).

2.5 Attention Based Architecture: Transformers

Transformers are the state-of-the art neural networks that are deployed as fundamental encoder-decoder based architecture for most of the major NLP and generative AI tasks due to their attention based workflow. Transformers were first presented by Vaswani et al. in 2017 [19]. Transformers fundamentally replaced RNNs, by eliminating the recurrence workflows completely and replacingit with a purely attention based mechanism. Transformers have evidently replaced all the NLP based tasks which were using RNN based neural networks. In a standard Transformer architecture, the input sequence is first transformed into a sequence of embedding, this represents inputs as high-dimensional tokens over a given space.

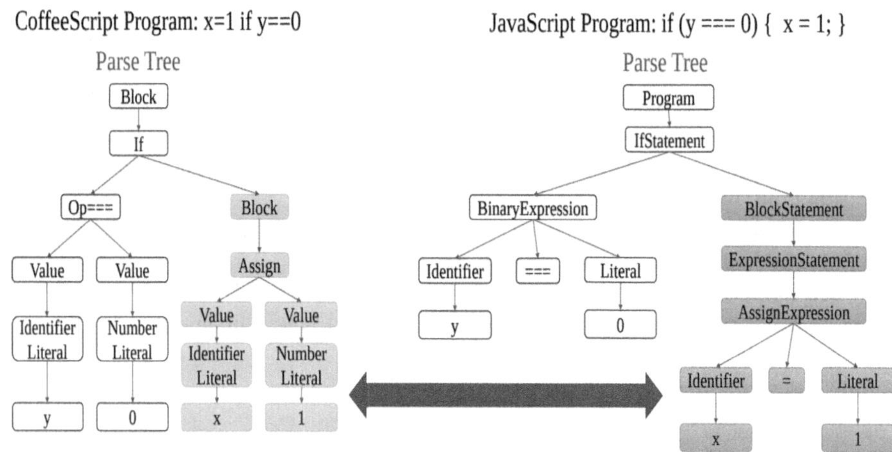

Fig. 2. Translating CoffeeScript to JavaScript. Highlighted sub-components are show in CoffeeScript and JavaScript which are translated

These embedding are then fed into a series of multiple self-attention layers, the training of the model takes place here and it learns to"attend" or give attention to various parts of the input sequence which is based on the relevance of each token to the task that currently being handled. After going through the self attention layers, the sequence of embedding is fed through a series of feed-forward network layers, that will generate the final output. Addition of such feed-forward layers improves the systems by refining the representation of input sequence and at the same time produce the final output for the given task.

2.6 Cross-Lingual Language Model Pretraining

Lample and Conneau [20] provide the basis for the model that is described in further sections for translation of programming languages. Authors proposed a new model where multiple languages were pretrained and the results generated were surpassing the then state-of-the-art translation models. There are two methods for cross-lingual language models (XLMs):

1. Unsupervised training based on monolingual corpus.
2. Supervised training based on parallel data with XLM based objective.

2.7 Unsupervised Translation of Programming Languages

Unsupervised learning over a large parallel corpus available on the internet can be used to train neural networks which then can be used to translate multiple program ming languages. Authors [21] introduced a new approach to solving the translation problem by using monolingual source code. The neural network used under the hood is XLM (cross-lingual modeling) and how unsupervised learning can be used to transpile programming languages. We consider a sequence-to-sequence model with attention heads [18], the core model architecture is kept the same for all the programming languages and is composed of an encoder-decoder with a transformer architecture. The three core unsupervised machine translation principles are initialization, language modeling, and back-translation as introduced by Lample et al. [19].

2.8 Deobfuscation Based Training Objective

Deobfuscation Pre-Training Objective for Programming Languages (DOBF) was introduced by Roziere et al. in 2021 [29]. It uses a new pre-training objective where model is trained on highly obfuscated source code to recover the original version of deobfuscated code by understanding of the structural aspect of the programming language. The pre-trained model gained around 12.2% more accuracy for programming language based downstream tasks and around 5.3% improvement in natural language code search.

Concept of code obfuscation was used for various aspects like reducing memory residue, intellectual property security and others. In code obfuscation developers usually write or modify the source code in such a way that it is hard to read for humans, without altering the internal logic of the programming language. Obfuscators are also used by compilers to optimize code on various levels.

3 Comparative Analysis of Programming Language Translation Techniques

The comparative analysis highlights various programming language translation techniques, ranging from deterministic transpilation to AI-driven approaches. Deterministic methods ensure precise code conversion but struggle with dynamic language features. Machine learning-based techniques, including attention-based transformers and unsupervised translation, improve adaptability but require extensive training data. Cross-lingual pretraining enhances multilingual translation, while deobfuscation based methods aid in security analysis. Each approach has unique advantages and trade-offs, making their applicability context dependent which is as shown in the Table 1.

Table 1. Comparative analysis of programming language translation techniques

Technique	Approach	Advantages	Limitations	Applications
Deterministic transpilation	Rule-based static analysis	Ensures precise code conversion with predictable output	Requires manually defined rules; struggles with dynamic features of modern languages	Legacy code migration, cross-platform code adaptation

(*continued*)

Table 1. (*continued*)

Technique	Approach	Advantages	Limitations	Applications
Statistical machine translation (SMT)	Probability-based translation using statistical models	Leverages large datasets to learn translation patterns	Requires parallel corpora; often struggles with long-range dependencies	Natural language processing, early programming language translation models
Attention-based transformers	Deep learning-based encoder-decoder with self-attention	Achieves state-of-the-art performance; captures contextual relationships effectively	Computationally expensive; requires large-scale training data	Neural code generation, AI-based code completion, advanced NLP tasks
Cross-lingual language model pretraining	Pretrained multilingual representations for translation	Improves generalization across multiple programming languages	Requires fine-tuning for optimal performance	Multilingual programming environments, AI-powered translation tools
Unsupervised translation of Programming languages	Self-learning via monolingual corpora with backtranslation	Does not require labeled datasets; highly adaptable to new languages	Can generate incorrect translations due to lack of explicit supervision	Code migration, automated software translation
Deobfuscation-based training objective	Learning to recover original code from obfuscated input	Improves robustness in handling obfuscated and minified code	Highly dependent on quality of training data	Security analysis, reverse engineering, malware detection

4 Recent Advances Over Original Model of Unsupervised Translation of Programming Language by Leveraging Automated Unit Tests

Roziere et al. [30] proposed a system which used the base system proposed in [1] to improve the performance of translation models presented in that paper. They made a fundamental change where they passed the input generated by the machine translation model through unit tests, during the back translation phase we either discard the translation or in case the a new parallel data is generated it is fed back to the translation model. This leads to major improvements over previous models, for Java to Python and Python to C + + translation, this model outperforms the stat-of-the-art model by more than 16% and 24% respectively [30]. Figure 3 represents the schematic work-flow of this model.

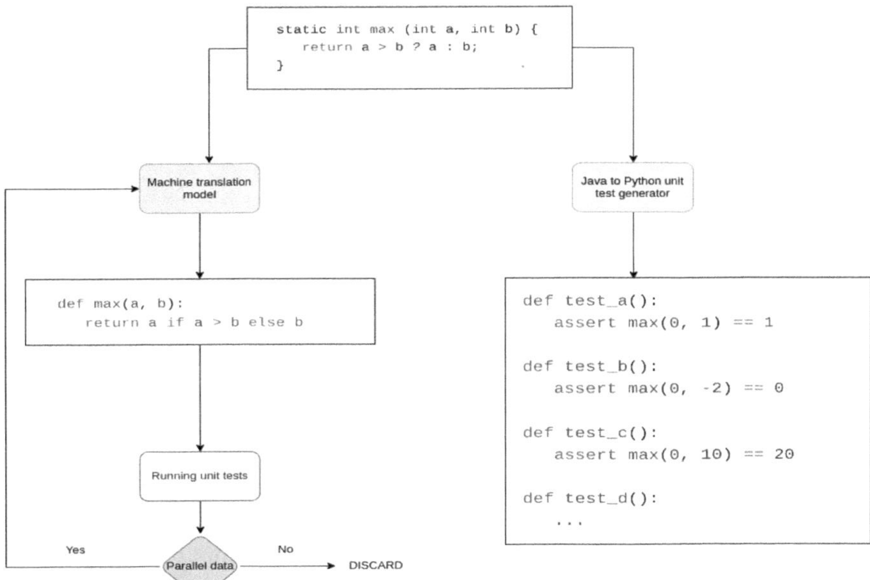

Fig. 3. Unsupervised machine translation with unit testing mechanism

4.1 Results

Table 2 shows the performance of TransCoder (unsupervised machine translation model given in [1]), DOBF (deobfuscation based mechanism given in [29]), offline and online self-training (ST) for translation between C + +, Java and Python.

Table 2. Performance comparison between various models. Offline ST represents finetuned training of the model until convergence and Online ST represents when parallel examples are created for input language on-the-fly while training, for online training authors have used a caching mechanism to improve quality of parallel results

	C++ → Ja	C++ → Py	Ja → C++	Ja → Py	Py → C++	Py → Ja	AVG
TransCoder DOBF	65.1%	47.1%	79.8%	49.0% 52.7%	32.6%	36.6% 45.7%	51.7%
Offline ST 1	65.5	56.2	81.6	61.8	46.8	55.1	61.1
Offline ST 2	65.5	58.3	83.7	63.3	46.4	52.2	61.6
Offline ST 3	66.5	56.2	**85.2**	66.3	48.1	56.6	63.1
Offline ST 4	65.3	48.2	81.1	58.1	48.9	54.7	59.4
Offline ST	**68.3**	**61.3**	84.6	**68.9**	**56.7**	**58.2**	**66.3**

5 Proposed Hybrid Deep Learning-Based Transpilation System

The proposed system as shown Fig. 4 on leverages a hybrid deep learning pipeline for deterministic transpilation, ensuring accurate and structured code translation between programming languages. The workflow begins with input tokenization using CodeBERT, which transforms source code into contextual embeddings while preserving syntax and semantics. Next, CodeT5 is employed as the primary translation model, utilizing an encoder-decoder architecture to generate accurate code in the target language. To ensure structural integrity, an Abstract Syntax Tree (AST) Transformer or Graph Neural Network (GNN) validates the translated code, ensuring compliance with syntactic rules and logical consistency. Finally, Codex/GPT-4 is utilized for final refinement, improving code readability, efficiency, and optimizing syntax. This hybrid approach effectively combines rule-based and deep learning methodologies, making it highly suitable for precise and automated code transpilation.

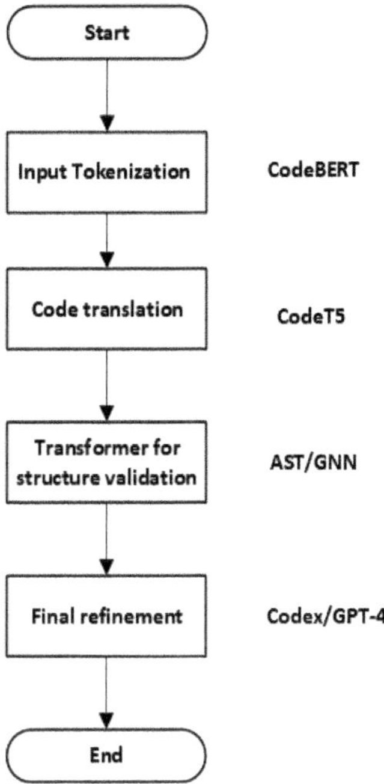

Fig. 4. Hybrid deep learning-based transpilation pipeline

6 Applications

Below are given a set of applications and usage of automated translation of programming languages:

- **High performance computing (HPC)**: HPC is still done in terse and hard to debug programming languages such as C and C + +, automating the translation process will help developers to write high performance code in a relatively safe higher level language and then translate to a fast low level language for production level deployment.
- **Solving mathematical equations**: There are NMTs that are designed to solve complex mathematical equations, but we can use the UNMT translation to give a generic symbolic mathematical representation which can be translated to a high performance language for production and research.
- **Learning programming language**: New developers can learn how a given algorithm can be represented in different programming languages giving them new perspectives on how different programming languages are structured.
- **Writing high performance code, critical code in a high level language**: Developers can write systems level code in a relatively higher level programming language which is easy to debug and does not have safety issues such as dangling pointer, or null pointer references, further these codes can be translated into faster lower level systems languages.

- **Converting/migrating legacy code**: A lot of banking infrastructure is still running on legacy code which is not efficient technically and financially. Automated code translation can help such code bases migrate to more safe and maintainable code bases.
- **Language interoperability**: Translating code between languages can help different systems communicate with each other more effectively. This is particularly useful in large-scale applications that use multiple programming languages [31, 32].

However this is not a definitive usage guide of automated programming language translation, the applications are usually endless, and are going to become more crucial in coming years.

7 Broader Impact and Future Work

Research in the field translation of programming languages will surely benefit the whole field of NLP and linguistics. We can also understand the logical learning pattern of neural networks in the pretraining phase and how they interpret the programming languages internally. If the field continues to work on this particular area, we can see improvements is formal linguistics as well, for example we can automate the construction of a compiler or even better, neural networks can discover new programming languages which are more intuitive and safe. We will clearly see a major impact on Theoretical Computer Science topics such as context free grammars, finite automata, push down automata to name a few, if neural networks get so robust that they can understand the formal logic of design and mathematics behind fundamental theoretical systems and grammars, they can have impact on almost all areas of computer science. Another major breakthrough can happen by training neural networks to translate programming languages is that they can learn how to argue logically this will surely help in a lot of pure mathematical related tasks, which in turn will produce new discoveries in various areas of research. The possibilities are endless and this area of research of translation of formal languages is still very new and at its initial phases, as new improvements are made the impact of deep learning will be in almost all the aspect of humanity, both positive and negative.

8 Conclusion

Programming language translation has advanced from rule-based deterministic methods to modern deep-learning approaches. Deterministic transpilation provides precise output but struggles with dynamically-typed languages. Statistical Machine Translation (SMT) introduced probabilistic models, improving translation accuracy but still lacking deep contextual understanding. The emergence of Transformers and Cross-lingual Language Models (XLMs) significantly improved translation quality, capturing long-range dependencies and enhancing multilingual adaptability. Unsupervised learning techniques further reduced reliance on labeled datasets, making translation scalable and efficient. Additionally, Deobfuscation-based training has strengthened security applications, aiding in reverse engineering and malware detection. Despite these advancements, deep-learning methods require extensive computational resources and training data.Future research should explore hybrid models that combine rule-based precision with neural adaptability to improve efficiency, scalability, and accuracy in programming language translation.

References

1. Lample, G., Chanussot, L., Lachaux, M.-A., Roziere, B.: Unsupervised translation of programming languages. NeurIPS (2020)
2. Lample, G., Conneau, A.: Cross-lingual language model pretraining. NeurIPS (2019)
3. Aho, A.V., Hopcroft, J. E., Ullman, J. D.: A general theory of translation. Springer (1969)
4. Schimpf, K.M., Gallier, J.H.: Tree pushdown automata. Elsevier (1985)
5. Fischer, G., Lusiardi, J., Wolff von Gudenberg, J.: Abstract syntax trees—and their role in model driven software development. In: International Conference on Software Engineering Advances (ICSEA 2007) (2007)
6. Prometeo, github.com/zanellia/prometeo/find/master
7. Lopez, A.: Static machine translation. ACM Computing Survey (2008)
8. Brown, P.E., Della Pietra, V.J., Della Pietra, S.A., Mercer, R.L.: The mathematics of statistical machine translation: parameter estimation. Computational Linguistics (1993)
9. Moon, T.K.: The expectation maximization algorithm
10. Ahmed, A., Hanneman, G.: Syntax based statistical machine translation (2023)
11. Bahdanau, D., Cho, K., Bengio, Y.: Neural machine translation by jointly learning to align and translate. ArXiv (2014)
12. Cho, K., van Merrienboer, B., Gulcehre, C., Bahdanau, D., Bougares, F., Schwenk, H., Bengio, Y.: Learning phrase representations using RNN encoder-decoder for statistical machine translation. EMNLP (2014)
13. Sutskever, I., Vinyals, O., Le, Q.V.: Sequence to sequence learning in neural networks. NeurIPS (2014)
14. Kalchbrenner, N., Blunsom, P.: Recurrent continuous translation models. In: EMNLP 2013—2013 Conference on Empirical Methods in Natural Language Processing, Proceedings of the Conference (2013)
15. Schuster, M., Paliwal, K.K.: Bidirectional recurrent neural networks. IEEE Trans. Signal Process. **45**(11), 2673–2681 (1997)
16. Chen, X., Liu, C., Song, D.: Tree-to-tree neural networks for program translation. NeurIPS (2018)
17. Sheng Tai, K., Socher, R., Manning, C.D.: Improved semantic representations from tree-structured long short-term memory networks. ACL Anthology (2017)
18. Sutskever, I., Vinyals, O., Le, Q.V.: Sequence to sequence learning with neural networks. NeurIPS (2014)
19. Vaswani, A., Shazeer, N., Parmar, N., Uszkoreit, J., Jones, L., Gomez, A.N., Kaiser, L., Polosukhin, I.: Attention is all you need. NeurIPS (2017)
20. Lample, G., Ott, M., Conneau, A., Denoyer, L., Ranzato, M.: Phrase-based and neural unsupervised machine translation. NeurIPS (2018)
21. Sennrich, R., Haddow, B., Birch, A.: Improving neural machine translation models with monolingual data. ACL Anthology (2016)
22. He, K., Zhang, X., Ren, S., Sun, J.: Deep residual learning for image recognition (2016)
23. Mikolov, T., Chen, K., Corrado, G., Dean, J.: Efficient estimation of word representations in vector space. In: Proceedings of Workshop at ICLR (2013)
24. Lewis, M., Liu, Y., Goyal, N., Ghazvininejad, M., Mohamed, A., Levy, O., Stoyanov, V., Zettlemoyer, L.: BART: Denoising sequence-to-sequence pre-training for natural language generation, translation, and comprehension. In: Proceedings of the 58th Annual Meeting of the Association for Computational Linguistics (2020)
25. Song, K., Tan, X., Qin, T., Lu, J., Liu, T.-Y.: MASS: Masked sequence to sequence pre-training for language generation. ICML (2019)

26. Lample, G., Conneau, A., Denoyer, L., Ranzato, M.A.: Unsupervised machine translation using monolingual corpora only. ICLR (2018)
27. Britz, D., Goldie, A., Luong, M.-T., Le, Q.: Massive exploration of neural machine translation architectures. In: Proceedings of the 2017 Conference on Empirical Methods in Natural Language Processing (2017)
28. Wang, Q., Li, B., Xiao, T., Zhu, J., Changliang, L., Wong, D.F., Chao, L.S.: Learning deep transformer models for machine translation. In: Proceedings of the 57th Annual Meeting of the Association for Computational Linguistics (2019)
29. Roziere, B., Lachaux, M.-A., Szafraniec, M., Lample, G.: DOBF: A deobfuscation pre-training objective for programming languages. NeurIPS (2021)
30. Roziere, B., Zhang, J.M., Charton, F., Harman, M., Synnaeve, G., Lample, G.: Leveraging automated unit test for unsupervised code translation. ICLR (2022)
31. Wang, B., Kolluri, A., Nikolic, I., Baluta, T., Saxena, P.: User-customizable transpilation of scripting languages. Proc. ACM Program. Lang. (2023)
32. Ramos, D., Lynce, I., Manquinho, V.M., Martins, R., Goues, C.L.: BatFix: repairing language model-based transpilation. ACM Trans. Softw. Eng. Methodol. (2024)

Author Index

J. Shreyas et al. (Eds.): CODE-AI 2025, CCIS 2690, pp. 389–391, 2026.
https://doi.org/10.1007/978-3-032-19321-6

BY

GPSR Compliance

The European Union's (EU) General Product Safety Regulation (GPSR) is a set of rules that requires consumer products to be safe and our obligations to ensure this.

If you have any concerns about our products, you can contact us on ProductSafety@springernature.com

In case Publisher is established outside the EU, the EU authorized representative is:

Springer Nature Customer Service Center GmbH
Europaplatz 3
69115 Heidelberg, Germany

Batch number: 10406159

Printed by Printforce, the Netherlands